ISBN 978-0-282-49415-5
PIBN 10853566

Air Navigation

BY

P. V. H. WEEMS
Lieutenant Commander, United States Navy, Retired

THIRD EDITION

McGRAW-HILL BOOK COMPANY, INC.
NEW YORK AND LONDON
1943

Preface to the Third Edition

This edition goes to press under critical war conditions with the purpose of assisting national defense. Material on civil airways and other phases of air navigation, which would have been treated more fully under normal conditions, has been reduced or omitted in order to make the text as suitable as possible for practical navigation under war conditions.

After fourteen years of evolution, the "Air Almanac" should remain free from radical changes for some years to come. Extracts giving the essential descriptive material of the almanac are included in Appendix B together with sufficient data to permit working the problems in the text.

The author is indebted to Mary Tornich for material from "Radius of Action of Aircraft," to Charles A. Zweng for material from "Instrument Flying," and to W. C. Konicek for help on radio. Link Aviation Devices, Inc., Sperry Gyroscope Company, Kollsman Instrument Company, Pioneer Instrument Company, and D. Van Nostrand Company, Inc., gave generous cooperation in the preparation of this book.

With need for an estimated 50,000 navigators for the 185,000-plane program for 1942–1943, navigation will provide a romantic career and a livelihood for hundreds of thousands of American youths within the next few years.

P. V. H. WEEMS.

ANNAPOLIS, MD.,
December, 1942.

Preface to the First Edition

In the earlier stages of aviation the aviator's efforts were almost completely absorbed in the problem of taking off and keeping his plane in the air. The increased knowledge of aerodynamics, the improvements in materials of construction, and the development of the air-cooled engine have done much to simplify this problem.

In the future the mechanics of flying will require less and less of the aviator's attention and he will be able to direct his efforts more and more toward finding his way in the air and overcoming such obstacles as blind flying and wind.

The sea-going "water-cooled" methods of navigation, while correct in principle, are too slow and cumbersome for use in the air. We must use "air-cooled" methods for pathfinding in flight. Only by practical work in the air can one realize how utterly impracticable it is to use any but the shortest and simplest methods for air work. It is my purpose to include in this book sufficient material for practical air navigation and to omit material which is of doubtful value in the cockpit of an airplane.

In addition to the commonly used methods of piloting and dead reckoning, this book is intended to stress the importance of position finding by means of celestial observations and by means of radio bearings. It is believed that the methods given in this book for finding positions from celestial observations are the simplest and fastest yet devised and that they are therefore especially suitable for use in the air. It must, however, be recognized that we have only scratched the surface of the art of celestial navigation in the air. There is not available at the present time an entirely satisfactory sextant or many handy tables and other simple devices needed in aerial navigation.

There are so many types of aircraft instruments that it is considered impracticable to include descriptions of all of them. Moreover, these instruments are continually being improved and new types developed so that such descriptions become out of date in a few months. Furthermore, each manufacturer sends out descriptive literature with the instruments which he sells. For these reasons detailed descriptions of aircraft instruments have been omitted.

In the preparation of this book, I am indebted to Lieutenant A. L. Danis, U. S. Navy, for part of the chapter on Theoretical Aerology and the entire chapter on Practical Aerology. I am especially indebted to

Lieutenant M. F. Schoeffel, U. S. Navy, for the chapter on the Earth Inductor Compass, for certain other material, and also for counsel on numerous occasions.

It was found desirable to revise most of the original chapters. The important and laborious work of making this revision was done by Lieutenant R. M. Watt, Jr., (CC), U. S. Navy. The material was rearranged in a more logical sequence, an index added, and much new material included, of which Lieutenant Watt is co-author.

Generous and valuable assistance was given in proofreading and indexing by Captain R. M. Watt, (CC), U. S. Navy; by my father-in-law, Mr. George E. Thackray; and by my nephew, W. M. Slayden.

Since this book is based largely on original material, credit is also due those who helped in this preliminary work: Lieutenant J. E. Gingrich, U. S. Navy, with the "Line of Position Book"; Lieutenant F. R. Dodge, U. S. Navy, Mr. Louis R. Johnson, and Harold C. Gatty, with the *Star Altitude Curves;* and Mr. J. F. Burke of the Navy Yard, New York, with Aviator's Dead-reckoning Tables. Also, the development of the *Star Altitude Curves* proved to be an extensive undertaking and I am indebted, for financial help and advice to my brother, Captain G. H. Weems, U. S. Army; to my friend, Lincoln Ellsworth; to my cousin, Andrew Gennett; to my uncle, F. A. M. Burrell; and to my friend, E. J. Willis.

In collecting data and making tests of instruments and methods discussed in this book, it was desirable to make repeated flights under various conditions. The following air lines gave generous assistance in this work: The Maddux Air Lines, The Western Air Express, West Coast Air Transport Company, and The Boeing Company.

In advocating new and untried methods of navigation, it has been a great encouragement to have the support of persons of recognized authority in this field. The author is especially indebted to Rear Admiral Moffett, chief of the Bureau of Aeronautics, U. S. Navy Department; Captain C. S. Freeman, superintendent, U. S. Naval Observatory; the Hydrographer, U. S. Navy; Captain H. A. Baldridge, formerly secretary-treasurer, U. S. Naval Institute; Colonel Charles A. Lindbergh; Rear Admiral R. E. Byrd; Rear Admiral Ridley McLean; Lieutenant-Commander D. C. Ramsey, in charge of flight training at the U. S. Naval Academy; and my friend, Lincoln Ellsworth.

P. V. H. Weems.

Annapolis, Md.,
July, 1931.

Contents

Lieutenant M. F. Schoefflel, U. S. Navy, for the chapter on the Earth Inductor Compass, for certain other material, and also for counsel on numerous occasions.

It was found desirable to revise most of the original chapters. The important and laborious work of making this revision was done by Lieutenant R. M. Watt, Jr., (CC), U. S. Navy. The material was rearranged in a more logical sequence, an index added, and much new material included, of which Lieutenant Watt is co-author.

Generous and valuable assistance was given in proofreading and indexing by Captain R. M. Watt, (CC), U. S. Navy; by my father-in-law, Mr. George E. Thackray; and by my nephew, W. M. Slayden.

Since this book is based largely on original material, credit is also due those who helped in this preliminary work: Lieutenant J. E. Gingrich, U. S. Navy, with the "Line of Position Book"; Lieutenant F. R. Dodge, U. S. Navy, Mr. Louis R. Johnson, and Harold C. Gatty, with the *Star Altitude Curves;* and Mr. J. F. Burke of the Navy Yard, New York, with Aviator's Dead-reckoning Tables. Also, the development of the *Star Altitude Curves* proved to be an extensive undertaking and I am indebted, for financial help and advice to my brother, Captain G. H. Weems, U. S. Army; to my friend, Lincoln Ellsworth; to my cousin, Andrew Gennett; to my uncle, F. A. M. Burrell; and to my friend, E. J. Willis.

In collecting data and making tests of instruments and methods discussed in this book, it was desirable to make repeated flights under various conditions. The following air lines gave generous assistance in this work: The Maddux Air Lines, The Western Air Express, West Coast Air Transport Company, and The Boeing Company.

In advocating new and untried methods of navigation, it has been a great encouragement to have the support of persons of recognized authority in this field. The author is especially indebted to Rear Admiral Moffett, chief of the Bureau of Aeronautics, U. S. Navy Department; Captain C. S. Freeman, superintendent, U. S. Naval Observatory; the Hydrographer, U. S. Navy; Captain H. A. Baldridge, formerly secretary-treasurer, U. S. Naval Institute; Colonel Charles A. Lindbergh; Rear Admiral R. E. Byrd; Rear Admiral Ridley McLean; Lieutenant-Commander D. C. Ramsey, in charge of flight training at the U. S. Naval Academy; and my friend, Lincoln Ellsworth.

P. V. H. Weems.

Annapolis, Md.,
July, 1931.

Contents

AIR NAVIGATION

CHAPTER I

INTRODUCTORY REMARKS

Navigation is defined as the science or art of conducting vessels on the water. It is derived from the Latin words, *navis* meaning "ship" plus the verb *agere* meaning "to move" or "direct." Since *avis* means "bird," it follows that the word *avigation* signifies the science of directing the man-made bird in flight. The term *avigation* is shorter and more distinctive, but it is new and not in general use. The purpose of navigation is to enable an aviator to determine the position of his aircraft on the earth's surface at any desired instant and the direction and distance from his position to the desired destination.

In studying the subject of navigation it is convenient to divide it into the four well-defined methods of *piloting, dead reckoning, radio position finding,* and *celestial navigation.*

Each of these methods may in turn be subdivided into three parts:

1. Definitions and theoretical principles.
2. Equipment and instruments used.
3. Actual practice of the method.

An aviator who wishes to be able to direct his plane on long and hazardous flights over unfamiliar territory in all sorts and conditions of wind and weather should have a good working knowledge of each of these four methods and in addition should be familiar with *applied meteorology* and *fog* or *blind flying.*

This book will treat first of a few general considerations that apply equally to all methods, such as the shape of the earth, position, direction and distance, charts, and the compass. Then each of the four methods will be studied in turn, considering under each method the theory, necessary equipment, and the actual practice of the method. Next there will be a brief study of applied meteorology.

Finally, a study will be made of the actual navigation of aircraft to show how an expert navigator would combine the use of all four methods with a knowledge of meteorology and blind flying in order to direct his plane with maximum effectiveness on long and difficult flights.

General Definitions.—*Piloting*, or air pilotage, is the method of directing aircraft from place to place by referring to visible landmarks on the earth's surface, such as church spires, lighthouses, beacons, railroads, rivers, mountains, and lakes.

Dead reckoning is the method of determining a position by keeping an account, or reckoning, of the track and distance from a previous known position called the *point of departure*. Some dead reckoning is ordinarily required whether or not other methods are used.

Radio position finding is the method of fixing the position by means of lines of position determined by the directional characteristics of radio waves.

Celestial navigation is the method of determining position by the observation of heavenly bodies—sun, moon, planets, and stars.

Meteorology is defined as the branch of physics treating of the atmosphere.

Applied meteorology is the application of this science to the problem of making the best use of the wind and weather in aircraft operations.

Blind flying is flying by means of instruments when all objects outside the plane are obscured by fog, darkness, rain, or snow. Fog gives so much more trouble than everything else combined that blind flying, as the practical pilot understands it, is almost equivalent to flying through fog or through clouds, which are nothing more than high fogs.

Uses of Each Method.—In a normal flight over land with a good compass and a suitable chart, the pilot guides his plane much as a motorist would direct his car from a road map; *i.e.*, by simple piloting, which means that objects observed from the plane are compared with objects shown on the chart.

If the flight is over water or over uncharted or unknown land, the navigator must have recourse to dead reckoning to keep track of his position. A track due east at a speed of 100 miles for 1 hr. from an airport shown on the chart may readily be plotted on the chart to show where the plane is at the end of the hour. If the plane changes course to the northeast this new run may be plotted similarly and, allowing for the wind as accurately as it can be estimated, a fairly exact account may be kept of the plane's position. With reasonable skill in following a compass course, the total course error will not exceed 3°. Also, let us assume that for *average* weather visibility is 10 miles. For a 3° course error, a 10-mile position error will be incurred on a flight of 200 miles. Since for flights shorter than 200 miles the position errors for the assumed conditions will be less than 10 miles, pilotage will meet ordinary requirements for this range.

By similar reasoning, the maximum dead-reckoning error for a flight of 400 miles is 20 miles. These arbitrary figures, 200 to 400 miles, are chosen as the limits within which dead reckoning exclusively may be employed. Where possible, piloting methods will be employed, of course, whether or not other methods are used.

For distances greater than 400 miles, errors greater than 20 miles will be encountered, and recourse is had periodically to radio position finding and celestial navigation to fix the plane's position, these being the only practical means, other than piloting, of determining definitely a plane's position in flight. These methods are discussed in later chapters.

We may visualize more readily the accumulative errors of dead reckoning by a study of Fig. 1.

Piloting and dead reckoning are usually combined, and both methods and more are needed. If, after flying several hundred miles by dead

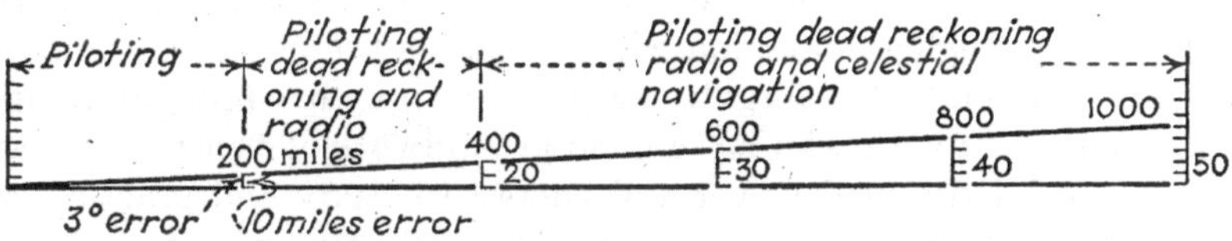

FIG. 1.—Distance limits for various methods of navigation.

reckoning, it is found that the wind has changed, and if for any reason it is impossible to estimate accurately the effect of the wind, the dead-reckoning position will be in error by an unknown amount. Any unknown error in the compass will also affect the accuracy of the *dead-reckoning* position. Furthermore, if, because of fog, darkness, or the lack of a good chart, it is impossible to recognize known objects on the earth, piloting, of course, cannot be used.

If the plane is equipped with a good radio set, the approximate position may be determined from two or more bearings from directional radio stations. This is of course dependent on the proper functioning of the radio and upon the proximity of radio beacons or radio-compass stations.

To date, most long-distance flying over water *which has succeeded* has made use of celestial navigation. This method and the radio are the only methods that determine the actual position regardless of the wind, compass error, or other similar difficulties. Prior to 1927, with the methods commonly used, about 10 to 15 min. were required to determine a position. Great progress has been made in this field in the last 10 years and with the latest equipment the time to determine a position has been reduced to approximately 5 min. for daywork and 2 min. for nightwork. The equipment required consists of a chart, a sextant, an accurate timepiece, and methods for converting the observed data into the position of the observer.

The greatest difficulty experienced by navigators at the present time is caused by fog. For position finding in a fog, radio alone can be used. In flying above a fog, celestial navigation or radio must be depended upon.

Preliminary Training Required.—It should be clearly understood at the start that a person does not need to be highly educated to become an expert navigator. It is desirable for the student to have a good foundation in the branches of mathematics commonly taught in high school. It is absolutely essential for him to have a reasonable amount of common sense and a systematic and orderly way of doing things. A navigator preparing for an important flight should think over and carefully plan his work well in advance. He should lay out the necessary equipment where it will not be forgotten and should be sure that he has a sharp pencil and the proper charts corrected to date. After starting an important flight, years of higher education will not enable the navigator to achieve practical results if he has forgotten a pencil or if he has brought along the wrong chart. Anyone who has a good groundwork of elementary mathematics, who has the will power and energy to study and practice faithfully, and who is willing to take great care and pains with his work, can easily become an expert navigator.

In addition to mastering this book, it is important to grasp every opportunity to take observations in flight with compass, drift indicator, and sextant, and actually to work out the positions from these data.

The idea of this book is to teach a prospective navigator what problems he will have to solve, what instruments he will need and how to use them, and, finally, how to apply the information he collects to reach the correct solution to his problem in the most direct manner.

As Kipling says, "There are nine and sixty ways of constructing tribal lays, and each and every one of them is right." There are just as many systems for accomplishing the four methods of navigation, but if, as comparative tests have demonstrated, some one system will do everything that all the others will do and as quickly and accurately, why bother about the other sixty-eight ways? Life is too short for a practical navigator to spend the best years of his life learning all the different ways of arriving at the same answer.

Importance of Mastering All Methods.—The ambitious student should understand clearly at the very start that it is not enough for him to be satisfied with a knowledge of piloting that will merely enable him to direct his plane from point to point on routine flights. He should launch out boldly and master all methods covered in this book.

Shape of the Earth.—The form of the earth's surface is approximately that of an ellipsoid of revolution whose shortest axis is the axis of revolution. The chief difference between the shape of the earth and an exact sphere is a bulge at the equator due to the action of centrifugal force. Thus the diameter extending from the north to the south pole is about 7,899.7 statute miles in length and the diameter of the equator is about 7,926.5 statute miles in length.

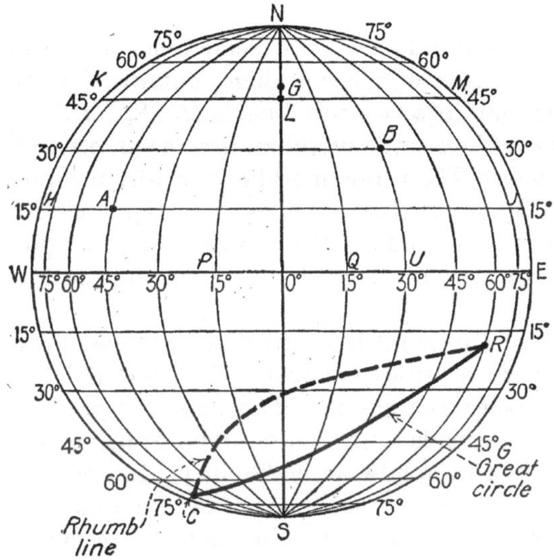

FIG. 2.—Sphere showing axis, equator, latitude, longitude, etc.

For the purposes of navigation these small departures from the exact spherical form may be neglected and the earth may be assumed to be a true sphere.

Definitions.—The following terms are in common use and should be thoroughly understood before proceeding to a discussion of latitude and longitude.

A *sphere* is a body bounded by a surface all points of which are equally distant from a point within called the center (see Fig. 2).

A *great circle* is a circle on the surface of the sphere, the plane of which passes through the center of the sphere and thus divides it into two equal hemispheres. It is important to remember that the shortest distance between any two points on the surface of a sphere is the arc of a great circle joining these points (see Fig. 2).

A *small circle* is a circle on the surface of a sphere, the plane of which does not pass through the center of the sphere (*KLM* in Fig. 2).

The *axis of the earth* is that diameter about which it rotates (*NS* in Fig. 2). The north end of the axis is the *north pole* of the earth, and the south end is the *south pole* (*N* and *S* in Fig. 2).

The *equator* is that great circle of the earth which lies midway between the poles. The plane of the equator is perpendicular to the axis of the earth and all points on the equator are 90° from the poles (*WE* in Fig. 2).

Parallels or *parallels of latitude* are small circles of earth's surface whose planes are parallel to the plane of the equator (*KLM* or *HAJ* in Fig. 2).

Meridians of longitude are great semicircles of the earth joining the poles. Thus the plane of the meridian contains the earth's axis which divides the meridian into two equal parts (*NGS*, *NBS*, *NAS* in Fig. 2).

The *prime meridian* is the meridian used as an origin for the measurement of longitude. The prime meridian used by most countries, including the United States, is that of Greenwich, England (*NGS* in Fig. 2).

Latitude and Longitude.—The position of any point on the surface of the earth may be defined by the latitude and longitude of that point.

The *latitude* of any point is its angular distance north or south of the equator. Latitude is measured from 0° to 90° north or south of the equator to the poles along a meridian and is expressed in degrees (°), minutes (′), and seconds (″). These units are known as units of arc. There are 360° in a complete circle, 60′ in 1°, and 60″ in 1′.

The *longitude* of any point is the arc of the equator intercepted between the meridian passing through the point and the prime meridian, usually that of Greenwich.

Longitude is measured from 0° to 180° east or west of the prime meridian along the equator and is also expressed in degrees, minutes, and seconds (units of arc). Longitude is sometimes expressed in units of time instead of in units of arc. This will be explained in detail in the chapter on Time.

Charts in the English language always use the meridian passing through the observatory of Greenwich, England, just outside London, as the prime meridian, but this is not universal for foreign charts. In Fig. 2 let *G* represent the position of Greenwich, let *NGS* be the prime meridian, and *WE* the equator. The position of the point *A* is Lat. 15°N., Long. 45°W. The position of the point *B* is Lat. 30°N., Long. 30°E.

Difference of Latitude and Difference of Longitude.—One position on the earth's surface is related to another by the difference of latitude and the difference of longitude between them. If an aircraft is proceeding from point *A* in Fig. 2 to point *B*, the difference of latitude is 15°N. because the parallel of the destination is 15°N. of the parallel of the departure point. The difference of longitude is 75°E. because the meridian of the destination is 75°E. of the meridian of the departure point. *Departure* is the east-west distance in nautical miles between two points as shown in Fig. 3, and is measured at mid-latitude.

Direction.—The most important, most used, and most troublesome definition in navigation is the general term *direction* and the related terms

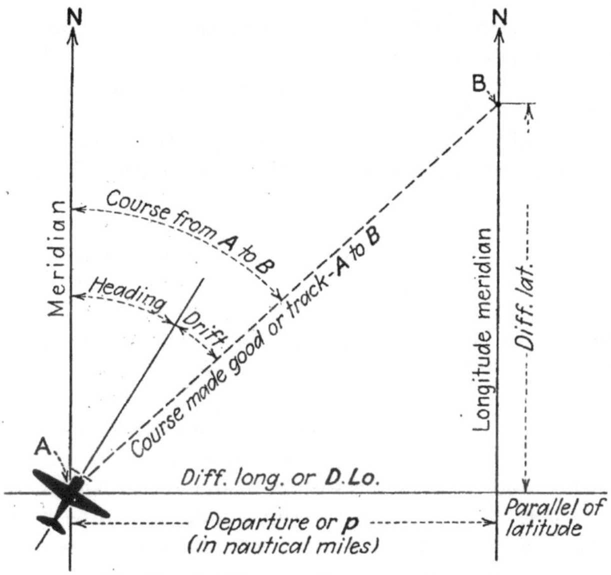

Fig. 3.—Position coordinates on the earth.

course, *heading*, *track*, *bearing*, and *azimuth*. Webster defines direction as "the line in which a body moves or to which its position is referred." The direction in which an aircraft is heading at any instant is the angle between the meridian and the longitudinal (or fore-and-aft) axis of the aircraft.

Since the meridians are only imaginary circles, which are marked on charts but not on the surface of the earth itself, the direction of the observer's meridian must be established before the bearing of a given point can be determined. The direction of the meridian is established by means of a compass.

All aircraft compasses (Chap. III) indicate the direction in which the plane is heading and many of them carry a device that enables the navigator to take bearings of objects outside the aircraft.

Aircraft compasses are of the magnetic type operating upon the principles that the mariner's compass has used for centuries. Compasses will be discussed in detail in later chapters.

The gyro compass, which is of such great value to mariners, is impracticable for use in aircraft at this time because of its weight, the space required, and the excessive vibration in a plane.

The official definitions approved by the Special Conference on Air Navigation Terms in 1935 will now be given, followed by a discussion of the ambiguous terms:

Azimuth.—The bearing of a celestial body measured as an arc on the horizon from the true meridian north or south to the east or west. Abbreviation: *Z.* Abbreviation Z_n is used where the azimuth has been changed to read from the north through east to 360°.

Bearing.—The direction of one object from another, expressed as an angle measured clockwise from true north. Bearing is true unless otherwise designated. Abbreviation: *B.*

> *Magnetic bearing.*—Bearing (true) with variation applied. Variation and deviation are discussed in Chap. III.
>
> *Compass bearing.*—The magnetic bearing with deviation applied.
>
> *Relative bearing.*—The direction of an object expressed as an angle measured clockwise from the heading of an aircraft.

Compass rose.—A small circle, graduated in degrees, 0 to 360, placed on maps or charts, as a reference to directions, true or magnetic.

Course.—The direction over the surface of the earth, expressed as an angle, with respect to true north, that an aircraft is intended to be flown. It is the course laid out on the chart or map and is always the true course unless otherwise designated. Abbreviation: *C.*

> *Magnetic course.*—The course (true) with variation applied. Abbreviation: MC.
>
> *Compass course.*—The magnetic course with deviation applied. Abbreviation: CC.
>
> *Course made good.*—The resultant true direction the aircraft bears from the point of departure. All courses are measured fom north through east to 360°.
>
> *Great-circle course.*—The route between any two places along the circumference of the great circle which joins them. It is the shortest distance between two points over the surface of the earth.

Heading.—The angular direction of the longitudinal axis of the aircraft with respect to true north. In other words it is the course with the drift correction applied. It is true heading unless otherwise designated.

Magnetic heading.—Heading with variation applied.
Compass heading.—Magnetic heading with deviation applied.

Intercept bearing.—The bearing that must be maintained in order to intercept another moving object.

Intercept heading.—The direction of the longitudinal axis of an aircraft to make good a given intercept course.

Intercept track.—The track flown by an aircraft over the earth's surface from a known position to a moving object.

Mercator course (rhumb line).—A line on the earth's surface which intersects all meridians at the same angle.

Track.—Actual path of an aircraft over the surface of the earth. Track is the path that has been flown. *Course (true)* is the path intended to be flown.

These fairly complete definitions still leave points to be cleared up. If a *bearing* is "the direction of one object from another," we must state whether the bearing angle is measured from north to the *great circle* through the two objects, or to the corresponding *rhumb* line. Since a bearing of an object is along the line of sight, it must be along the great circle between the observer and the object observed. Since an aircraft can fly a *rhumb* or *Mercator course* only by compass, the course must be along the rhumb line. This may be stated, "a *bearing* is the direction of a great circle; a *course* is the direction of the rhumb line." Through carelessness or ignorance these terms are often misused or interchanged. The beginner should make every effort to avoid confusion by getting a complete understanding of these related terms.

The *course line* is drawn on a chart by connecting the point of departure with the destination. The navigator then estimates the *course to be steered*, or *heading* to make good the course line. This heading seldom results in the plane's making good the course line, due to combined unavoidable errors in evaluating the following factors:

1. Variation (Chap. III).
2. Deviation (Chap. III).
3. Winds.
4. Steering.
5. Plotting course line.

Distance.—It is well known that on a plane surface the shortest distance between two points is the straight line joining them. On the surface of a sphere such as the earth, the shortest distance between two points is the lesser arc of the great circle passing through the two points.

Thus the shortest distance between two points, both on the equator, is the distance between them measured along the equator. Similarly, the shortest distance between two points, both on the same meridian, is the distance between them measured along that meridian. In the case of an aircraft flying between the two points on the equator (*P* and *Q* in Fig. 2), the shortest course, which is along the equator, will make the same angle (90°) with each meridian crossed. In flying between two points on the same meridian (*U* to *B* in Fig. 2) the course will be north along the meridian or 0°.

In Fig. 2, the solid line joining points *C* and *R* represents the arc of a great circle so that an aircraft seeking the shortest possible route from *C* to *R* would follow this line. It should be noted that this great circle cuts each meridian at a different angle. Thus at the point of departure *C*, the course is about 120° and this gradually changes to about 30° at the point of destination. This does not show up so clearly on a flat sketch, but if the student will look at an ordinary globe such as is used to teach geography in grammar school, the principle will at once be apparent. The shortest distance between two points may be determined by means of a piece of thread. The thread, when stretched between the two points, traces the great circle that passes through both points, and it will be seen that, except in the two cases noted above, it makes a different angle with each meridian. Thus, if an aircraft is to fly the shortest route from one point on the earth to another, its track must be a great circle, and the direction of this track is constantly changing, except in the two special cases noted. In following such a track in practice, the course is changed at regular intervals so that the aircraft follows a series of rhumb lines that approximate the great circle.

A *rhumb line* is a line on the earth's surface which cuts all meridians at the same angle. Thus an aircraft flying a steady true course is following a rhumb line. In Fig. 2 the dotted line from *C* to *R* cuts every meridian it meets at the same angle (60°) and represents the rhumb line through these two points. It should be noted that if this rhumb line is followed indefinitely the navigator will travel in a long spiral and finally end up at the pole. By referring to the globe again it is clear that the rhumb line is not the shortest distance between the points. However, the rhumb line joining any two points offers a great advantage in that it may be followed by flying a constant course.

The comparisons shown in the accompanying table will help the navigator to decide when to follow a great-circle course and when to follow a rhumb-line course. In the first case (New York to Boston), it will be noted that about 0.6 per cent of the rhumb-line distance is saved by following the great circle and in the last case (New York to Tokyo) about 15.5 per cent of the distance is saved.

From	To	Rhumb-line distance, nautical miles	Great-circle distance, nautical miles
New York, N. Y.	Boston, Mass.	166.4	165.4
New York, N. Y.	Chicago, Ill.	621.9	618.8
New York, N. Y.	Calshot, Eng.	3,088.0	2,976.0
New York, N. Y.	Paris, France	3,290.4	3,149.4
Tokio, Japan	Calshot, Eng.	6,182.0	5,219.0
New York, N. Y.	Tokio, Japan	6,932.3	5,855.9

Although the great-circle track offers the shorter distance, it has the disadvantage of making it necessary to alter course frequently in order to fly along a great-circle track.

For flights of less than 1,000 miles the saving in distance by great circle is small and the convenience of a single course makes the rhumb line preferable. For flights in excess of 1,000 miles the saving in distance will generally outweigh the inconvenience of altering course, and the great-circle track is to be preferred provided it does not lead into high latitudes where dangerous cold will be encountered.

Units of Distance.—The navigator should be familiar with the following units which are in common use:

a. The *statute mile* is 5,280 ft. This arbitrary unit of length has been adopted as the standard in the English-speaking countries. A statute mile is approximately 0.87 nautical mile.

b. The *nautical mile* is 6,080.27 ft. in the United States and 6,080 ft. (called the "admiralty mile") in Great Britain. This length was chosen because it was thought to be the length of 1′ of latitude or of 1′ of arc on the equator. The nautical mile is approximately one-seventh longer than the statute mile. Nautical miles may be converted into statute miles by multiplying the number of nautical miles by eight-sevenths, or by 1.15. It is the standard unit of measure for marine navigation and for work with the Mercator chart. It is sometimes called the "geographical mile," although the latter is slightly different, being 6087.1 ft.

c. A *knot* is a speed of 1 nautical mile per hour. It is the standard unit of speed for marine navigation and is much used by seaplanes and military planes.

d. A *meter* is 39.37 in., or 3.281 ft. The meter is the unit of length in the metric system. The meter occurs frequently on foreign charts.

e. A *kilometer* is 3,280.8 ft., or 1,000 meters, and is one ten-thousandth part of the distance from equator to pole.

f. A *fathom* is 6 ft. Depths of water on charts are usually expressed in fathoms, sometimes in feet.

g. 1 kilometer = 0.62 statute mile = 0.54 nautical mile.
1 nautical mile = 1.15 statute mile = 1.86 kilometers.
1 statute mile = 0.87 nautical mile = 1.61 kilometers.

Notation.—The abbreviations and symbols used in this book will be accurately defined in the chapter where they are first introduced, but for reference, the most common abbreviations and symbols and those which have the same meaning throughout the book are here briefly defined:

Abbreviations

a	Altitude difference between the observed and the computed altitudes.
A/C	Altered course.
Alt. (or h)	Altitude.
A.M.	Ante meridiem or forenoon.
AS	Air speed.
Az (or Z)	Azimuth.
C (or Co.)	Course.
C_N	Course measured from north around to the right.
CC	Compass course; also changed course.
Corr. (or C)	Correction.
d (or dec.)	Declination.
D Lo.	Difference of longitude.
Dep (or p)	Departure.
Dev.	Deviation.
Diff.	Difference.
Dist. (or D)	Distance.
DL (or l)	Difference of latitude.
D.R.	Dead reckoning.
E.	East, or error.
E.P.	Estimated position.
E.T.A.	Estimated time of arrival.
ft.	Feet.
G (or Gr.)	Greenwich.
GCT	Greenwich civil time.
GHA	Greenwich hour angle.
GS	Ground speed.
H (or Alt.)	Altitude of a celestial body.
H_c	Computed altitude.
H_o	Observed altitude, or the sextant altitude with all known corrections applied.
H_s	Sextant altitude.
HA (or t)	Hour angle.
H.D.	Hourly difference.

H.P.	Horizontal parallax.
hr. (or h)	Hour.
Ht. Eye	Height of eye above sea level.
IC	Index correction.
Int.	Interval.
L (or Lat.)	Latitude.
L_m	Middle latitude.
LCT	Local civil time.
LHA	Local hour angle.
Long. (or λ)	Longitude.
LPB	"Line of Position Book."
Mag.	Magnetic.
Merid.	Meridian.
min. (or m)	Minutes.
m.p.h.	Miles per hour.
N.	North.
Naut.	Nautical.
p	Departure.
P	Pole; or polar distance.
P_n	North pole.
P_s	South pole.
p.c.	Per compass.
P.M.	Post meridiem or afternoon.
P.M.	Prime meridian.
Ref.	Refraction.
S.	South; also speed.
S.D.	Semidiameter.
SAC	"Star Altitude Curves."
t	Local hour angle.
Var.	Variation.
W.	West.
W.T.	Watch time.
z.	Zenith distance.
Z	Azimuth.
Z_n	Azimuth measured from north pole to the right through 360°.
ZT	Zone time.

Symbols

$\odot$	Sun.
$\overline{\odot}$	Sun, altitude of upper limb.
$\ominus$	Sun, altitude of center.
$\underline{\odot}$	Sun, altitude of lower limb.
☾	Moon.
$\overline{☾}$	Moon, altitude of upper limb.
☾̶	Moon, altitude of center.
$\underline{☾}$	Moon, altitude of lower limb.
♃	Jupiter.
♀	Venus.
♂	Mars.
$^\circ$	Degrees of arc.
$'$	Minutes of arc.
$''$	Seconds of arc.
h	Hours.
m	Minutes of time.
s	Seconds of time.
*	Star or planet.
♈	Vernal equinox, or First Point of Aries.
λ	Longitude.

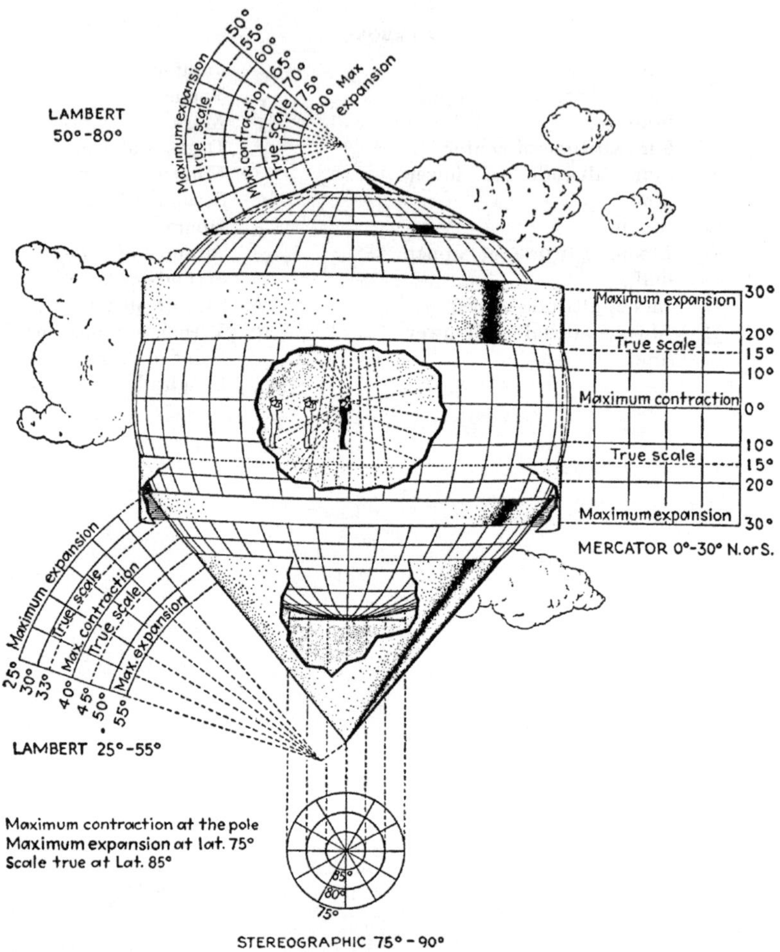

FIG. 4.—Illustration of projections used in the construction of WSN navigation charts.

CHAPTER II

CHARTS

A *chart* or *map* is a representation of the earth's surface or a part of it on a flat piece of paper. The term "chart" is generally used where aids to navigation are indicated. Except for short flights, an accurate chart of the route to be traversed is probably more important than any other item of equipment. If the destination is not in sight at the start, the navigator's first act is to look at the chart to find the direction in which he must go to reach the desired destination. In the course of the flight, as the navigator fixes his position by any or all the various methods, he plots his fix or determined position on the chart and checks the direction from this fix to the destination. If there is available a chart made up especially for aerial use, the navigator may, by referring to it, lay out his course so as to take advantage of airports and avoid dangerous terrain such as mountain peaks. Thus, whatever method he may use, the navigator's first requirement is a chart on which he plans his journey before starting and records his progress in flight. As intermediate positions are determined, he sets his course from these positions by the most advantageous route to the desired destination.

Systems of Projection.—Any boy who has knocked the cover off a baseball and then tried to spread it out flat knows that it cannot be done without a good bit of stretching and wrinkling. Similarly the spherical surface of the earth cannot be represented on a flat piece of paper without some distortion of certain features. In an ideal chart, distances, bearings, shapes, and areas would be shown in their true relation and the shortest distance between two points would be represented by a straight line. It is possible to preserve one of these properties by sacrificing some of the others and charts are constructed by various methods of projection so as to preserve that feature which will be most useful for the particular purpose a chart is to serve. The three principal systems of projection in use in the United States are:

1. The Mercator.
2. The conformal conic (Lambert).
3. The polar stereographic.

Mercator Projection.—The Mercator projection, so called after Gerhard Mercator, the Flemish geographer who invented it, shows the parallels of latitude as horizontal parallel lines and the meridians of longitude as vertical parallel lines. On the earth's surface the length of 1° of longitude is 60 nautical miles at the equator, 30 nautical miles in Lat. 60°N. or S., and zero at the poles where all the meridians converge. On a Mercator chart, however, where the meridians are parallel, the length of a degree of longitude is the same in any latitude. To balance this east and west distortion Mercator conceived the idea of expanding the north and south dimensions, *i.e.*, the latitude scale, in just the same proportion that the longitude scale is expanded. This method shows the compass courses as straight lines and the shapes of areas are shown correctly, although areas in high latitudes appear larger

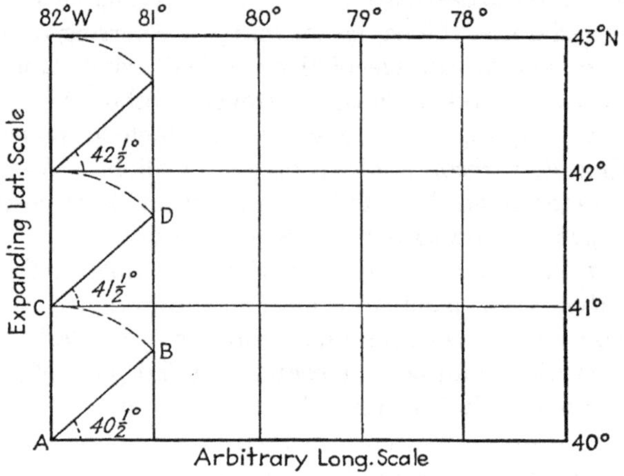

FIG. 5.—Practical construction of a Mercator chart.

than they really are. On a Mercator chart the correct distance between two points is measured by the latitude scale at the mid-latitude. Distance should *never* be measured by the longitude scale, which is to be used only for picking off the longitude.

Figure 4 illustrates the Mercator projection. Although the cylinder for it is tangent to the earth at the equator, the illustration shows the cylinder cutting the sphere at two latitudes in order to reduce distortion.

Practical Construction of a Mercator Chart.—The principles of construction of a Mercator chart have been frequently and incorrectly illustrated as the development of a sphere on a cylinder. However, for practical purposes, a Mercator chart may be constructed easily and with sufficient accuracy for any desired area and scale. Assuming that the earth is a sphere, which is true for practical navigation, a Mercator chart may be constructed as follows:

1. Draw parallel vertical lines to represent the longitude to the desired scale (see Fig. 5).

2. For the lowest latitude, say 40°N. or S., draw a horizontal line for the bottom of the chart.

3. From the intersection of the left vertical line and Lat. 40°, draw a line *AB* making an angle of 40½° with the 40° parallel of latitude.

4. The portion of *AB* intercepted between two adjacent longitude meridians is the scale of the latitude between 40° and 41°.

5. Repeat the procedure for 41° and 42° of latitude by drawing *CD* at angle of 41½° to the 41st parallel of latitude, and continue the construction as desired. Number the longitude and latitude parallels as required for the area covered.

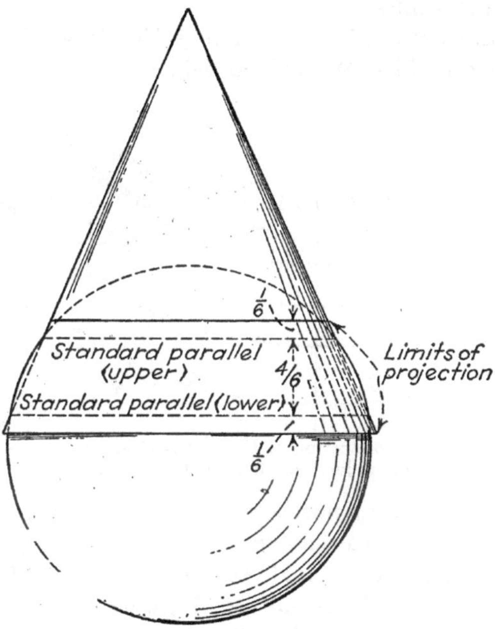

Fig. 6.—Lambert conformal conic projection. Diagram illustrating the intersection of a cone and sphere along two standard parallels. The elements of the projection are calculated for the tangent cone and afterwards reduced in scale so as to produce the effect of a secant cone. The parallels that are true to scale do not exactly coincide with those of the earth, since they are spaced in such a way as to produce conformality.

For areas covering only 1° to 3° of latitude, a fixed latitude scale may be used as described under Universal Plotting Chart.

In practice, the navigator buys his charts. If the student is interested in the method of construction used by the cartographer for making Mercator charts, see Bowditch's "American Practical Navigator."

Summary.—The important features of a Mercator chart are:

1. Rhumb-line courses appear as straight lines.

2. The scale of distance is taken directly from the latitude scale at the mid-latitude.

3. The chart is easy to construct.

4. Limited areas shown on the chart have their true shape but are distorted in size.

5. A straight line, called a rhumb line, connecting two points is the true course between them.

The principal advantage of this system of projection is having the compass courses appear as straight lines.

The principal disadvantage is the distortion of areas as the chart gets up into high latitudes.

Lambert Conformal Conic Projection.—This type of chart came into prominence during the First World War, when it was adopted by the allied

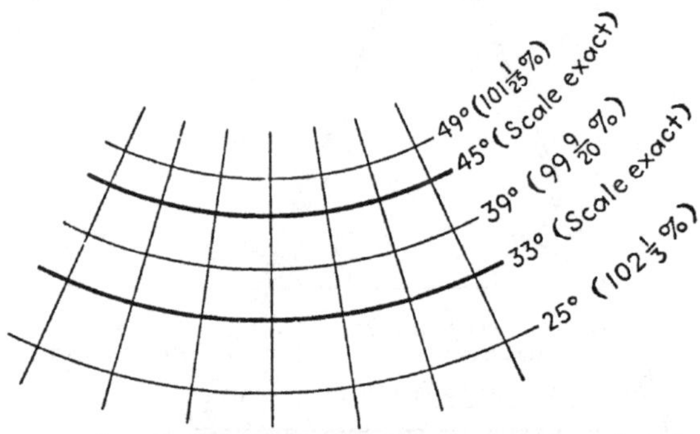

FIG. 7.—Appearance and scale distortion of the Lambert conformal conic projection with the standard parallels at 33° and 45°.

powers. Since then India, northern Canada, the north Atlantic, the United States, and other areas have been charted on this projection. The charted area is developed on a single cone cutting the surface of the earth at two standard parallels of latitude (see Figs. 6 and 7). This projection has the following features and advantages:

1. Areas appear in the proper perspective.

2. Scale distortion is small.

3. Straight lines represent very nearly the great circle between two points; hence radio bearings may be plotted directly on the chart.

4. A large area may be charted in sections so that adjacent sections fit.

The principal disadvantage is that the rhumb line or compass course does not appear as a straight line. In practice, the course measured at the mid-longitude meridian gives a close approximation to the plotted course.

Polar Stereographic Projection.—In the polar stereographic projection the plane of projection is tangent to the earth at one of the poles; the eye of the observer is located at the other pole of the earth, where it is in the plane of every meridian, and therefore all meridians are projected as straight lines. The straight line between any two points on this stereographic projection closely represents the arc of a great circle and is therefore the shortest distance between those points. On this chart the parallels of latitude are projected as circles whose center is the pole at point of tangency, and the meridians appear as straight lines radiating from this pole. This projection would give a distorted chart for low latitudes but is excellent for use in latitudes above 80° where the distortion of the Mercator construction becomes excessive.

The *gnomonic projection* used in the construction of *great-circle charts* (published by the Hydrographic Office) is the development of the surface of the earth on a plane tangent to the earth's surface as it would be projected with the observer's eye at the center of the earth.

Polar stereographic charts are similar to, but have less distortion than, polar gnomonic charts. The one advantage of the latter is the fact that a great-circle course is shown as a straight line. This advantage is offset by greater advantages in the three projections already discussed.

Figure 4 illustrates the Mercator, Lambert, and stereographic projections, on which about 99 per cent of United States aeronautical charts are constructed.

Both the Mercator and stereographic projections may be considered as special cases of the Lambert projection. In the former the apex of the cone has receded to an infinite distance; in the latter the apex of the cone has been brought down into the plane of the projection itself.

If the reader studies Fig. 4 carefully, he should understand the projections used for aviation charts of the United States.

Foreign Charts.—Foreign and international charts, using various projections and scales, will undoubtedly come more and more into use in this country and will therefore be discussed briefly at this point.

International Chart of the World.—In 1909 an Official International Conference met in London and organized The International Map Committee, which adopted rules for the construction of the 1:1,000,000 chart of the world. In 1913 a second conference was held in Paris, and further rules were adopted, so that the earth's surface is being charted by international agreement. These charts, sometimes called 1/M charts, are to the scale of 1:1,000,000, and use international standard map symbols. Aeronautical features have been added to these charts to make the 1/M aeronautical charts discussed below.

Up to latitude 60° the separate sheets are to include 6° of longitude and 4° of latitude; from latitude 60° to the pole they are to include 12° of longitude. The sheets are designated in longitude by consecutive numbers to the east from 1 to 60, beginning at 180°W., and in latitude alphabetically from *A* to *Y*, beginning at the equator and extending both north and south. Thus the sheet for Ireland is *N* 29 North (see Fig. 10).

The Lallemand projection is adopted for these charts in which the meridians are straight lines and meridional errors are lessened and distributed somewhat (except in an opposite direction) as in the Lambert projection described above. In other words, it provides for a distribution of scale error by having two standard meridians 2° on each side of the

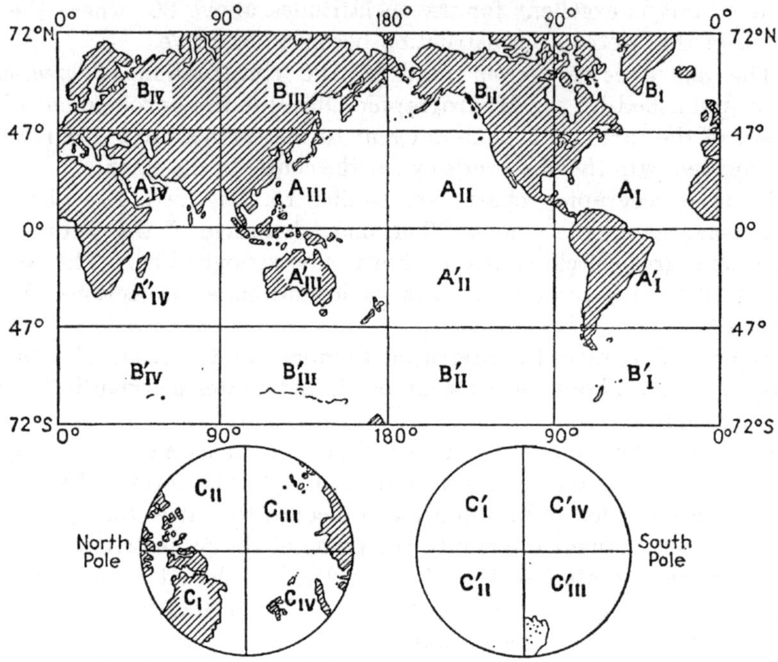

FIG. 8.—Index map of Basic Aeronautical Map of the World.

center. Meridians appear as straight lines, parallels as circles, and any sheet fits its four neighboring sheets exactly along its margins.

International Aeronautical Charts.—The Convention Relating to the Regulation of Aerial Navigation met at Paris on Oct. 13, 1919, and the International Commission for Air Navigation (I.C.A.N.) came into force in 1922; it is permanently organized with headquarters at 15 *bis*, rue Georges-Bizet, Paris. The I.C.A.N. has decided to produce the following international aeronautical charts:

I. *Basic Aeronautical Charts:*
   Scale: 1:10,000,000 at equator.
   Mercator projection to latitude 72°.
   Stereographic projection above latitude 72°.
   Index chart shown in Fig. 8.

II. *Itinerary or Route Charts:*
   Scale: 1:10,000,000.
   Oblique true projection on cylinder tangent along great-circle route. (Kahn charts, etc.)

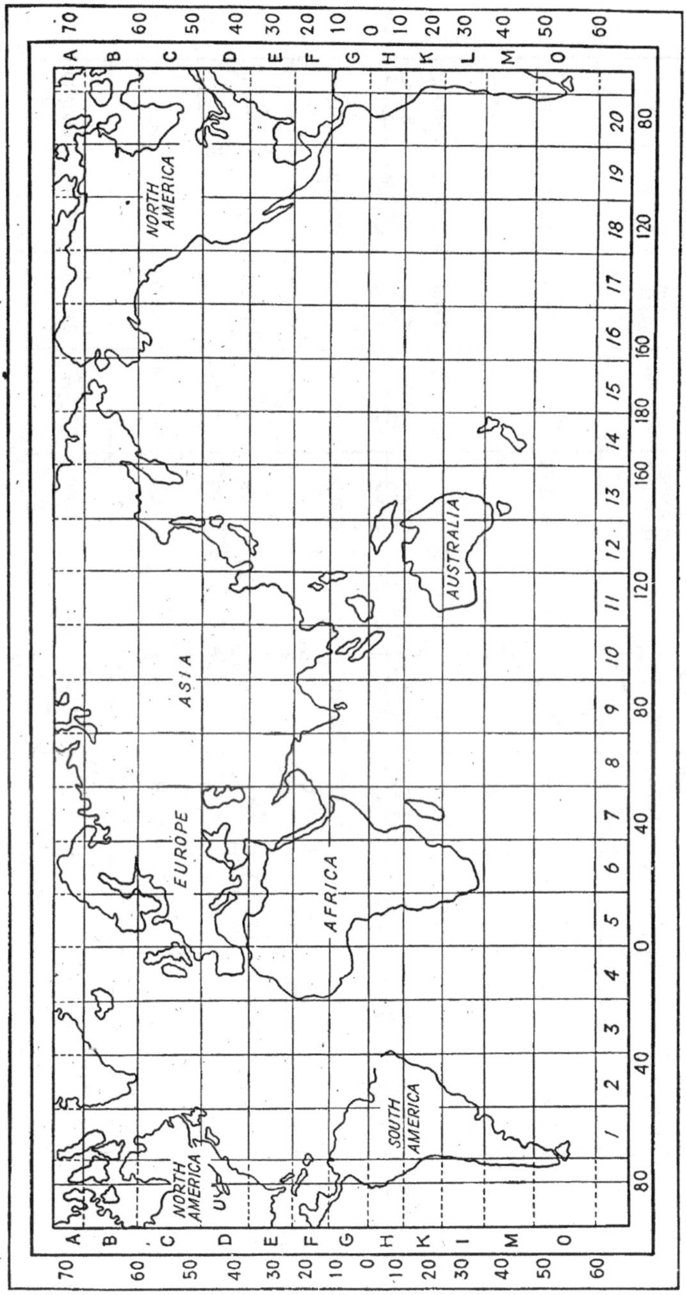

Fig. 9.—Index map of General Aeronautical Map.

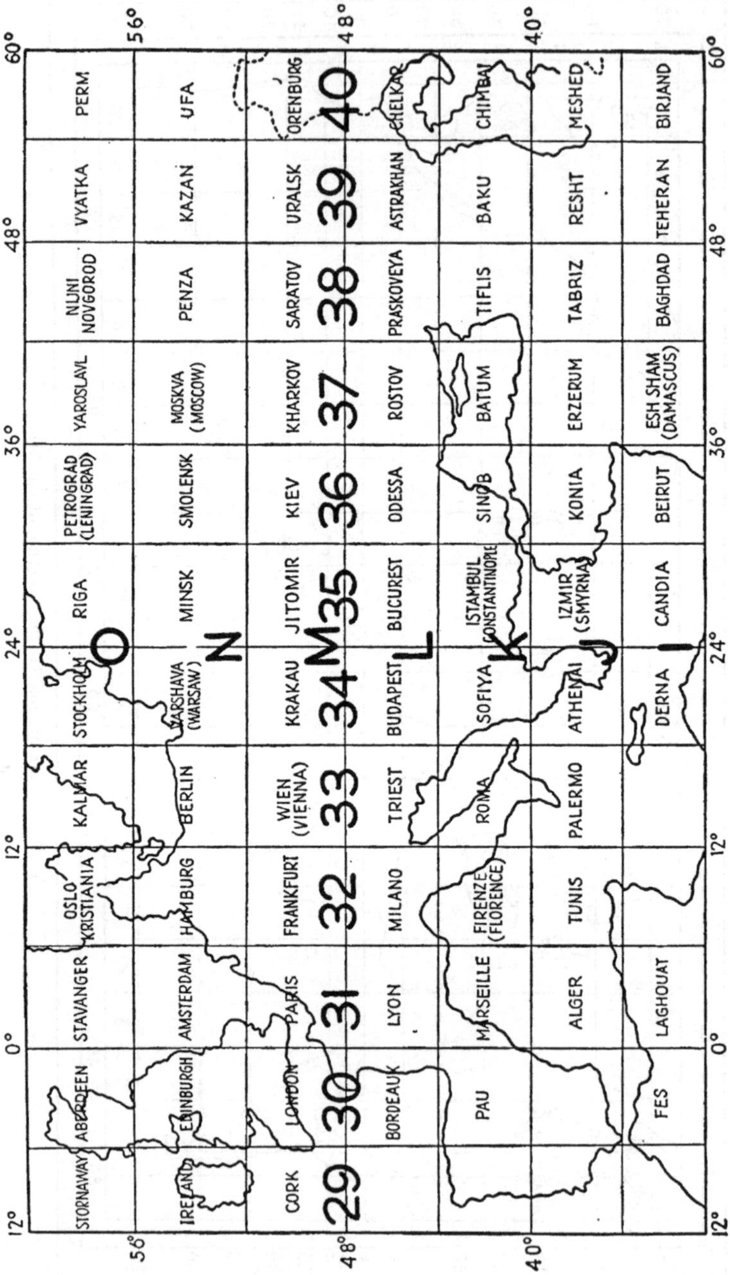

FIG. 10.—Index map of Europe for Local Aeronautical 1/M Map.

III. *General Aeronautical Chart:*
Scale: 1:3,000,000 at equator.
Mercator projection extending to 68°N. and S.
Index chart shown in Fig. 9.

IV. *Local Aeronautical Chart:*
Scale: 1:1,000,000.
Projection: modified polyconic (Lallemand).
Same as International Chart of the World with aeronautical features added.
Index chart for Europe shown in Fig. 10.

Skeleton Navigation Charts of the World.—The speed and range of modern planes have created a demand for special skeleton navigation charts to a relatively small scale. Such charts are not issued by either the Hydrographic Office or the U. S. Coast and Geodetic Survey. To meet this need the author has published the WSN series of five charts, scale 1:5,000,000 as shown in Fig. 4, which also shows the principles of the projections used. The following table shows the projections selected as best suited for the latitudes covered:

SCALE DISTORTION AND OTHER DATA ON WSN CHART SERIES

Chart	Latitude N. or S.	Projection used	Standard parallels	Latitude	Scale distortion, %
1	0°–30°	Mercator	15°	0°	− 3.4
				15	0.0
				25	+ 6.6
				30	+11.2
2	25°–55°	Lambert	33°	25°	+ 2.3
			45°	33	0.0
				39	− 0.5
				45	0.0
				55	+ 3.8
3	50°–80°	Lambert	55°	50°	+ 1.7
			75°	55	0.0
				65	− 1.5
				75	0.0
				80	+ 2.7
4	75°–90°	Stereographic	85°	75°	+ 1.5
				85	0.0
				90	− 0.2
5	30°–53°	Mercator	15°	30°	+11.2
				50	+20.6

These charts cover sufficient areas to plan transoceanic flights, and yet both courses and distances may be measured with sufficient accuracy

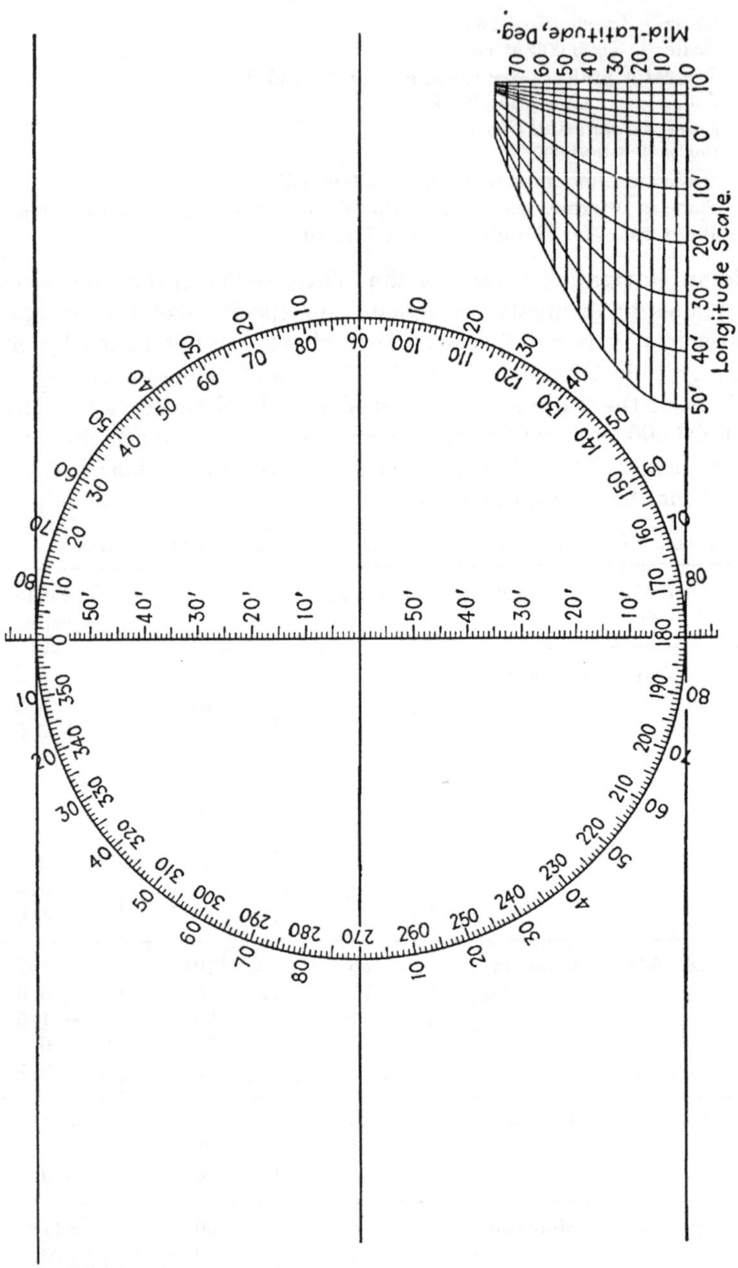

FIG. 11*a*.—Universal plotting chart.

TIME	POSITION		COMPASS	TRUE	DRIFT or			I.A.S. or				
	LAT.	LONG.	HEADING	HEADING	LEEWAY	TRACK	G/S	D.R. SP.				

Fig. 11*b*.—Form for entering navigational data and extension of universal plotting chart.

for practical purposes. The aircraft plotter, Fig. 23, may be used to measure distances on charts 2, 3, and 4 by adding a zero to the distances shown on the 1:500,000 scale. Chart 5 is an extension of chart 1 to cover continental United States on the Mercator projection.

Universal Plotting Chart (Fig. 11).—This is a partly constructed blank Mercator chart for small areas, with a fixed latitude and distance scale, a compass rose, and a longitude scale. The construction is based on the fact that the longitude scale is the cosine of the latitude scale, and assumes that the earth is a sphere, which, for practical purposes, is true. It is constructed on a scale, approximately 16 nautical miles to the inch, such that the scale of nautical miles is equal to the 1:1,000,000 miles scale of the Mark II plotter.

This chart may be used for any latitude and longitude, covers an area of 175 by 260 nautical miles, and is supplementary to the WSN chart series. To complete the skeleton chart, mark the angular distance of the mid-latitude on the compass rose above and below the mid-latitude parallel and draw a vertical line between the two points; this gives 1° of longitude at mid-latitude. Name the parallels and meridians to suit the problem. For minutes of longitude, measure along the mid-latitude line on the longitude scale in the lower right-hand corner of the sheet. Alternatively the plotter may be placed diagonally between two adjacent meridians with the 0 of the distance scale on one meridian and the 60 on the next meridian; the minutes of longitude can then be read off the plotter.

For 2° or 3° of latitude, there will be practically no difference between distances on the Universal plotting chart and distances on a Mercator chart. Where the latitude changes more than 2° or 3° it is only necessary to construct a new longitude scale for the latitude used.

Figure 11*a* shows the Universal plotting chart, which is used as the left-hand page in the "Navigation Note Book." Figure 11*b* shows the extension used as the right-hand pages, on which navigational data are recorded.

"Navigation Note Book" (Weems).—For keeping a record of dead reckoning, and for use for celestial navigation, a "Navigation Note Book" has been published, having on the left pages the Universal plotting sheet and on the right pages vertical columns for arranging the work. A pocket is provided on the front cover page to contain the Aircraft Plotter, Mark II.

Aircraft Plotting Sheets.—The Aircraft Plotting Sheets published by the Hydrographic Office are laid out on a Mercator projection to the scale of 1° of latitude equals 3 in. at the middle latitude of the sheet, resulting in a uniform scale of 1 in. equals 20 miles for the middle degree of latitude of all sheets.

Each degree is divided into divisions of 10′ each to assist the navigator in quickly estimating his approximate latitude and longitude without the use of dividers.

U. S. Aeronautical Charts.—Three principal series of aeronautical charts are now being published by the Coast and Geodetic Survey. The limits of these charts and their relative size and extent are shown in Fig. 12. The series are as follows:

Sectional charts, of the entire United States, in 87 sheets, at a scale of 1:500,000, or about 8 miles to the inch.

Regional charts, to cover the whole country, in 17 sheets, at a scale of 1:1,000,000, or about 16 miles to the inch.

Radio direction-finding charts, of the entire United States, in 6 sheets, at a scale of 1:2,000,000, or about 32 miles to the inch.

In addition to these, the following special charts are also available:

Aeronautical planning chart of the United States (chart 3060*b*), at a scale of 1:5,000,000, or about 80 miles to the inch.

Great-circle chart of the United States (chart 3074) at approximately the same scale as chart 3060*b*.

Magnetic chart of the United States (chart 3077), showing lines of equal magnetic variation, at a scale of approximately 1:7,500,000, or about 115 miles to the inch.

Kenai, and *St. Elias,* Alaska, at a scale of 1:1,000,000, or about 16 miles to the inch. These are the first two of a series of charts intended to cover the entire Territory of Alaska.

Because of the larger scale and the more complete information of the sectional charts they are necessary supplements to the regional series. They will always be required for detailed studies of an area and should generally be used whenever piloting is employed.

The regional charts are designed particularly for air navigation, as contrasted with piloting. They are more convenient than the sectional charts for comparatively long flights, with faster planes, since pilots do not need to change charts as often while in the air.

Figure 13 shows a portion of Chart 9MN. The symbols, colors, spacing, scales, etc., illustrate the text on the corresponding subjects. The navigator should develop the habit of making a careful study of charts to be used.

The radio direction-finding charts have been designed especially for use in the plotting of radio bearings. Their smaller scale and wider extent make it possible to plot bearings from radio stations that would frequently be outside the limits of the local chart when using either of the larger-scale series.

The aeronautical planning chart of the United States is very useful in planning routes between distant points, and for selecting the proper sectional or regional charts along the route.

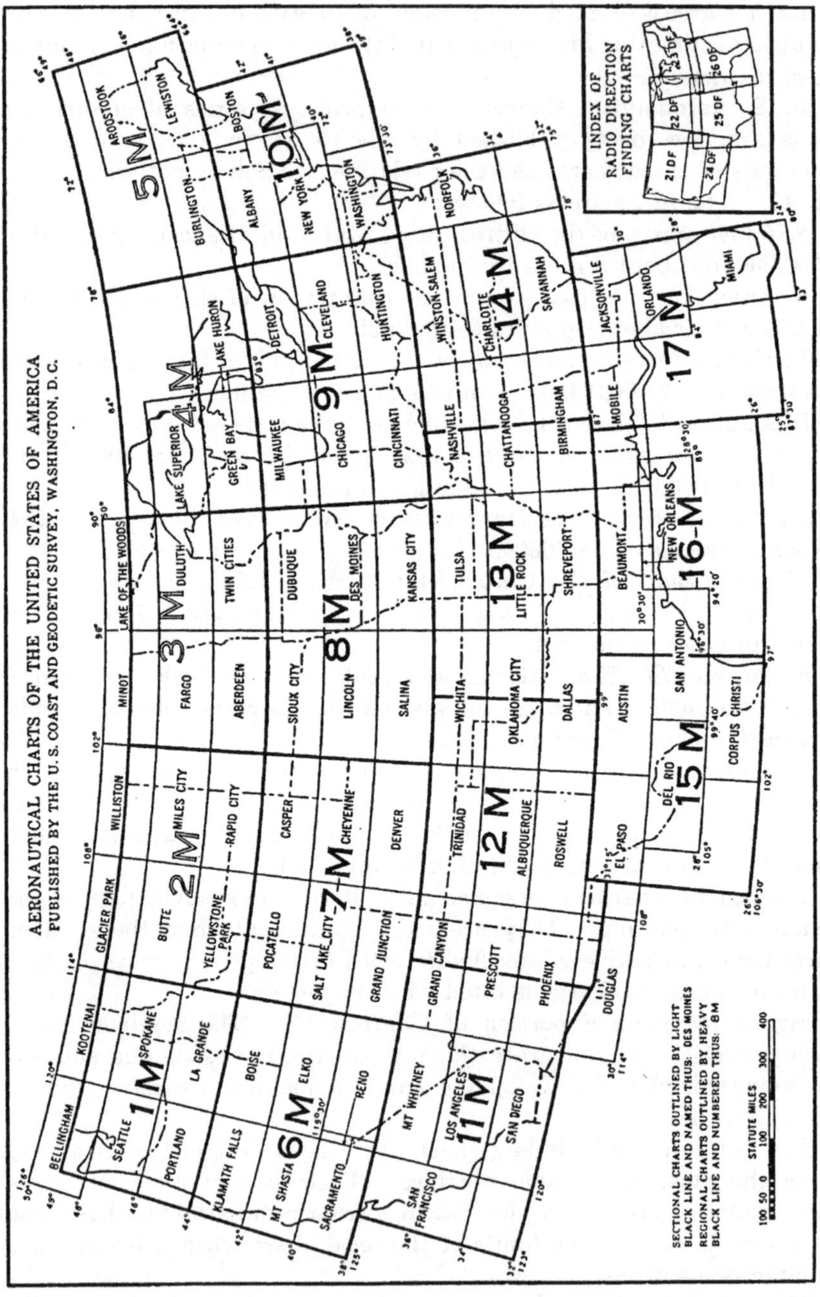

FIG. 12.—Index chart of W. S. sectional, regional, and radio direction-finding charts.

The great-circle chart is of value for one special purpose only, namely, the easy determination of the exact great-circle route between any two points. It cannot be used directly for the scaling of courses or distances. Its use is limited, then, to the most exacting record flights and to comparative studies.

The Coast and Geodetic Survey also publishes marine harbor charts, coast pilots, and other items of possible interest to the air navigator. See Appendix C.

Navy Strip Charts.—Navy strip charts cover principally the coastal areas of Alaska and Central America and are convenient for their special areas. They are constructed on the Mercator projection and conform more to the naval and marine type of chart.

Figure 14 shows the areas covered by the Navy strip charts.

In addition to strip charts, the Hydrographic Office publishes marine charts, great-circle charts, Notices to Aviators, pilot charts of the upper air, aircraft plotting sheets, and other special charts and publications of interest to the air navigator. See Appendix C.

Chart Reading and Handling Charts in the Air.—The ability to translate into useful knowledge the symbols and other features of a chart is of extreme importance to the navigator. There is not sufficient time in the air to study the characteristics of a chart; all this should be done before taking off.

In studying a chart, the first step is to read the legend, which is usually found in the lower right-hand corner and should show the title, area covered, the survey date, date of last correction, the scale used, and the meanings of the symbols used.

The area along the course to be flown should be scrutinized and particular stress should be placed on the location of possible emergency landing fields and the aids and dangers to navigation. The magnetic variation (discussed in a later chapter) and the changes in the variation along the course should be noted. The elevation of the terrain passed over should be studied with a view to choosing the best course to fly, and to make certain that the altitude along the course is not greater than the ceiling of the plane.

Chart Coordinates.—For ordinary navigation the coordinates of latitude and longitude, called the *graticule*, are preferred; for military use, however, a *grid* of rectangular coordinates is superposed.

The Chart Scale of Distance.—Every chart should have plainly marked on it a scale of distances. There are three distinct ways of showing a scale of distance on a chart. The chart scale may be given as:

1. A statement in words, as 1 in. equals 8 miles, meaning that a distance of 1 in. on the map represents an actual distance of 8 miles on the earth.

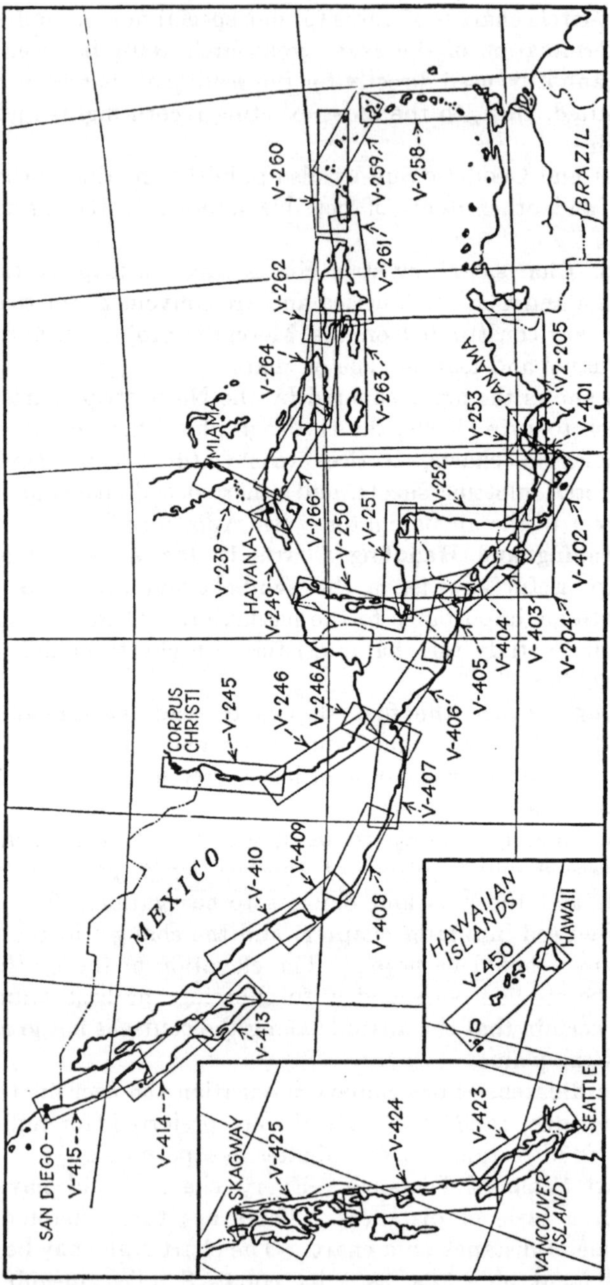

FIG. 14.—Index map of Hydrographic Office aviation strip charts.

2. A representation fraction, as 1/1,000,000, 1:1,000,000, or 1/M, which means that one unit of distance on the chart represents an actual distance of 1,000,000 of these units on the surface of the earth.

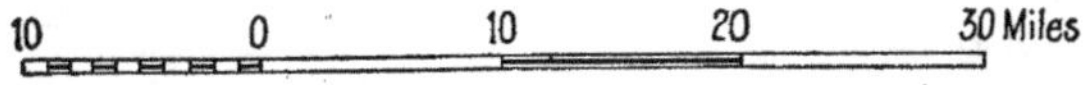

Fig. 15.—Graduated line for scale of distance on map.

3. A graduated line as shown in Fig. 15. The graduated line may and usually does show the scale of distances for the entire chart. However, on a Mercator chart, a graduated scale in nautical or geographical miles is given at the side, as shown in Fig. 16.

The margins of these charts are subdivided into minutes of latitude and minutes of longitude. These scales provide convenient means of plotting points when their geographic coordinates (latitude and longitude) are known, or of determining the geographic coordinates of points from their positions on the charts.

Chart Symbols.—The international chart symbols shown in Fig. 17 are those adopted by the Board of Survey and Maps, of the United States of America, corrected to Oct. 1, 1932. It will be found that the symbols for most charts will conform closely to this system of symbols.

Use of Colored Charts.—Most air-navigation charts are shown in several colors—green, brown, red, etc. The colors add to the clearness of the aids and dangers to navigation and make it easier to visualize the nature of the terrain.

Water Features.—Water features are represented on the aeronautical charts in blue, the smaller streams and canals by single blue lines, the larger streams and other bodies of water by blue tint within the solid blue lines outlining their extent (see Fig. 18).

Cultural Features.—Cultural features are generally indicated in black. Towns of less than 1,000 in population are indicated by a conventional black circle. Towns between 1,000 and 5,000 are shown by a yellow square outlined by purple; the actual shapes of larger cities are shown in yellow within a purple outline (see Fig. 18).

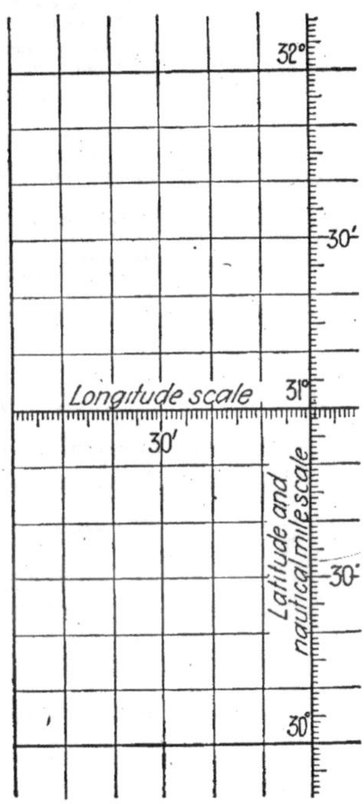

Fig. 16.—Section of a Mercator plotting sheet. Both the unit of latitude and the scale of nautical or geographical miles are shown on the meridian, or vertical scale. The scale of longitude corresponding to the latitude is shown on the horizontal scale. The distance between 31° and 32° is slightly greater than the distance between 30° and 31°, owing to expansion of the latitude.

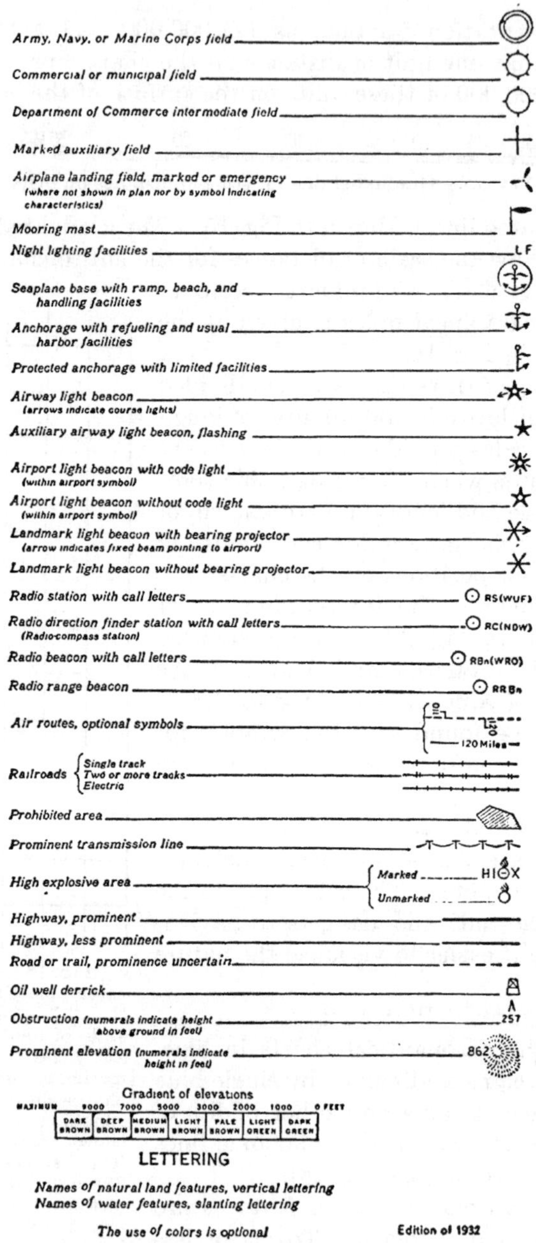

Fig. 17.—Standard symbols for maps.

Reliefs and Contours.—The symbols and standards used in the construction of charts establish easily remembered relations between the graphic representations on the charts and the physical features on the

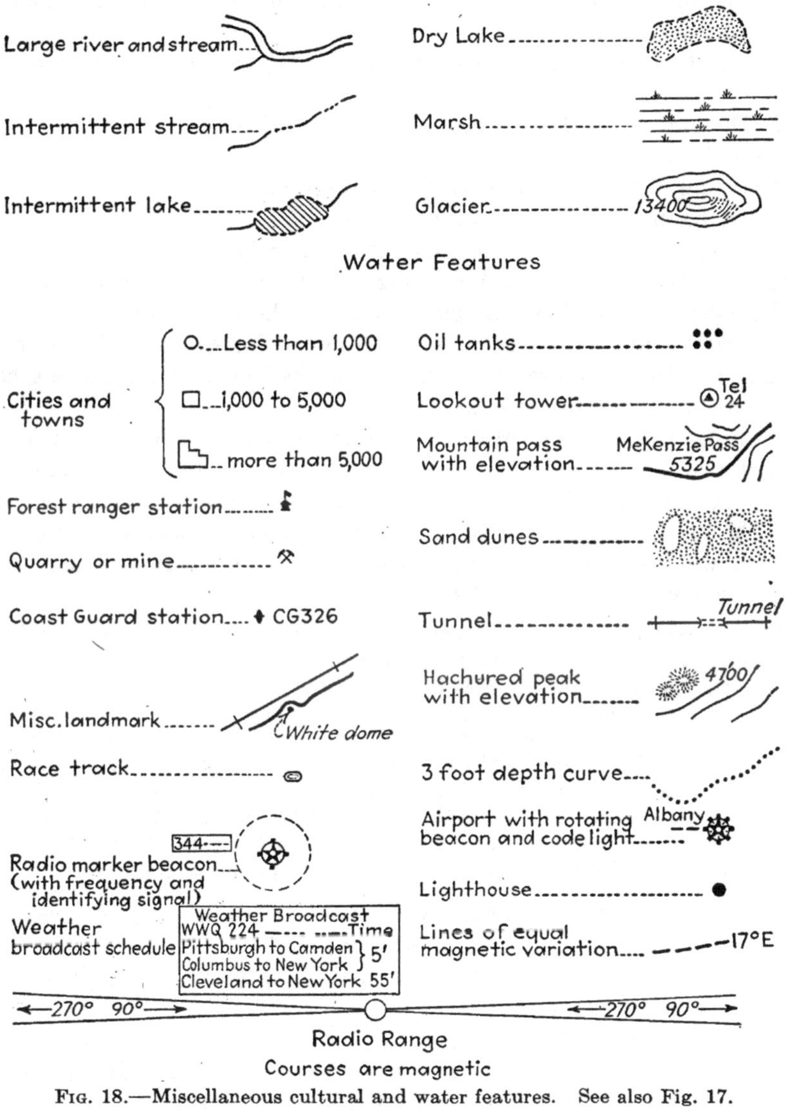

FIG. 18.—Miscellaneous cultural and water features. See also Fig. 17.

earth's surface which they represent. The features of the terrain are shown by hachures, which are lines or shading, to indicate hills or elevations and contours, which are lines representing equal altitudes. For instance, a cone-shaped hill would be indicated by a series of lines resembling concentric circles with shading between the lines. The steepness of the hill would be indicated by the closeness of the lines, and any irregularity in the shape of the hill would be indicated by a corresponding curvature in the contour lines. The different altitudes are indicated by one or more of three ways:

1. By the contour lines as explained above, the spacing showing the *gradients*.
2. By colors, or layer tinting, the dark brown representing high altitudes.
3. By numbers that indicate the altitude in feet.
4. By hachures or hill shading.

Marshy ground, trees, and many other special features may be represented by the elaborate symbols utilized by cartographers (chart makers). In the construction of air-navigation charts, the scale, aids and dangers to navigation, and other features are designed for the special use of the navigator. As many useful details as are possible without causing confusion are worked into the chart.

Plotting Courses on Mercator Charts.—The methodical, efficient navigator will ordinarily lay down the courses to be steered and will record such data as will assist in the navigation of the plane. In order to show the details of plotting courses and distances, a sample elementary problem will be worked out on a section of an aircraft plotting sheet, using true courses only (Fig. 19).

Example.—Plot the true courses and distances on the closed course over Puget Sound starting from Seattle and flying over Bremerton, Port Angeles, Victoria, Bellingham, and thence back to Seattle. Also tabulate the true course and distance for each leg of the route.

Solution: The cities on the route are first connected by straight lines. Each line or leg of the flight is then referred by means of the aircraft plotter or drafting machine to the compass rose, or to any meridian, to find the true course. This may be done by any of several methods. The use of the aircraft plotter (see Fig. 23) is a convenient way of doing it. Measure the length of each leg on the vertical latitude scale at approximately the middle latitude. Tabulate the results for each leg.

Only the true courses are given. The variation being 24° east for the locality, this amount subtracted from the true courses will give the magnetic courses. The subject of variation and deviation will be treated in the chapter on Compasses.

In describing in detail the operations for setting a course, we make it seem harder than it really is. It is important to have handy equipment and by practice to be able to do this quickly and accurately.

Use of Sectional and Regional Charts.—On the sectional and regional charts the meridians converge as shown in Fig. 20. Since the compass course cuts successive meridians at the same angle, it is obvious that a course measured at the starting point at the left of Fig. 20 would not be correct for the entire distance to the destination. However, the *average* course for the entire distance may be measured at the mid-meridian.

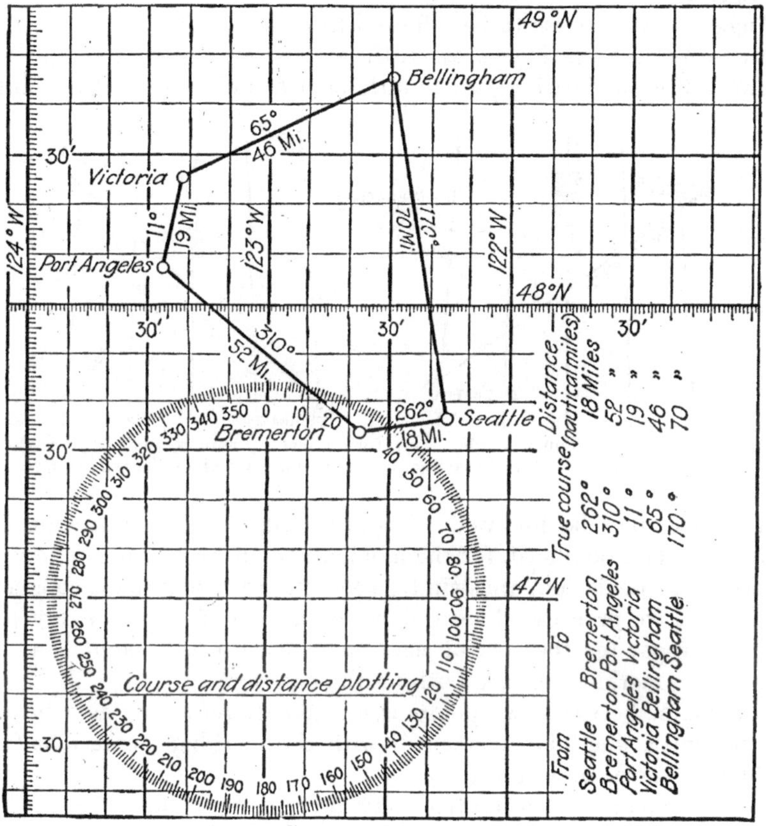

FIG. 19.—Course and distance plotting.

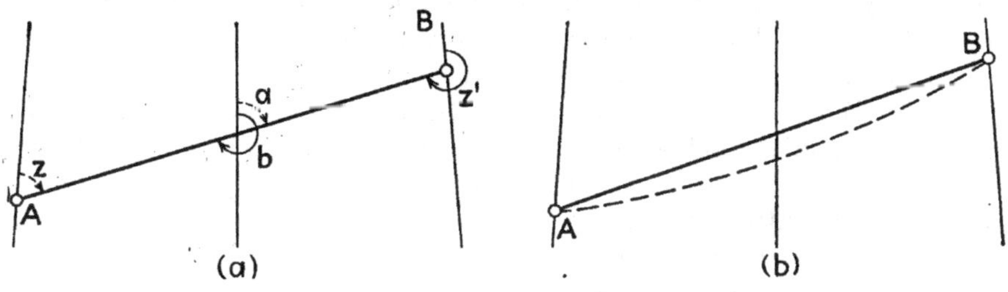

FIG. 20.—(*a*) Course and bearings; (*b*) course and track.

In order to understand clearly problems involving directions, it is important to distinguish carefully between a course and a bearing (or azimuth). Figure 20 illustrates the difference between these terms, as well as the methods of measuring courses and bearings on the Lambert projection, between any two points, A and B. Referring to the figure,

Angle a is the course to be followed from A to B.

Angle b is the course to be followed from B to A.

Angle Z is the bearing, or azimuth, of B as measured at the point A.

Angle Z' is the bearing, or azimuth, of A as measured at the point B.

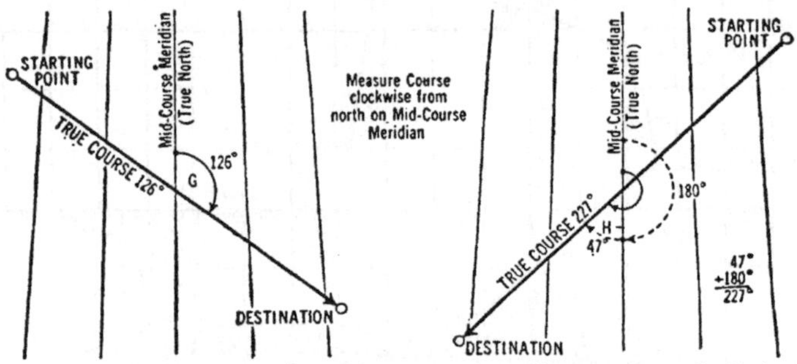

FIG. 21.—Measuring the course angle. *Left*, for an easterly course, angle G gives the true course. *Right*, for a westerly course, angle H plus 180° gives the true course.

A course may be followed without change for the entire distance between the two points (if, for the moment, we disregard magnetic variation, compass deviation, and wind); a bearing (or azimuth) is constantly changing as we progress along the route and is different at every point thereon (except for the special cases in which the two points are both on the same meridian, or are both on the equator).

If a course between two points is mistakenly measured as a bearing, with the initial meridian instead of with the meridian nearest halfway, considerable error may result.

To clear up any confusion that may yet remain, it should be explained that when the course is measured with the meridian nearest halfway (as the angle a, Fig. 20a), a plane following that course will not exactly follow the straight line AB on the chart, but will slightly depart therefrom near the middle of the route, as indicated by the light broken line (greatly exaggerated) in Fig. 20b. However, when courses are measured as recommended in the following paragraphs, the departure is so slight that it may be considered that the plane does exactly track the straight line throughout its entire length.

A course measured with the true geographic meridian printed on the chart is the true course.

When the two points are separated by not more than 3° or 4° of longitude, the true course may be measured with the meridian nearest halfway, as described above and as illustrated in Fig. 21, and the entire distance flown as one course.

When the difference of longitude between the two points is more than 3° or 4°, the straight line on the chart should be divided into sections crossing approximately 2° of longitude each, and the true course to be flown for each section should be measured with the middle meridian of that section.

For example, Fig. 22 illustrates the method of determining the series of true courses to be flown between St. Louis and Minot. The distance is 862.7 miles, and the difference of longitude is nearly 12°, which is too great to be flown satisfactorily in one course. The route is therefore divided into five sections crossing approximately 2° of longitude each, the two end sections being slightly longer than the others. The true course to be flown throughout the total length of each section should be measured with the middle meridian of that section, and the course should be changed in flight as the end of each succeeding section is reached.

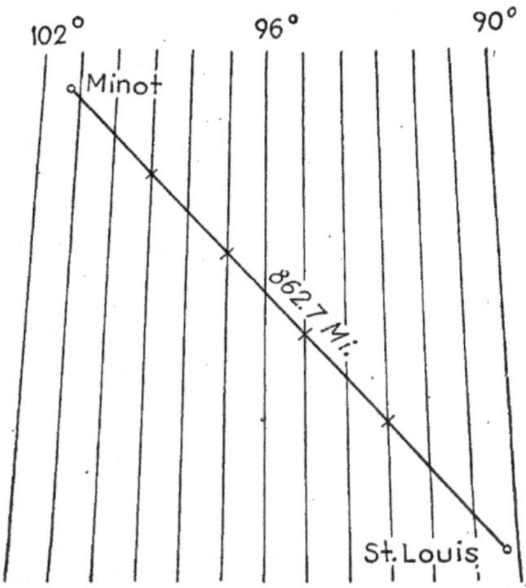

FIG. 22.—Subdividing a long route.

On the Lambert projection, for all practical purposes, a straight line is the great-circle route (shortest possible distance) between its extremities. The method just outlined makes it possible to fly the great-circle route by a series of short courses (rhumb lines).

Aircraft Plotter, Mark II (Department of Commerce Type).—This type of plotter is designed especially for use with the Department of Commerce aeronautical charts and replaces the protractor, dividers, and parallel rulers (see Fig. 23).

To Measure a Course.—*a.* Place the distance scale along the course line with the center over the intersection of the course line and the mid-course meridian. A pencil point held on the course line helps to do this.

b. Read easterly courses on the outer semicircle, and westerly courses on the inner semicircle. In other words, read the course on the semicircular scale adjacent to the arrow that points in the direction of flight.

To Lay a Course.—*a.* Mark the north point with reference to the starting point. This may be done as follows: (1) by orienting the plotter at right angles to a latitude line, (2) by orienting the plotter parallel to the nearest meridian, or (3) by orienting the plotter with the 90° line along a parallel and with the distance scale line through the starting point.

b. Place the center of the plotter over the starting point with the semicircle to the left and with the distance scale to the north.

c. Rotate the plotter to the right, with a pencil point as a pivot, until the desired easterly (westerly) course is indicated on the outer (inner) semicircle.

d. Draw the course line along the distance scale.

To Measure Distance.—*a.* Note the scale of the chart and choose the proper scale on the plotter.

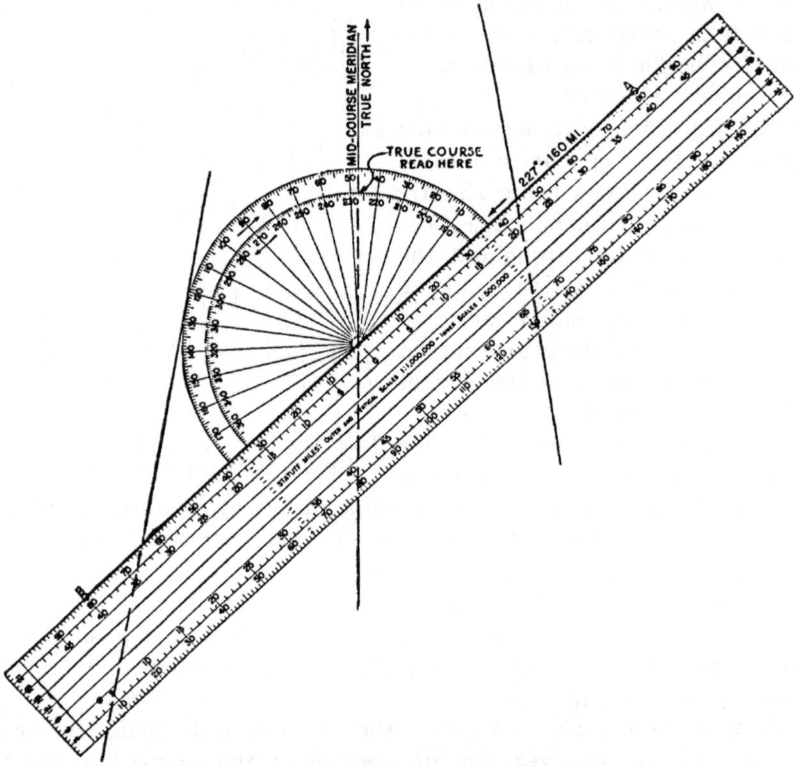

FIG. 23.—Aircraft Plotter, Mark II (Department of Commerce type).

b. Place the straightedge along the course line with 0 over the starting point. Or, read the distances from the mid-meridian to the starting point and to the destination and take the sum.

c. Read the distance in miles directly from the proper scale.

To Parallel Any Line.—*a.* Align the parallel lines of the plotter with the original line and at the desired distance from it.

b. Mark a line along the straightedge of the plotter.

c. Or, place the center and the 90° point of the plotter along the original line and then make dots on the scales of both edges of the plotter and at the desired distance from the original line and draw the required parallel line through the two dots.

The plotter is used in the same way with Mercator charts, or with Universal plotting charts, except that the courses may be measured from any meridian, and the distance shown by the plotter is referred to the latitude scale at the mid-latitude to get the correct distance in nautical miles.

Bearings from or to any point, wind problems, and, in general, all chart problems may be accomplished with this type of plotter, by following the same procedure as outlined above for each step; the one exception is that a bearing should always be measured at the meridian passing through the place at which it is determined, instead of the mid-course meridian.

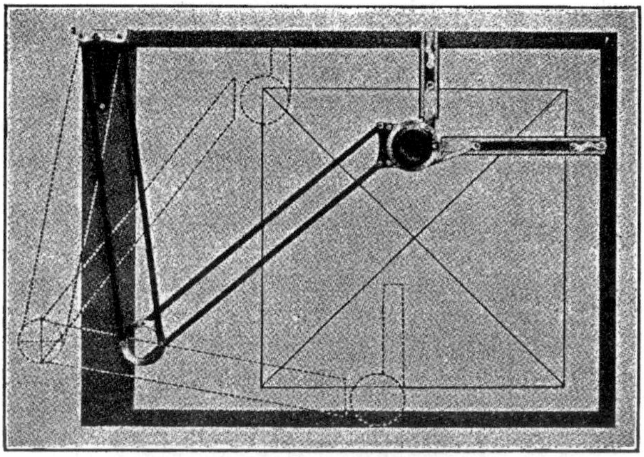

FIG. 24.—Universal drafting machine.

Other types of plotters are the Warner plotter, which is similar to the Mark II plotter except that it has a spinner for the protractor; and the WSN course protractor with a large spinner pivoted to a long arm. The latter is especially convenient for accurate work over long distances and for use with marine charts.

Universal Drafting Machine.—This machine offers a ready means for plotting courses and distances on a chart laid out on a board.

It consists of two parallelograms, a protractor, and a square having graduated ruling edges as shown in Fig. 24. The two parallelograms joined together constitute an arm which, anchored to the board, gives the protractor and square an accurate parallel motion about the drawing. This form of parallel motion permits either zero on the ruling edges to be instantly placed at any point on the drawing by a single direct movement, owing to the fact that the arm is similar to the human arm, and the action is just as free and direct as when the hand is moved to any position.

The Universal drafting machine is probably the most efficient means of plotting courses and distances on a chart, especially on Mercator charts.

Folding Charts.—Charts should be folded once, back to back, along the line AB (Fig. 25), then in four or six "accordion folds" in the other direction, along the vertical broken lines indicated in the figure. In

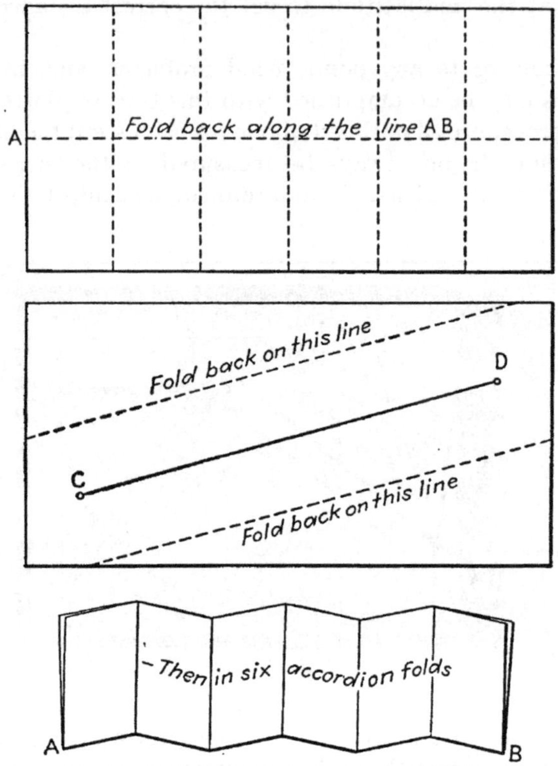

Fig. 25.—Folding the map for use in flight.

this way the entire chart may be consulted merely by turning over the accordion folds. If it is desired to make a strip chart covering a certain route (as the route CD), fold the chart so as to leave the route in the center of a strip 10 or 12 in. wide; then fold the strip in the accordion fold illustrated. It should be so arranged that necessary information, such as longitude, latitude, and distance scales, is readily accessible.

CHAPTER III

COMPASSES

Each of the methods of navigation requires its own particular set of instruments which are essential to secure proper results with that method, but *all methods* require a proper chart and a *reliable compass.* Aircraft compasses may be of the magnetic, the earth-inductor, the sun, or the gyroscopic types. Most aircraft compasses are of the magnetic type because it is the simplest, cheapest, and easiest to keep in working order.

The magnetic compass is an instrument by means of which the directive force of that great magnet, the *earth,* upon a freely suspended needle is used to determine direction upon the surface of the earth. The Chinese, the Arabs, and the Greeks are all supposed to have used the magnetic compass as early as the thirteenth century. The modern mariner's compass is very accurate and quite reliable. Because of the limitations of weight and space, and the vibrations and accelerations to which it is subjected in service, the aircraft magnetic compass is less accurate than the standard magnetic compass found on a modern ship. But throughout the ages and in spite of numerous improvements, the fundamental principle has remained the same.

The Molecular Theory of Magnetism.—A great many substances have magnetic properties, for example, iron, nickel, cobalt, aluminum, water, and bismuth. The first three are vastly more magnetic than any other known substances. For an understanding of the magnetic compass it is necessary to consider briefly the magnetic properties of iron and steel only.

If a bar magnet is dipped into a pile of iron filings, it will be found that they adhere more strongly at or near the ends of the magnet, which are called *poles.* A magnet suspended horizontally and free to turn about a vertical axis will always take up a position with one end pointing to magnetic north. (Hereafter in this chapter, when we speak of *north pole, northern hemisphere,* etc., we refer to the *north magnetic pole, north magnetic hemisphere,* etc.) This pole is called the north-seeking or *red pole* and the other pole is called the south-seeking or *blue pole.* The law always holds good that red poles repel each other and blue poles repel each other, and that a red and a blue pole attract each other. This is usually expressed by stating that (1) like poles repel each other, and (2) unlike poles attract each other.

The molecular theory of magnetism is that every substance capable of being magnetized consists of smaller parts, which are themselves magnets. These small magnets are termed molecular magnets; they

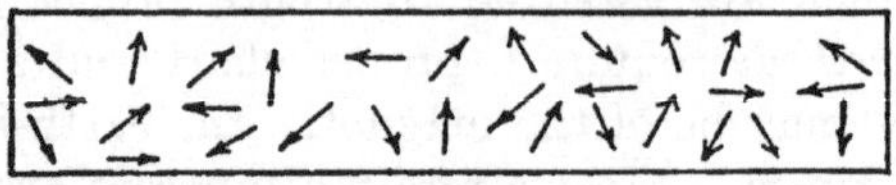

FIG. 26.—An iron bar before being magnetized.

should not be confused with the term molecule as used in chemistry. Before the substance is magnetized these minute magnets are arranged indiscriminately in all directions as shown in Fig. 26. The arrows represent molecular magnets and the arrowheads are red poles.

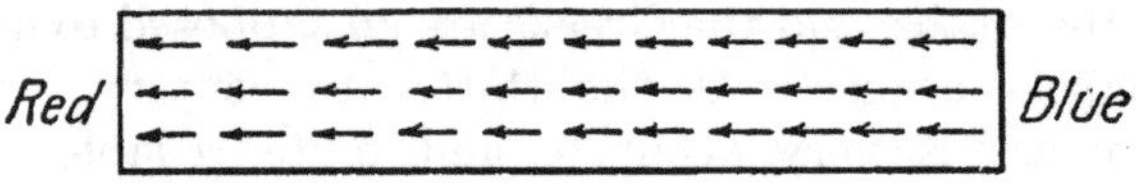

FIG. 27.—An iron bar after being magnetized.

The effect of magnetizing a bar is to arrange all these small magnets in the same direction. Figure 27 shows the bar after being magnetized. At one end there are free red poles, which account for the red pole of the magnet, and at the other end there are free blue poles, which account for the blue pole of the magnet.

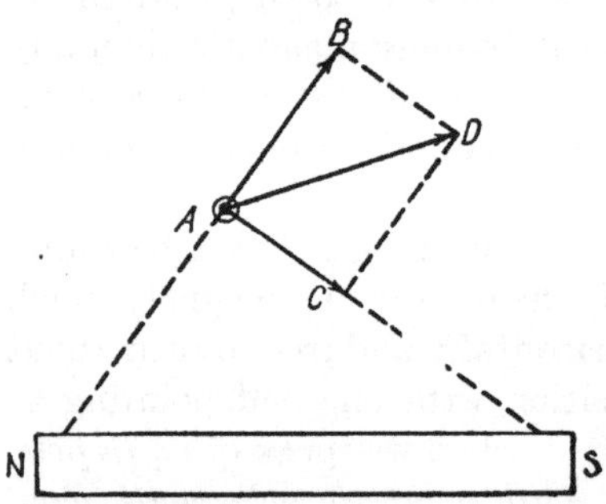

FIG. 28.—Forces in a magnetic field.

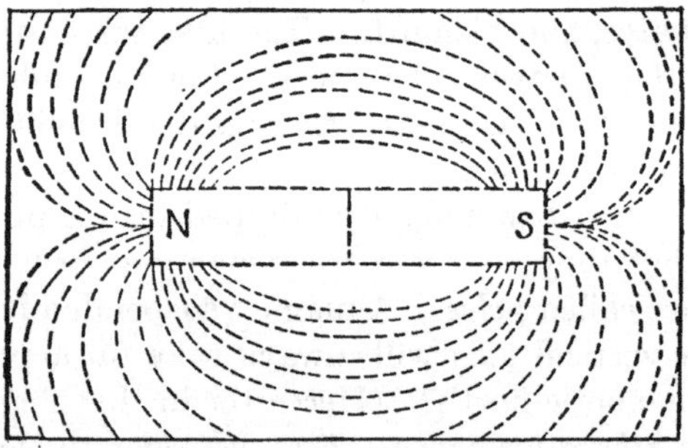

FIG. 29.—Magnetic lines of force.

Magnetic Lines of Force.—The space surrounding a magnet throughout which its influence extends is called its *magnetic field*. Imagine that a free red pole is situated at the point A in the field of the bar magnet NS as shown in Fig. 28. Then the red pole would be subjected to a force of repulsion AB from the red pole N and a force of attraction AC toward the blue pole S. The resultant AD of these two forces would indicate the direction of the magnetic field at the point A which is the direction in which the free pole would begin to move. The force AB is greater than AC because the isolated pole is nearer to pole N than to pole S. As soon, however, as the free pole has started to move, the distances NA and SA

change and the forces due to N and S would likewise change. By plotting a number of positions of the free pole A, it will be found that they lie on a curve running from N to S. This curve indicates the direction of the *magnetic line of force*, which is defined as a line, the direction of which at each point is the direction of the magnetic field at that point.

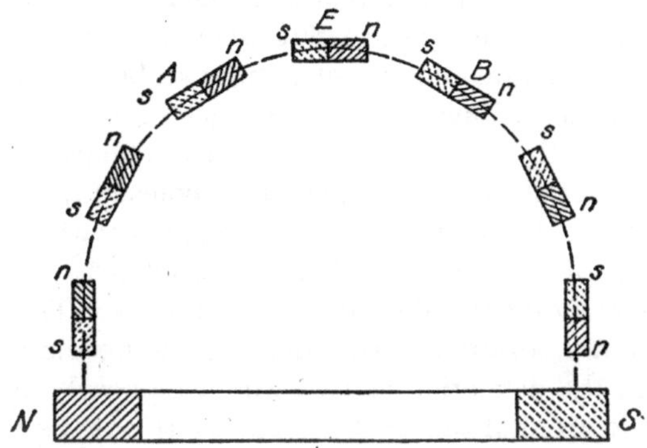

FIG. 30.—Action of bar magnet on small magnet.

Magnetic lines of force never meet and never cross each other. Their directions may be indicated by placing a piece of drawing paper over a bar magnet and sprinkling iron filings over the paper. If the paper is tapped, the filings will arrange themselves in fairly definite lines as shown in Fig. 29.

Suppose a small freely suspended magnet *ns* were placed in the field of the bar magnet *NS*. The small magnet would take up the various positions as indicated in Fig. 30.

Induction in Hard and Soft Iron.—When a piece of unmagnetized iron or steel is brought within the influence of a magnet, certain magnetic

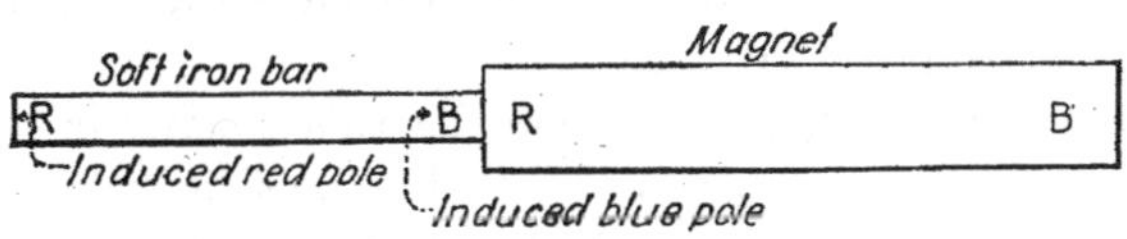

FIG. 31.—Polarity of induced magnetism.

properties are at once imparted to the iron or steel, which itself becomes magnetic and continues to remain so as long as it is within the field of the permanent magnet. The magnetism acquired in this way is said to be induced, and the properties of induction are such that the end or region which is nearest the pole of the influencing magnet will take up a polarity opposite to that of the influencing pole as shown in Fig. 31. If the magnet is withdrawn, the induced magnetism is soon dissipated. If the magnet is again brought near the iron or steel, but with the opposite pole nearer, magnetism will again be induced but this time the polarity will be reversed. If a piece of magnetically hard iron or steel, while temporarily magnetized through induction, is subjected to repeated blows, twisting, or mechanical work of any sort, the induced magnetism is thus made permanent. Soft iron is easily magnetized and readily demagnetized when the magnetizing influence is withdrawn. Hard iron is more difficult to magnetize but when the magnetizing influence is withdrawn it retains magnetism of a more stable nature.

The Inverse Square Law.—The force exerted between two magnetic poles of unit strength varies inversely as the square of their distance apart. It will generally be found that both poles of a magnet are acting and when this is true the total force of the magnet varies nearly inversely as the cube of the distance apart. The rule applies to the correction of a magnetic compass by means of permanent magnets. Thus if a correcting magnet 4 in. from a compass needle deflects the needle through 5°, at a distance of 2 in., the deflection caused would be nearly 40°.

The Magnetic Field of the Earth.—The earth acts as a great spherical magnet having the characteristic properties of a magnet as described in preceding paragraphs. Many theories have been advanced to account for this magnetism of the earth but each in turn has been discarded as wider knowledge indicated it to be untrue. However, the earth's magnetism has been under observation for more than 300 years and, although there is no acceptable theory to account for it, there is available a great fund of information about it.

The earth's magnetism at any place is measured by determining the direction and intensity of the field at that place. Many years of observation have shown that both the direction and intensity change from time to time. These changes in the intensity of the earth's field are so small that they need not be considered in navigation. The changes in direction, however, must be carefully considered as will be explained later in the discussion of Variation.

The Earth's Magnetic Poles.—The earth's magnetic poles do not coincide with the geographical poles. One is situated in Baffinland, *i.e.*, in about Lat. 73°N. and Long. 96°W., and the other in Victoria Land, in Lat. 72°S. and Long. 155°E. It should be clearly understood that the north-seeking end of a magnetic compass needle points to this north magnetic pole and not to the true geographical north. These magnetic poles have not the same properties as the poles of a bar magnet for if they did there would be an enormous increase in the intensity of the earth's magnetic field in approaching them. It is definitely known that this is not so.

Direction of the Earth's Magnetic Field.—A small bar magnet suspended so that it may turn freely about its center of gravity will

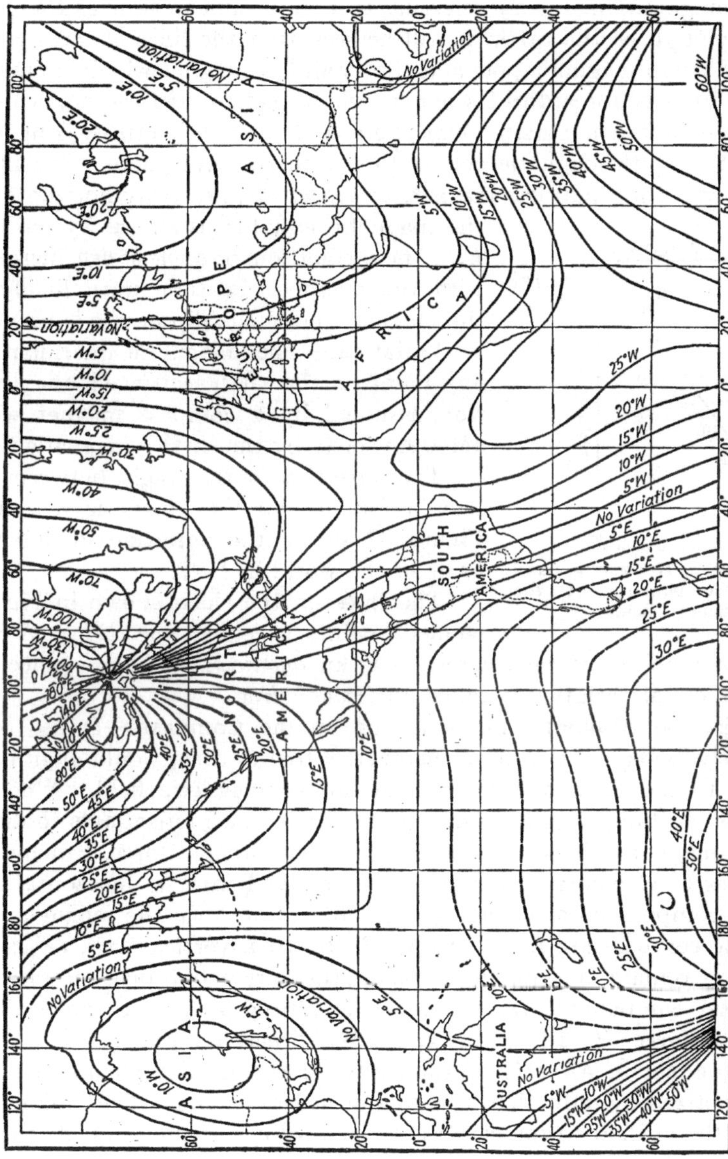

FIG. 32.—Lines of equal magnetic declination (variation) for 1942.

take a position with its magnetic axis parallel to the lines of force of the earth's magnetic field. By referring to Fig. 30 and imagining the large bar magnet to be the earth, it will be seen that the lines of force curve in to the north magnetic pole and the small bar magnet which is parallel to these lines of force dips its north-seeking end toward the north magnetic pole in the northern hemisphere and dips its south-seeking end toward the south magnetic pole in the southern hemisphere. It will be evident why the small bar magnet, or *dip needle*, lies truly horizontal at the magnetic equator as shown by position *E*. The inclination, or *angle of dip*, increases with the magnetic latitude until finally at the magnetic poles the dip needle is vertical.

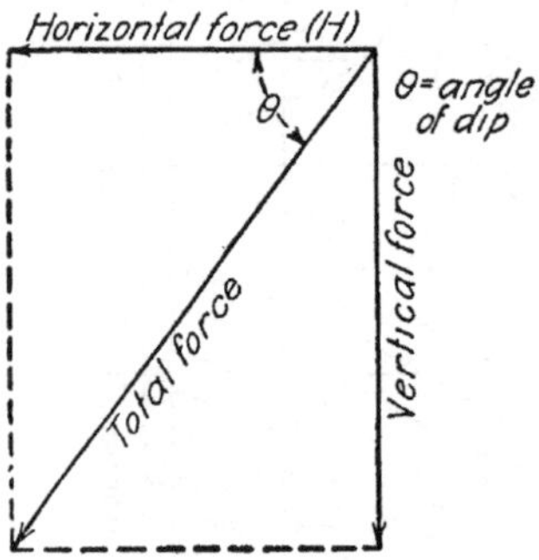

FIG. 33.—Horizontal and vertical components of the earth's magnetic field.

Variation.—The magnetic meridian at any point on the earth's surface is the direction assumed by a compass magnet when acted on solely by the earth's magnetic field.

The *variation* is the angle between the true meridian and the magnetic meridian. Figure 32 shows the variation in different parts of the world in 1942.

The lines on the chart shown in Fig. 32 connect points of equal variation. A line connecting points of zero variation is an *agonic line;* a line connecting points of equal variation is called an *isogonic line.*

Horizontal Magnetic Force Acting on the Compass.—It has been previously explained that a bar magnet, if free to turn, would align itself parallel to the lines of force of the earth's magnetic field. When it is in this position, the total force of the earth's magnetism acts to direct the bar magnet. This force, like any other directed force, can by the principles of elementary mechanics be resolved into its horizontal and vertical components as shown in Fig. 33. Now since the magnets of a magnetic compass are free to turn only about a vertical axis, *i.e.*, since they must remain in a horizontal plane, they are acted upon only by the horizontal component, *H*, of the earth's total magnetic force. This horizontal force, *H*, is therefore called the directive force.

Magnetism in Aircraft.—Practically all the iron in an aircraft will be magnetized by induction to a greater or less extent. The bulk of the steel in the ordinary type of plane is forward of the pilot and it is there that most of the magnetism will be found. If an airplane were built on a heading of (*magnetic*) north, a fore-and-aft bar of hard iron would be magnetized so that its forward end would be a red pole. If built on a heading of south, the bar would be magnetized so that it would have a blue pole forward. If the plane were built on an east-and-west heading, the bar would have practically no magnetism, and what little magnetism was induced would be red on the north side and blue on the south side of the bar. Thus, if the position of the plane in building is known, it is possible to know in a general way where its red and blue poles will be.

The conditions in an airplane cockpit are unfavorable for a magnetic compass. The compass should therefore be mounted with great care and checked at frequent intervals.

Description of the Magnetic Compass.—The purpose of the compass is to show the heading and to measure bearings. The pilot's compass usually indicates only the heading; the navigator's compass not only shows the heading but is also adapted to measure bearings.

The essentials of a modern aircraft compass are:

1. Magnetic needles that align themselves with the earth's magnetic field.

2. These needles are attached or referred to a *card* on which are marked the directions N., E., S., W., from 0° to 360°.

3. The card (in card-type compasses) and needles are supported on a pivot by a sharp point at the center of the card. The card thus turns on this pivot under the action of the needles.

4. The card, needles, and pivot are contained in a liquid that partly floats them, thus reducing the weight and friction on the pivot.

5. The whole of the above assembly is contained in a bowl that has a glass window permitting the card to be seen. The bowl is connected directly to the airplane and the pivot is connected directly to the bowl.

6. In the window, and secured to the bowl, is a *lubber's line* or square mark, which is set accurately in the fore-and-aft line of the plane. The heading of the aircraft by compass is read by noting the marking on the card, which is under or behind this lubber's line.

7. A *compensating* or correcting device.

In addition to the above essentials practically all compasses of standard make have the following:

8. A float chamber, on types other than the aperiodic. This is an empty capsule attached to the card, giving the card assembly (card, needles, and pivot) almost but not quite enough buoyancy to float it.

9. An expansion chamber, the function of which is to prevent the expansion and contraction of the liquid from either cracking the bowl or permitting bubbles to form as the temperature changes.

10. A light for reading the course at night.

11. An antivibration mounting to reduce the effect of the airplane's vibration.

The liquid in the compass has another extremely important function, which is to "damp" the oscillations of the magnetic element. If there were no liquid and the magnetic element for any reason rotated to one side, it would oscillate back and forth a great many times before coming to rest. The damping action of the liquid brings the needles to rest quickly.

Magnetic Needles.—The magnetic elements of the compass which supply the directive force are small needles or rods, about six in number. They are made of alloy steel, specially hardened.

Marking of the Card.—Airplane magnetic compasses of the card type are usually mounted forward of the pilot and near the level of the

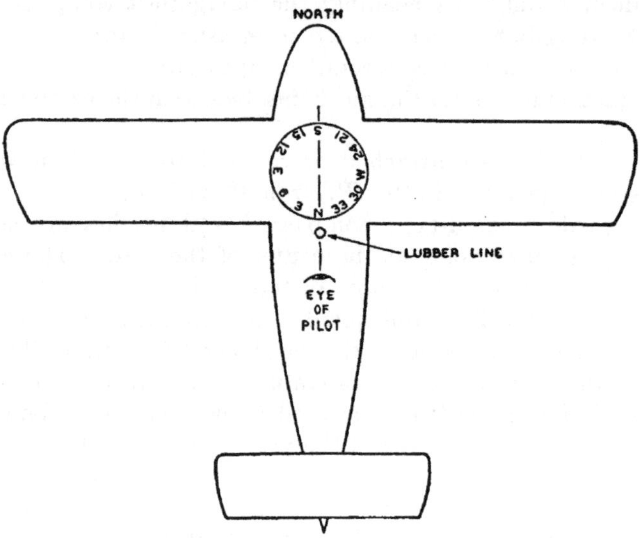

FIG. 34.—Typical compass mounted in plane headed north.

eye; for this reason the scale graduations are made on the vertical circumference of the card as well as on top of it. For this reason, also, there is usually a back lubber's line 180° from the front one, and the scale graduations are marked with the 0° aft and the 180° forward so that the heading of the aircraft can be read directly from the back lubber's line.

Direction is measured from 0° at the north to the right or clockwise through 360°. Steering compass cards are usually marked with the numbers every 30°. To save space, the final zero is omitted from each heading so that 30° is marked "3," 120° is marked "12," 270° is marked "27," and so on. Intermediate marks are placed every 5° so that an experienced pilot should be able to steer a course within 2°. It is desirable for these scale markings to be coated with luminous paint so that they can be clearly seen at night in case of failure of the illuminating device. Figure 34 shows a typical marked compass mounted in a plane headed north. Figure 35 shows the markings of a standard compass card as seen from behind by the pilot. The best modern compass cards are marked for each degree and numbered every 10°.

As previously explained magnetic dip depresses the north-seeking pole of a magnet in the northern hemisphere. For this reason a small weight is usually attached to the southern side of the card (in north latitude) to hold it horizontal.

Pivot and Cup.—At the center of the rotating system is a pivot of a hard material such as agate or iridium. This pivot rests in a socket or

FIG. 35.—Magnetic compass Mark IX. (*U. S. Navy, official photograph.*)

cup of an even harder material such as sapphire, so that the pivot will not wear it into irregularity. The pivot and cup are a most important part of the compass. They must stand routine wear and also withstand violent shocks.

Liquid.—The liquid reduces the weight of the rotating system on the pivot by partly floating it, lessens the friction, diminishes the harmful effect of shocks and vibration, and, most important of all, damps the oscillations of the rotating members. Extremes of temperature make it difficult to select a suitable liquid. Specifications for a modern aircraft compass provide for a range of temperature from −50°C. to +50°C. Pure alcohol is the most suitable liquid and will not freeze at the low temperatures. Unfortunately, it has been very difficult to find a paint or other coating that is not gradually dissolved by the alcohol. When this dissolution of the paint takes place, the liquid becomes discolored.

For this reason the pure alcohol is usually diluted with distilled water. Pure white kerosene that is free from acid is now being frequently used in place of the alcohol and water mixture.

The Bowl.—The compass bowl is usually cylindrical or spherical in shape. Some clearance is allowed between the inside of the bowl and the edge of the card so as to reduce to a minimum the errors due to swirling of the liquid. The bowl is directly connected to the airplane. Its primary purpose is to hold the damping fluid and support the bearing in which the pivot rests.

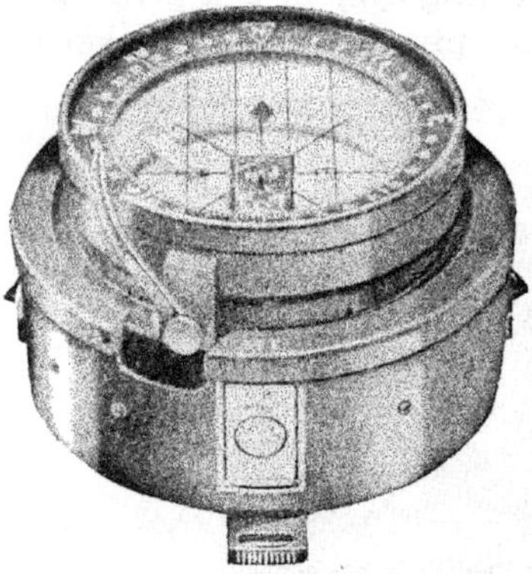

Fig. 36*a*.—Late Type Campbell-Bennett aperiodic compass, Mark IIIA. (*Courtesy Henry Hughes and Sons, Ltd.*)

The Lubber's Line.—A vertical line called the lubber's line is marked on the inner surface of the bowl and the compass is so mounted that a line joining its pivot with the lubber's line is parallel to the fore-and-aft line of the plane. Thus the lubber's line represents the head of the airplane and the scale marking on the compass card directly behind the lubber's line is the compass direction in which the aircraft is heading. The compass is generally mounted in front of the pilot and about on a level with his eye. When such is the case, a second lubber's line is marked on the after side of the compass bowl and the card is so marked that the course or heading of the aircraft is read from this second or after lubber's line.

Compensating or Correcting Device.—It has been previously explained that a considerable magnetic field exists in the cockpit of an airplane because of the magnetic metals used in construction. This is further modified because of the magnetic fields set up by the operation of the generator and magneto. These undesirable magnetic fields are compensated by small bar magnets, one set of which is placed in a fore-and-aft position and the other athwartships in a suitable carrier attached to the bottom of the compass. There are several patented types of compensating devices, such as the *microadjuster*, for conveniently correcting the compass.

Types of Compasses.—Typical aircraft compasses are shown in Figs. 35 to 37. Figure 36*b* shows a type of compass for observing bearings, which is compact and serviceable and has proved very popular. Pilot compasses are used primarily for steering; observers' compasses are equipped for taking bearings.

Aperiodic Compass.—A type of magnetic compass worthy of special note is the aperiodic compass shown in Fig. 36*a*.

The *aperiodic magnetic compass* is a compass without a "period"; that is to say, it returns, after being displaced from its equilibrium position, by one direct movement to the north-pointing position, instead of executing a series of oscillations.

The aperiodic compass has no card. The azimuth degree marks are shown on a rotatable grid ring that carries a set of parallel grid lines running in the north and south direction. The pilot steers his course by keeping the grid lines parallel to the long north and south pointers of the needle system. The grid ring is previously set for the desired course, turning it till the degree mark of the course comes against the forward lubber line and then locking it.

FIG. 36b.—0.6 aperiodic observer's compass.

Owing to this method of reading, the compass may be placed in any convenient position in front or on one side of the pilot, and he has only to set his course by the grid ring and then turn his airplane till the red *NS* wire appears parallel to any one of the four parallel wires. The pilot is thus relieved of any need to remember the figures of the course he is steering, and his head is free to occupy any position without causing any parallax error in the reading.

This compass was specially designed for aircraft work and to fulfill the following conditions:

1. Minimum weight of moving part (under 3 grams).

2. The magnetic element shall quickly take up its correct position, but shall not be liable to oscillation; it shall be aperiodic.

3. The magnetic element shall not set up any current in the liquid which would cause it to give a false reading.

This has been done by using very small but strong magnets, combined with eight fine radial damping wires. If the card is deflected, these wires, as they move through the liquid, do not drag the liquid with them or set up eddies. In the case of the ordinary compass card, the liquid is dragged round by skin resistance and a current produced in the liquid, which is almost entirely absent in this compass.

The eddy resistance is very great and the card does not appreciably overswing even when using such large deflections as 90°.

Vertical Dial Steering Compasses.—It is inconvenient to steer a plane by a horizontal reading compass. To overcome this, readings may be made from vertical cards as shown in Figs. 34 and 35. A new development is the vertical dial magnetic compass with *stationary dial*, *indicating pointer*, and *reference index*, shown in Fig. 37. The reference index may be set to any desired heading by turning the knob at the bottom of the dial. Compass observation over a long period of time is made simple by setting the index to the desired heading and steering the airplane so that the pointer remains parallel to the reference lines. Eyestrain is greatly reduced, since individual course numbers are neglected after the reference lines are set.

FIG. 37.—Vertical-dial compass.

Notes on Care and Maintenance.—Magnetic compasses are very delicate instruments and should always be handled with special care.

The appearance of bubbles means a leaking compass. Bubbles the size of a quarter dollar generally have little effect; if they become much larger, the effect will be bad and the compass should be refilled through the filling plug that is always provided. Some compasses are filled with alcohol-water mixtures and some with petroleum derivatives. If the compass to be refilled requires alcohol water and only bubbles have appeared, add distilled water. If the compass has leaked to such an extent that a free liquid surface has formed, a mixture of 50 per cent of alcohol and 50 per cent of water should be added. If the compass is filled with a petroleum liquid, any leakage should be made up by adding water-white kerosene. Do not put kerosene in an alcohol-water compass, or vice versa, for the result is certain to be bad.

Mounting the Compass.—Whenever possible, the compass should be installed on the fore-and-aft center line of the airplane. The lubber's lines must be accurately lined up with the fore-and-aft center line.

No definite rules can be laid down for the distance of a compass from magnetic parts to ensure given deviation, as the deviation depends upon the degree of magnetization which the parts have accidentally acquired. However, if the airplane designer keeps the following recommendations in mind, he may confidently expect that the magnetic conditions surrounding the compass will be good.

The following distance should be allowed between the compass and parts, if they are magnetic:

a. Structural rods and wires, 15 in.

b. Fire walls, 24 in.

c. Tanks or floor plates above or below the compass, 15 in.

d. Tanks on the center line ahead of or behind the compass, 24 in.

e. Tanks not on the center line, 30 in.
f. Cowling or instrument board, 24 in.
g. Movable parts such as the control column, 18 in.
h. Untwisted direct-current wiring, 36 in.
i. Essential removable parts, such as cranks and tool kits, 36 in.
j. Engines, as far as is consistent with good visibility of the compass.

If direct-current wires are twisted one over the other, they have no effect. Alternating-current wires have no effect.

Doubling the distance from a magnetic part decreases the effect of that part to approximately one-eighth of its previous value, since the force between two magnets varies inversely as the cube of their distance apart, although single magnetic poles obey the inverse-square law.

The compass should be secured to a surface that is sufficiently rigid not to magnify the airplane's vibration.

Compass Errors.—The most common reasons why the airplane compass does not indicate true north are:

1. Magnetic variation.
2. Magnetic deviation.
3. Acceleration errors.

Magnetic Variation.—It has been previously explained that a magnetic compass, if operating perfectly and undisturbed by outside forces, will point to the magnetic north pole. Since the magnetic pole does not coincide with the true north pole, the magnetic meridian thus established does not coincide with the true meridian. *Magnetic variation* is the angle between the true meridian and the magnetic meridian.

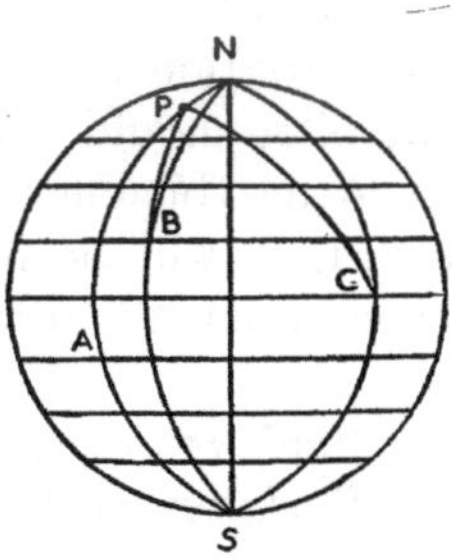

FIG. 38.—Difference in variation from place to place.

Variation differs from place to place on the surface of the earth for the reason shown in Fig. 38. Let *N* represent the geographical north pole which is called true north; let *P* represent the north magnetic pole toward which magnetic compasses point, called magnetic north; let *A*, *B*, *C* represent three different points on the earth's surface. Point *A* is on the same true meridian as the magnetic pole *P*. Therefore, the compass in pointing to the magnetic pole *P* also points to true north *N* and the variation is zero. Point *B* is on a different meridian and has a variation equal in value to the angle *PBN*. Point *C* has a greater variation equal to the angle *PCN*.

Variation also changes from time to time because of the slow movement of the earth's magnetic poles relative to the geographical or true poles, and it is imperative for the navigator to know the variation for the time and place in question. In general, in the United States, variation will not change more than about 1° in a 100-mile east-west flight and much less in a north-south flight. The variation is given on charts for the year in which they were published and a statement of the annual change is given so that the variation to date may be figured.

It is important to remember that compass markings are on the card and that these markings indicate direction by a lubber's line representing the head of the aircraft moving past them. Keeping this in mind, it is obvious that if a compass card swings to the left or west, the reading at the lubber's line is increased. If the compass card swings to the right or east, the reading at the lubber's line is decreased.

Suppose it is desired to steer a course of 20° true. If the compass is pointed to true north, the problem would be solved by simply swinging the plane's head until the 20° mark coincided with the lubber's line. *Actually, however, the magnetic compass seldom points to true north because of variation.* If the compass card is pulled 10° to the left or west by variation, the plane heading 20° true would show a reading of 30° by the magnetic compass. (For the purposes of this discussion neglect deviation so that the compass indicates magnetic north.) Thus by subtracting the variation of 10° west from the magnetic course of 30°, the true course is found to be 20°.

It is seen likewise that easterly variation should be added to the magnetic course to obtain the true course; and westerly variation should be subtracted. There is a great possibility of a serious error by applying the variation in the wrong direction, and it is therefore imperative that the pilot should understand this rule.

After this effect of variation upon the compass card is clearly understood, it will be found helpful to remember the following jingle as a check upon the direction in which the variation is applied:

Variation east, magnetic least. (Meaning smaller numerically.)
Variation west, magnetic best. (Meaning greater numerically.)

For example, if the variation is 5° east and the true course is 350°, the magnetic course will be least, or 345°. If the magnetic course is 40° and the variation is 10° west, the magnetic course would be best and the true course would be 40° less 10°, or 30°. Thus, starting with the magnetic course, the variation is applied in accordance with the signs, east (+) and west (−). That is, the easterly variations are added to the magnetic readings to obtain true readings, and the westerly variations are subtracted from the magnetic readings to obtain the true readings.

The effect of variation on the compass card is shown graphically in Fig. 39.

Magnetic Deviation.—It has been explained in discussing variation that the magnetic compass seldom points to true north but that when it is *undisturbed* by outside influences it points to magnetic north. In practice the compass is seldom undisturbed but is usually affected by local magnetism within the plane. The angle by which the compass needle is deflected from the magnetic meridian by this local magnetism is known as the *deviation.* The deviation varies with each direction in which the airplane is headed, but it can be corrected so the compass will read within 2° or 3° of the proper magnetic headings. This local magnetism is due to magnetic induction in hard and soft iron in the plane, and to the magnetic fields set up by the currents in electric wiring.

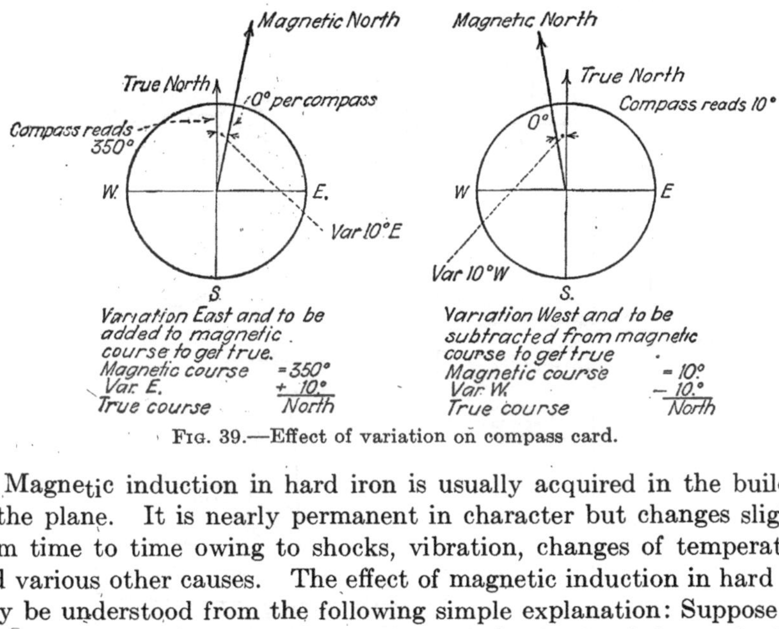

FIG. 39.—Effect of variation on compass card.

Magnetic induction in hard iron is usually acquired in the building of the plane. It is nearly permanent in character but changes slightly from time to time owing to shocks, vibration, changes of temperature, and various other causes. The effect of magnetic induction in hard iron may be understood from the following simple explanation: Suppose one of the longerons (fore-and-aft members of the fuselage) forward of the compass has become magnetized with a north pole near the nose and a south pole near the compass. This might be the case in a plane built with its head north.

By referring to Fig. 40 it will be seen that when the airplane is headed north, this south pole in the longeron will attract the north pole of the compass needle and hold the compass on north more strongly than usual. When the airplane is headed east, the compass magnets should lie directly athwartships, but the south pole in the longeron will pull the north end of the magnets to the east and cause an easterly deviation, so called because the compass needles have been pulled to the east. When the airplane is headed south, the south pole in the longeron will tend to repel the south pole of the compass needles, which pole is now toward the nose of the airplane and will weaken the effect of the earth's magnetism and cause sluggishness of the compass, but will not directly cause any deviation. When the airplane is headed west, the compass needles should again lie athwartships but the south pole in the longeron will pull the

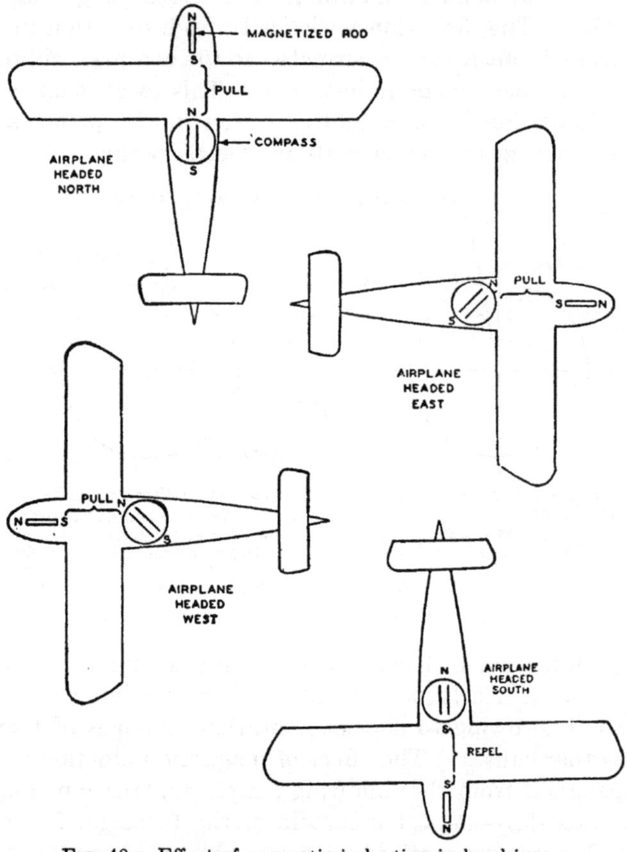

FIG. 40.—Effect of magnetic induction in hard iron.

north pole of the compass needles to the west this time and cause a westerly deviation, so called because the needles have been pulled to the west of north.

If the longeron had become magnetized with the south pole forward, the effect would be just the opposite.

Although errors due to the temporary induced magnetism in soft iron are at times troublesome, there is usually so little soft iron in the modern plane that they are small in amount and difficult to correct because of their variableness. Therefore, it is customary to make no compensation for the error caused by this type of magnetism in aircraft magnetic compasses.

The controlling factor in deviation is the magnetic induction in hard iron or "subpermanent magnetism," which is nearly permanent in character. By referring again to Fig. 40 it will be seen that as the plane is swung to the right, the deviation caused by this magnetized longeron starts at zero on a heading of north, increases to a maximum on east, and falls to zero on south. The deviation rises to a maximum in the

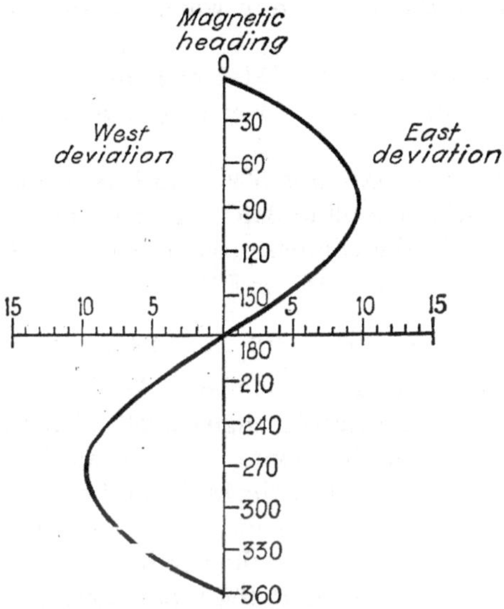

Fig. 41.—Curve of deviation due to magnetic induction in hard iron.

opposite direction on west and again falls to zero on north. Therefore, this is called semicircular deviation. A curve for this type of deviation alone, before the compass has been corrected, is shown in Fig. 41. This curve is for a plane built with its head pointing to magnetic north, and only fore-and-aft magnets would be required. Usually, deviation is found on north-south headings.

To compensate for this type of deviation (the only kind considered in compensating an airplane compass), small bar magnets are placed in receptacles under the compass bowl, both in the fore-and-aft line and athwartships. The correction is effected by placing these magnets so that they give an effect equal and opposite to that of the iron and steel in the plane. These small magnets close to the needles compensate large disturbances farther away.

In applying the error due to deviation it is important to remember, as with variation, that the compass markings are on the card, and that these markings indicate direction by a lubber's line representing the head of the aircraft moving past them. If the compass card is pulled 10° to the left or west by deviation, a plane heading 20° magnetic would show a compass reading of 30°.

Hence, it is seen that easterly deviation should be added to the compass course to obtain the magnetic course, and westerly deviation should be subtracted.

After this effect of deviation upon the compass card is clearly understood it will be helpful to remember the jingle that was used with variation, as a check upon the direction in which the deviation is applied.

Deviation east, compass least. (Meaning smaller numerically.)
Deviation west, compass best. (Meaning greater numerically.)

For example if the deviation is 5°E. and the magnetic course is 350°, the compass course will be least or 345°. If the compass course is 40° and the deviation is 10°W., the compass course would be best and the magnetic course would be 40° less 10° or 30°.

ACCELERATION AND TURNING ERRORS

Restatement of General Principles.—As previously described, a freely suspended magnetic needle aligns itself with the earth's lines of force, which dip below the horizontal from 0° at the magnetic equator to 90° at the magnetic pole. Only the horizontal component of the earth's total magnetic force at any place is utilized to indicate direction, the vertical component being overcome by the pendulous effect of gravity on the compass magnetic element, so that the latter normally remains approximately horizontal. For simplicity, we shall consider a compass with a single needle mounted with the center of gravity below the pivot, as shown in Fig. 42.

Action of Compass When Not Horizontal.—If the compass needle is placed in a vertical plane, the horizontal directional force would be overcome and only the vertical component of the total force would affect the compass, which in this vertical position would become a dip needle instead of a compass. If the compass needle should be placed in a plane, say, at 45° to the horizontal, both the horizontal and the vertical component of the earth's magnetic force would affect the compass, so that it would give neither the correct direction nor the correct dip, but a combination of the two. For any angle of tilt of the plane of the compass needle, the general position of the needle may be predicted, neglecting friction, inertia, and other factors. An understanding of this principle will make clear the action of the compass under acceleration and in turns.

Turning Errors.—This error was first noted for turns made on north-south courses, and is called the *northerly turning error*. Actually it is only one case of the general problem just mentioned. We shall first discuss turns in north latitude. If on a northerly course a plane turns short right and banks normally, the acceleration is to the east and the compass will also be banked with the plane, so that the plane of the needle is tilted, say 60°, with the horizontal. In this position of tilt the compass needle will be acted on by the vertical as well as the horizontal component of the earth's total magnetic force, while at the same time the needle is mechanically restricted to effective motion in the tilted

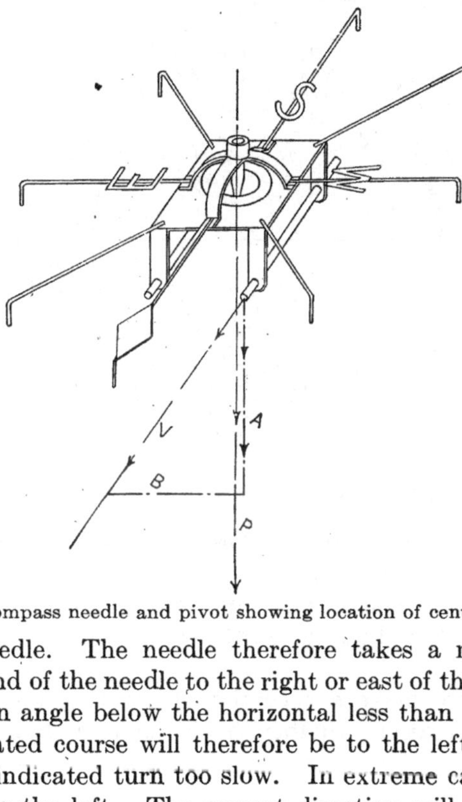

FIG. 42.—Compass needle and pivot showing location of center of gravity.

plane of the needle. The needle therefore takes a resultant position with the north end of the needle to the right or east of the correct compass course, and at an angle below the horizontal less than the total angle of dip. The indicated course will therefore be to the left of the magnetic course, and the indicated turn too slow. In extreme cases the indicated turn might be to the left. The correct direction will obviously not be indicated under these conditions, and the plane must be leveled off and the compass permitted to settle down before it will do so.

Acceleration Errors.—In turning right from north it is the acceleration to the east that tilts the plane of the needle and causes the easterly deviation. Acceleration on an easterly course sufficient to give this same angle of tile would cause the same deviation. In other words, turning, slowing, or speeding up causes accelerations, which in turn tilt the compass. It should be remembered that slowing is "deceleration," or acceleration in the opposite direction.

In turning left from north, or accelerating to the west, by the same reasoning, we find that the needle is deflected to the west, giving a westerly deviation. In the case of northerly accelerations, the plane of the needle is tilted up aft and the needle is more nearly aligned with the earth's lines of force so that the directive force of the needle is made stronger, without causing deviation. For southerly accelerations the needle becomes more nearly perpendicular to the earth's lines of force

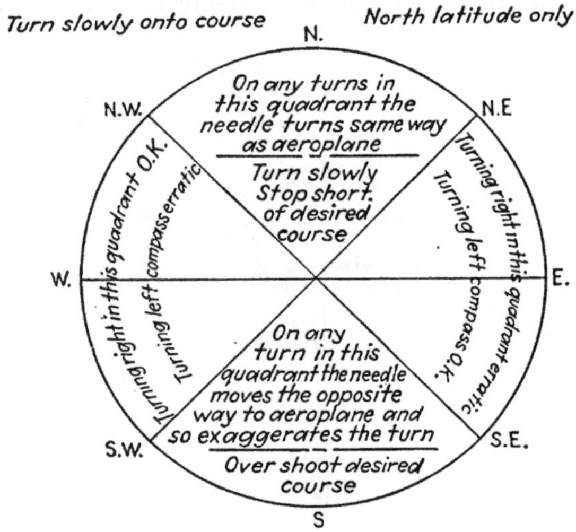

FIG. 43.—Compass turning error diagram.

and loses all or part of the directive force. If the tilt due to southerly acceleration reaches about 25° in the United States, the compass loses all directive force and, while in that condition, is inoperative. The same reasoning will make clear the action of the compass on other headings and latitudes.

We may therefore make a general rule covering the action of the compass under acceleration, whether this is due to a change of course or speed.

North latitude: Easterly acceleration causes *easterly* deviation.
Westerly acceleration causes *westerly* deviation.
South latitude: Easterly acceleration causes *westerly* deviation.
Westerly acceleration causes *easterly* deviation.

Figure 43 is a diagram designed to show what effect might be expected when turning from any course.

Other Disturbances in Turning.—Whereas the preceding discussion refers to an ideal compass without inertia or friction, the magnetic element has a moment of inertia as well as friction. The theoretical deviations will therefore not be attained since the deviation in the first part of the turn would be less (due to inertia) and at the end more than the theoretical deviation.

In addition to the pendulous effect of the magnetic element, its south end is weighted (in north latitude), and this unbalanced condition is also effected by acceleration. Therefore, the complete picture of the action of a compass in flight cannot be painted, nor exact values assigned, since these values depend on the angle of tilt, the period of the compass, the angle of dip, the course, the rate of turn, etc.

In practice, however, turning and acceleration errors cause little trouble since the compass is used principally to indicate direction on straight, level flight.

The importance of knowing the exact nature of the compass errors is so that the navigator will know its limitations, and so that he will not lose confidence in the compass because of apparently erratic actions under special conditions.

ANALYSIS OF DEVIATIONS

For the purpose of correcting marine compasses, the deviations have been analyzed into five coefficients, and a brief description of these is now given, omitting those parts which are of no interest to the air navigator.

The deviations, as has already been explained, are due to the presence of hard and soft iron in the structure of the plane; in general, the hard-iron effect, or permanent magnetism, predominates, the effect due to soft iron usually being quite small.

In a plane we generally ignore soft iron, as the effect is small, but it must always be remembered, and if the effects are serious it will be necessary to call in an expert to correct them as far as can be done.

Coefficient A is the constant deviation due to the lubber line not being truly fore and aft, together with another small constant deviation due to horizontal induction in unsymmetrical soft iron. Coefficient *A* is + when giving easterly deviation; it is corrected by rotating the compass by a suitable amount. It is independent of latitude.

Coefficient B is the fore-and-aft component of the semicircular deviation caused by the permanent magnetism of the airplane structure together with an effect due to vertical induction in soft iron; it is corrected by fore-and-aft permanent magnets, and the correction changes with the latitude to some extent. Coefficient *B* is + when the deviation is easterly in the eastern semicircle and westerly in the western semicircle.

Coefficient C is the athwartship component of the semicircular deviation caused by the permanent magnetism of the airplane's structure; *C* is corrected by athwartship permanent magnets; it is called + when the deviation is easterly in the northern semicircle and westerly in the southern semicircle. C does not vary with latitude when properly corrected, but an uncorrected C does so.

Coefficient D is due to induction in horizontal soft iron and is not usually taken into account in the adjustment of airplane compasses. The correction would be made by fitting soft-iron correctors to right and left of the compass in the plane of the magnets of the magnetic element. D is called + when giving easterly deviation in the northeast and southwest quadrants, and westerly deviation in southeast and northwest quadrants. It is not affected by latitude.

Coefficient E is due to induction in unsymmetrically disturbed soft iron and is corrected in conjunction with coefficient D by placing the soft-iron correctors at an angle to the athwartships line instead of athwartships. E is called + when giving easterly deviation in north and south quadrants, and westerly in the east and west quadrants, and like D, is not affected by latitude.

The five coefficients are derived from the deviations on the north, northeast, east . . . west, northwest points as follows, where north, northeast . . . represent the deviations on those points regarding easterly deviation as positive and westerly as negative.

Coefficient A = the sum of all 8 divided by 8. Averaging errors on N. E. S. and W. is enough for practical purposes.

$$\text{Coefficient } B = \frac{\text{E.} - \text{W.}}{2}$$

$$\text{Coefficient } C = \frac{\text{N.} - \text{S.}}{2}$$

These are all that are in general use on airplanes.

It is frequently of interest to take out the values of D and E just to see that they are small and that no trouble is present.

$$\text{Coefficient } D = \frac{\text{N.E.} - \text{S.E.} + \text{S.W.} - \text{N.W.}}{4}$$

$$\text{Coefficient } E = \frac{\text{N.} - \text{E.} + \text{S.} - \text{W.}}{4}$$

PRACTICAL ADJUSTMENT OF AIRCRAFT COMPASSES

Testing Ground.—A site should be chosen well away from all magnetic disturbance such as electric generators, motors, cables, iron, and steelwork. On the site a large circle of diameter not less than the length of the airplane whose compasses will require adjusting is drawn, and the eight points—north, northeast, east, southeast, south, southwest, west, northwest—marked out accurately from the center by means of a good prismatic compass. In doing this, and at all times when adjusting, the operator must take special care that he has previously removed all iron and steel articles from his person, as these will cause errors in the work. Penknives, keys, eyeglasses with steel rims or springs, and notebooks with steel clips are especially to be guarded against; as many such articles are of steel and nickel plates, it is clear that the greatest suspicion is necessary with regard to all metal articles.

In marking out the eight points, their directions are to be "magnetic," not "true." Opposite points are to be connected by painted lines or lengths of string for a temporary job, and the four lines will intersect at the center.

The testing ground is now ready for the aircraft.

First Adjustment.—The airplane is prepared for the operation by having plumb lines hung from its nose and tail and having all *tools and equipment* placed in their permanent positions on board; this is important, as tools are steel.

The machine is now placed on the north-south line, nose to north, and trued to the line by means of the plumb lines that indicate its fore-and-aft axis. The machine should be chocked up into flying trim with controls in flying position, and if possible the engine should be running.

The compass should now read 0° or 360°, but it probably will not do so; it should be made to do so by rotating the athwartships corrector-operating head (using *microadjuster*) till the compass does read 0° or 360°.

It will be found that a clockwise movement of the corrector head results in a clockwise movement of the compass element, and vice versa.

Having made the reading on north correct, or very nearly so, the machine is turned and trued up to face east. The fore-and-aft operating head must now be turned to bring the compass reading to 90°.

The compass should now be completely adjusted if the compass position is a good one in a type of machine that is not difficult to adjust. The corrector box is now closed, and the compass readings taken on each of the eight directions in turn, to check the adjustment and, if this is sufficiently good, to prepare a deviation table.

In the majority of light aircraft the adjustment will be sufficiently good so that the deviation table will be considered next. Those cases that require further adjustment will be considered later.

Preparing the Deviation Table.—From the compass readings on the eight specified fixed magnetic courses obtained on the testing ground a deviation table for any number of magnetic courses is readily prepared by interpolation. If the deviations are tabulated for intervals of 45°, this is quite close enough in practice. The interpolation is, by simple proportion, applied to corresponding intervals.

Thus, if 45° magnetic corresponds to 49° compass reading and 90° magnetic corresponds to 91° compass reading, then the interval of 45°

from 45° to 90° magnetic corresponds to the interval from 49° compass reading to 91° compass reading, *i.e.*, an interval of 42°.

Hence our compass reading corresponding to, say, 60° magnetic will be 49°, corresponding to 45° magnetic, plus

$$\tfrac{15}{45} \text{ of } 42°, \textit{ i.e.}, 49° + 14° = 63°$$

Thus, 60° magnetic corresponds to 63° compass reading.

Readjusting or Improving the First Adjustment.—This is not often necessary on light aircraft, but is essential in difficult cases. The airplane is set up as before, and the readings taken on the eight readings of the machine and noted. There are then three corrections to be dealt with. First make a table of the compass readings for each heading, and in a third column put the deviation. For example, let the readings be as follows:

Magnetic		Compass	Deviation
	°	°	°
N.	0	348	+ 12 (E.)
N.E.	45	41	+ 4 (E.)
E.	90	93	− 3 (W.)
S.E.	135	142	− 7 (W.)
S.	180	186	− 6 (W.)
S.W.	225	226	− 1 (W.)
W.	270	262	+ 8 (E.)
N.W.	315	302	+ 13 (E.)

The three coefficients A, B, and C are now computed as follows, paying attention to signs:

$$A = \tfrac{1}{8} \text{ (sum of deviations)} = \tfrac{1}{8}\,(+20°) = 2\tfrac{1}{2}°$$
$$B = \tfrac{1}{2}(\text{E.} - \text{W.}) = \tfrac{1}{2}[-3° - (+8°)] = \tfrac{1}{2}(-11°) = -5\tfrac{1}{2}°$$
$$C = \tfrac{1}{2}(\text{N.} - \text{S.}) = \tfrac{1}{2}[+12° - (-6°)] = \tfrac{1}{2}(+18°) = +9°$$

The three alterations required are:

1. Rotate the lubber's line through an angle A (2½°), in a counterclockwise direction.

2. Place the plane on heading east, note compass heading, and operate fore-and-aft corrector magnets until the reading has been altered by B (−5½°).

3. Place the plane on heading north, note compass heading, and operate athwartship corrector magnets until the reading has been altered by C (+9°).

When doing this remember that a clockwise movement of the key results in a clockwise movement of the compass element.

These operations may be carried out in any order desired, and the machine swung again to get the new improved deviation table, which will then be analyzed as above to see whether any further improvement is likely, remembering that we can alter the differences between the north and south and the east and west pairs of readings only by altering the setting of the microadjuster, and that the compass adjuster has to choose the nearest he can to his ideal. The example given above is, of course, deliberately chosen with large errors.

Recording and Using Compass Deviations.—A convenient table is to give the compass readings for the corresponding magnetic headings, and after the deviation is determined on the eight equidistant headings, a curve of deviation should be plotted on cross-section paper and a more complete table made out by picking off from the curve the deviation for, say, every 10°.

Converting Compass Readings.—The conversion of compass readings to magnetic or true readings, and vice versa, is a most important and frequent duty of the navigator. In addition to the mental operation of visualizing the card as being drawn to the right by easterly deviation, with the compass markings at the lubber's line becoming smaller, the pilot may have recourse to convenient memory ticklers in the form of the jingle previously mentioned, *deviation east, compass least*, etc.

There are numerous rules for this purpose. One of the favorite ones with the midshipmen at Annapolis is

"*C*-an *D*-ead *M*-en *V*-ote *T*-wice?"

Using the first letters we have *C, D, M, V, T*, which are the initial letters for

Compass Course, Deviation, Magnetic Course, Variation, True Course

Now apply the rule that in going from a compass reading to a magnetic reading, or to a true reading, variation and deviation are applied additive if east, and subtractive if west. The mental operation is to remember the sentence, then to write down the form with an arrow pointing to the right and marked "east (+)." The best part of this arrangement is the facility with which any conversion problem may be worked. A table is given below with a couple of conversions completed and with several to be worked out, if desired, by the reader. The form is shown as written down by the student.

EAST (+) ⟶

C	D	M	V	T
40°	5° W.	35°	10° E.	45°
30°	10° E.	40°	10° E.	50°
40°	5° W.	..	5° E.	
60°	....	70°	....	80°
225°	20° E.	..	10° W.	
320°	5° W.	..	....	300°
350°	10° E.	..	10° E.	

Summary of Compass Compensation.—The deviation due to sub-permanent magnetism is the only part of the deviation compensated. This is corrected by the use of small permanent magnets. Stow the tool kit, crank, etc., where they are usually carried and proceed to correct this deviation as follows:

a. Head the plane on magnetic north.

b. Place athwartship magnets so as to make the compass read 0°. Whenever the compass reading is being taken, the fuselage and controls should be approximately in the position of level flight and the compass should be tapped with the finger.

c. Head magnetic east.

d. Place fore-and-aft magnets so as to make the compass read 90°.

e. Head 180° magnetic. If the compass is out more than 2° or 3°, remove half the remaining deviation by athwartship magnets.

f. Head 270° magnetic. If the compass is out more than 2° or 3°, remove half the remaining deviation with fore-and-aft magnets.

g. Now head 0° magnetic and note the deviation of the compass. Head 45°, 90°, 135°, 180°, 225°, 270°, and 315° magnetic and note the deviation in the same way and make out a table of the deviation for all headings on which it was noted.

h. With the table of deviation as made out, plot a curve of magnetic-course deviations, then pick off from this curve the deviation on each 10° magnetic heading and complete a compass card or course converter for handy reference in the plane.

To check the compass, occasionally compare the compass reading, when the airplane is on a known magnetic heading, with the value given on the table.

The small amount of work required in compensating a compass will be well repaid by the resulting ability to fly across country without following railroads.

Finally, after the compass is properly compensated and the deviation table is made out, we are ready to take compass bearings or to set compass courses, and by applying the deviation and variation we can convert these compass courses and bearings to true courses and bearings.

Checking Compass Deviation in the Air.—Swinging the aircraft on the ground is the best method of compensating the compass, but it will be found in practically all cases that the deviation for various headings determined on the ground differ from those obtained for the same headings in the air. Swinging for deviation in the air is therefore a necessity for aircraft, particularly those engaged in long-range flying.

There are two methods available for swinging in the air:

1. By using magnetic azimuths of the sun.
2. By using ranges.

In the first case a curve of magnetic azimuths of the sun is determined for the time of swinging, and by using a pelorus or a drift meter the deviation is found on different headings. In the second case a suitable range (such as a straight stretch of highway, railroad track, etc.) is selected and then its magnetic direction is determined, and the deviation for various headings is obtained by using a drift meter. In both cases the compass azimuth or bearing, obtained by applying the angle found by drift meter to the compass, is compared to the curve of magnetic bearing of the range and the deviation obtained.

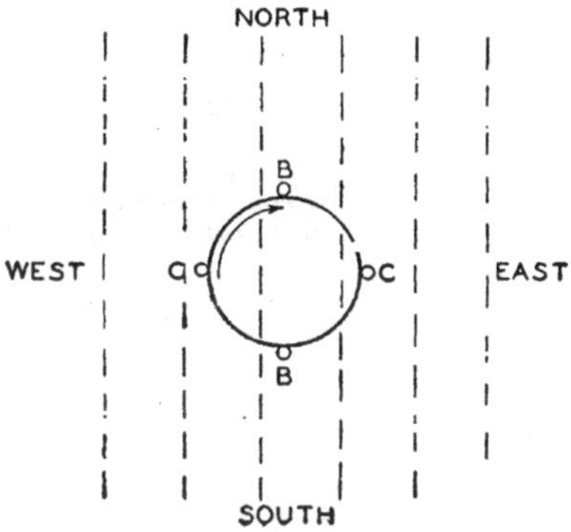

FIG. 44.—Principle of the earth-inductor compass.

The Earth-inductor Compass.—The principle of this compass is briefly as shown in Fig. 44. By the basic theory of the electric dynamo, an electromotive force (e.m.f. or voltage) is generated when a conductor cuts a magnetic line of force. In the diagram, the conductor *BB* is cutting the lines of force in the position shown, and an electromotive force is being generated. *CC* is, however, at the instant shown, moving parallel to the lines of force, and is not generating electromotive force. Since *BB* is cutting the lines at right angles, the position shown is the one of maximum electromotive force.

Suppose a commutator, brushes, and a voltmeter are added, and, further, that the magnetic field is the earth's field; then

a. If the brushes are placed to commutate the various conductors as they swing over the east-west line, the voltmeter will read zero.

b. If the brushes are placed to commutate the conductors as they swing over the north-south line, the voltmeter will read a maximum value.

The possibility of using such a circuit as a compass is at once apparent (see Fig. 45). Actually, commutation on the east-west line is used, as zero is a definite value and, further, for the extremely important reason that the direction of the electromotive force changes as the line of

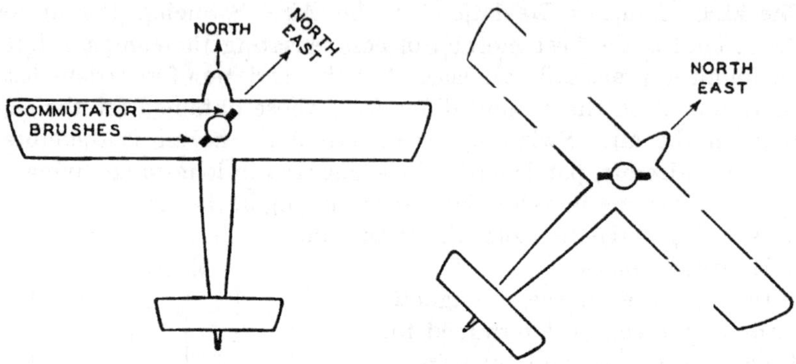

(a) (b)

FIG. 45 (a)—Setting course northeast by the earth-inductor compass—first step. (b) Setting course northeast by the earth-inductor compass—second step.

commutation crosses the east-west line. In the actual compass the brushes can be rotated easily by the navigator through a controller, which shows him at what angle they are set with reference to the fore-and-aft line of the plane.

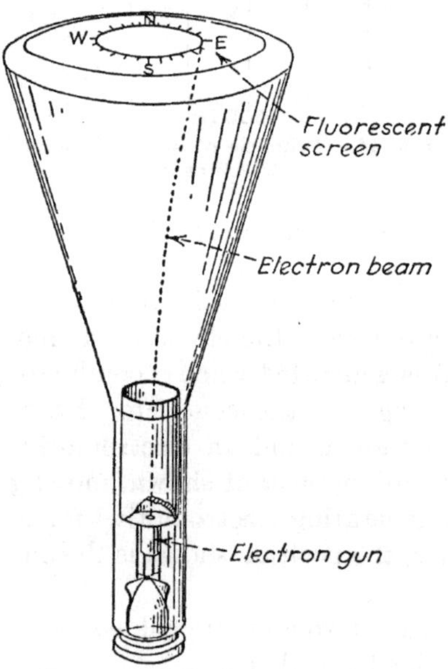

FIG. 46.—Cathode-ray compass.

The Cathode-ray Compass.—Another type of magnetic compass that has recently been suggested and is now undergoing development is the cathode-ray compass. This compass has not yet been developed far enough for tests in the air, but when brought sufficiently near perfection will offer definite advantages in its freedom from acceleration effects, although it will have other sources of error.

An ordinary cathode-ray tube, consisting of an electron "gun" at one end of an evacuated tube and a fluorescent screen at the other, may be used as a magnetic compass (see Fig. 46). If the electron beam is projected vertically upward, the horizontal component of the earth's field deflects it toward the east, in accordance with the usual laws governing the interaction between an electric current and a magnetic field. This means that the electron beam shows on the fluorescent screen as a bright spot to the east of the center of the screen. If the tube is rotated about a vertical axis, the spot of light traces a circle on the screen, and each position of the spot on the screen corresponds to a definite angular position of the tube. Hence the provision of a compass rose on the end of the tube is all that is required to turn a cathode-ray tube into a compass, remembering, of course, that the spot points east, not north.

Non-magnetic Compasses.—All the compasses hitherto mentioned have made use of the earth's magnetic field in some way or other. There

FIG. 47.—Bumstead sun compass.

is still another means of obtaining direction relative to the direction of the earth's axis which defines geographical north and south, and that is by utilizing the rotation of the earth about that axis. Consequently, compasses constructed to utilize the rotation of the earth to show direction will show true or geographical and not magnetic north. Up to the present there are two types utilizing this principle—the astronomical type, making use of the position of some celestial object, and the gyroscopic type.

The Sun Compass.—The astronomical type generally uses the sun, which by virtue of its brilliancy can throw a shadow and so dispense with sighting and viewing telescopes. There are two known examples, the Goerz and the Bumstead, the latter being shown in Fig. 47. In order to understand the theory of this instrument, the reader is advised to read Chap. XIII and then return to the following description.

The compass consists of a mean-time clock with a 24-hr. dial, the hour hand being replaced by a diametral bar carrying a pin and translucent screen at the ends. The pin and screen are set parallel to the earth's axis by tilting the clock, which is mounted on horizontal trunnions, and clamping it by means of the latitude arc. The brackets carrying the trunnions, etc., are mounted on a horizontal azimuth dial that can be set to the course desired. In operation the clock is wound, the pin and screen set to the local apparent time, the latitude arc set to the latitude and the azimuth dial set to the required true course; the airplane is then steered so that the shadow of the pin falls on the center of the screen.

The sun compass can give correct indications only when:

1. The sun shines.
2. The local apparent time is known. This entails a knowledge of the longitude.
3. The latitude is known.

Its use is therefore very restricted, and in practice it has been used only in polar flights and then only to a limited extent.

Gyroscopic Compasses.—Up to the present there is no gyroscopic compass that will give satisfactory performance in airplanes, although gyroscopic compasses give good service at sea. The reason for this failure is that the disturbance due to acceleration is so much larger in the air than at sea. The weight of the marine-type gyroscopic compass is also prohibitive in the air.

Many people have the mistaken idea that the directional gyroscope discussed in Chap. IX is a compass and call it a gyro compass. This is not the case, as the directional gyro has no azimuth-seeking property whatever and so cannot be a compass, for which this property is essential.

Problems

1. A pilot wishing to fly from Los Angeles, Calif., to San Francisco finds from his chart that his initial magnetic course is 338°. He knows his deviation on this heading to be 12°W. (*a*) What compass course should he steer to make good the given magnetic course? (*b*) The variation shown on the chart for this locality is 16°E. What is the corresponding true course? *Ans.* (*a*) 350°; (*b*) 354°.

2. The compass course is 274°, the deviation is 4°W., the variation is 3°W. (*a*) What is the true course? (*b*) What is the magnetic course? *Ans.* (*a*) 267°; (*b*) 270°.

3. The compass course is 74°, deviation 6°E., variation 6°W. (*a*) What is the magnetic course? (*b*) What is the true course? *Ans.* (*a*) 80°; (*b*) 74°.

4. A pilot finds from his chart that the true course to his objective is 356°. The variation is 7°E., the deviation 4°W. (*a*) What is his compass course? (*b*) What is his magnetic course? *Ans.* (*a*) 353°; (*b*) 349°.

5. The true course is 235°, variation 7°E. and deviation 12°W. (*a*) What is the compass course? (*b*) What is the magnetic course? *Ans.* (*a*) 240°; (*b*) 228°.

6. The compass course is 40°, deviation 5°W., variation 17°W. (*a*) What is the true course? (*b*) What is the magnetic course? *Ans.* (*a*) 18°; (*b*) 35°.

7. The compass course is 90°, true course 85°, magnetic course 100°. (*a*) What is the deviation? (*b*) What is the variation? *Ans.* (*a*) 10°E.; (*b*) 15°W.

8. The compass course is 349°, true course 357°, magnetic course 338°. (*a*) What is the deviation? (*b*) What is the variation? *Ans.* (*a*) 11°W.; (*b*) 19°E.

9. The compass course is 317°, deviation 11°W., variation 17°E. (*a*) What is the magnetic course? (*b*) What is the true course? *Ans.* (*a*) 306°; (*b*) 323°.

10. The true course is 291°, deviation 9°E., variation 14°W. (*a*) What is the compass course? (*b*) What is the magnetic course? *Ans.* (*a*) 296°; (*b*) 305°.

CHAPTER IV

AIR PILOTAGE

Air pilotage, or *piloting*, is the method of directing aircraft from place to place by referring to visible landmarks on the earth's surface such as church spires, lighthouses, beacons, railroads, rivers, mountains, and lakes. In normal flight over land, the pilot guides his plane much as a tourist directs his automobile, *i.e.*, by comparing objects observed from the plane with objects shown on the chart.

Since radio bearings are taken on terrestrial objects, they may be considered a part of piloting; however, the important subject of radio is discussed separately in the next chapter.

An aviator flying from St. Paul, Minn., to New Orleans, La., would, by following the course of the Mississippi River have almost continuous landmarks to guide him. It would not be necessary to follow the meanderings of the river. The pilot would take short cuts so as to make his path as nearly as possible a straight line and when these short cuts took him to the west of the river he could pick up the river again by altering his course to the left; or, if to the east, he could check his position by altering his course to the right.

Preparation for Air Pilotage.—Air pilotage is extremely restricted as a method of navigation when compared with the other three methods. In practice, however, both dead reckoning and air pilotage are used on all flights, and the use of radio and celestial navigation is becoming more common.

Before the take off, even for a comparatively short and simple flight, the proper charts should be studied, the course or courses laid down, the route subdivided either by distances or by time of flight, and the necessary data plainly marked on the chart. The route should be divided into 20-min. runs, 50-mile intervals, or other convenient divisions. Use a soft pencil and make heavy black lines that will be legible. The charts should be carefully folded and laid out in the order of use.

Once in the air, *alertness* is the most important part of air pilotage. By continually referring landmarks sighted to the chart, and charted landmarks to the ground in view, there should be little difficulty in recording the track of the plane on the chart as the flight progresses.

General Rules for Overland Flying.—Some good general rules for cross-country flying are:

1. Make careful preparations in advance.

2. Keep a continuous check on the plane's position.

3. Carry a good compass, and when in doubt steer a steady compass course. Avoid frequent changes in course when the position is uncertain, as these changes only add to the uncertainty.

4. Practice chart reading until objects seen on the earth may be readily identified on a good chart.

5. When lost, use every means to get a fix as quickly as possible. The more time that is lost in getting a fix, the more difficult it becomes to do so.

Procedure in Flight.—An experienced navigator instructs his pupils as follows: Lay off from the point of departure courses 10° to the right and left of the track as an aid to pilotage. After taking off and reaching altitude, set the compass course that has been computed to allow for deviation, variation, and wind drift; hold this course till a definite landmark is identified.

Suppose a landmark is identified beneath the plane after flying 10 miles on the computed compass course. The position of the landmark, as shown on the flight chart, is 1 mile to the right of the course. One mile in 10 is 6 miles in 60, which gives a drift angle of 6°. This drift may be corrected by any of several methods, as follows:

1. The plane may be turned 90° to the left till it has covered 1 mile, after which the course is set 6° to the left of the course originally steered.

2. The drift angle may be doubled and the plane steered 12° to the left of the original course for the distance traveled, namely, 10 miles. A course 6° to the left of the original course is then steered till another landmark is identified.

3. A new course may be set to the destination, closely paralleling the original course, by steering 6° to the left of the original course.

The choice of the procedure to be followed depends on local conditions and on the pilot's preference. In any case, each leg of the flight should be on a steady compass course till a definite landmark indicates that a change in course is required. The lines drawn 10° to the left and right of the original course will help the pilot to estimate the amount he is off course.

On Getting Lost.—Once the position of the plane becomes uncertain because of inattention or from other causes, the work of the navigator increases rapidly. As all experienced navigators know, it is extremely easy to get lost. Often it is difficult to fix a position even over well-charted areas once the air-pilotage and dead-reckoning sequence of navigation is broken.

The navigator should, by force of habit, frequently note the plane's position and record it on the chart or elsewhere, giving in abbreviated form the time and place and other items of possible use later. The importance of such procedure is at once apparent to anyone who has been enveloped suddenly in a fog or runs into a rain squall without knowing his exact position.

Finally, it may be remarked that there is no disgrace in being lost in the air. This happens to the best navigators. The important thing is to reduce the periods of being lost or uncertain of position to the lowest limit humanly possible. There is an increased hazard to flying the moment the position of the plane becomes unknown or uncertain.

The Use of Ranges.—The frequent use of fixed ranges affords a check on the drift and assures the pilot what true course is being followed.

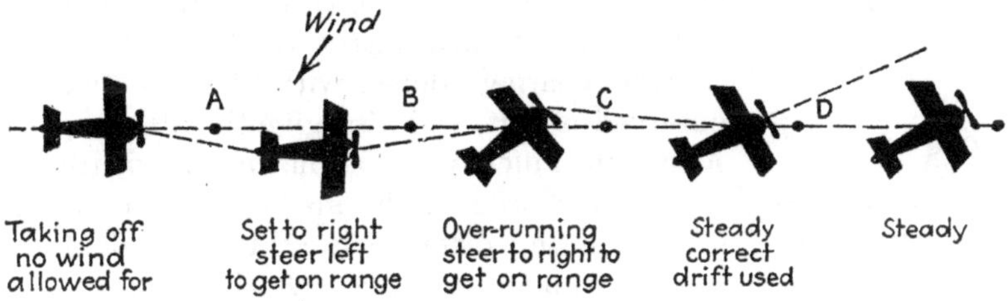

Fig. 48.—Use of ranges.

If on taking off, the pilot sets a compass course and notes any two objects ahead in line, it is a simple matter to keep on the course by keeping the objects in line. The objects need not be especially prominent. Any two objects which may be seen clearly, and which lie directly on the course to be made good, will suffice. The objects may be a smokestack, a green field, a hill, a tree, or a highway. Before the front or nearest range is passed over, another back or more distant range should be picked out, thus keeping, as far as possible, a continuous series of ranges. Theoretically, if the true course is set on taking off and ranges are carefully followed, the true course would be followed indefinitely. In practice, however, it is not always possible to select two definite, clear-cut objects exactly on the course. Therefore, although ranges are valuable aids and should be used whenever possible, sole dependence should not be placed on them. Figure 48 illustrates the use of ranges.

If, while in the air, two objects are seen and known to be on the true course, or a road or railroad is seen running directly toward the destination, this range may be used to check the compass error and the proper course to steer.

The Use of Bearings.—When two or more known objects are in sight, the navigator may locate his position definitely by observing and plotting on the chart the bearing of each. Each bearing gives a position line, *i.e.*, a line along which the plane's position is known to be. And the intersection of these lines gives a definite position or fix.

Suppose that in piloting along the seacoast, the compass bearings are taken of two lighthouses A and B that are recognized by their characteristics as given on the chart. Since it is more handy to work with true bearings on the chart, the two compass bearings are converted to magnetic bearings by applying the compass deviation on the given heading, and from magnetic to true by applying the variation. Next, through the position of each lighthouse on the chart draw the true line of bearing as shown in Fig. 49. The true bearing of lighthouse A is 210° and that of lighthouse B is 310° and the lines of position intersect at point C which is the true position of the plane at the time the bearings were observed.

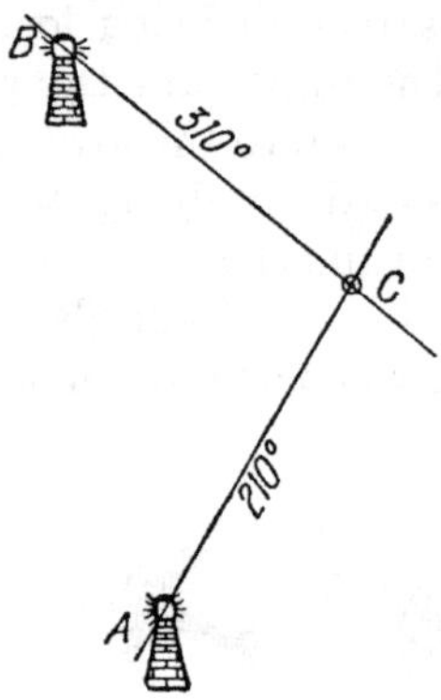

FIG. 49.—Fixing position by cross bearings.

One bearing alone, with the distance to the observed object, will also give the plane's position. Although much used in marine navigation, this case is less used in the air because an instrument for measuring distance is seldom available in an airplane. If, when a bow angle is observed the course is held until the bow angle is *doubled*. The run between bearings is the distance to the observed object at the time of the second bearing.

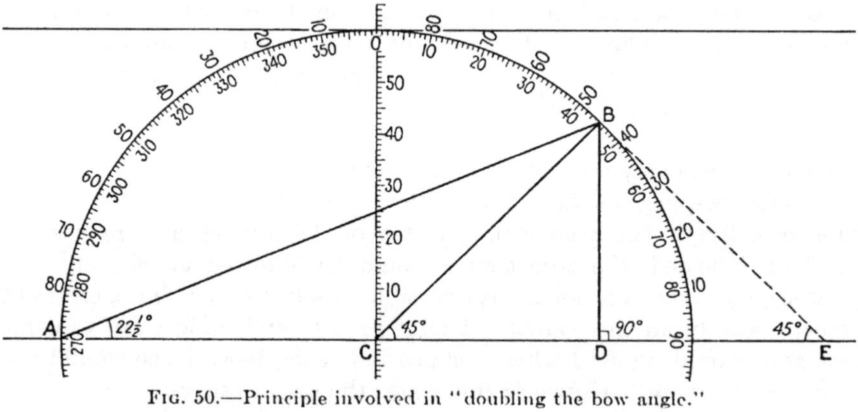

FIG. 50.—Principle involved in "doubling the bow angle."

Bow and beam bearings, etc., are special cases of *doubling the angle on the bow*. Figure 50 shows the results of observing angles 22½° and 45°; and of 45° and 90° on the object at B. The beam and quarter bearings are shown at D and E, the run DE being equal to DB.

Frequently, however, the distance from an object may be judged by the eye with sufficient accuracy for practical purposes. In this case the navigator may simply make a small circle with a cross in it,

⊕, on the line of bearing, the approximate distance away from the object, and note alongside this the time of observation.

Another method of fixing position is to take several bearings of the same object and to note the run between. For example, a plane sights the lighthouse L on the starboard bow (Fig. 51). The bearing is taken and found to be 60° and the exact time is noted. Fifteen minutes later the bearing is found to be 120°. These bearings are plotted on the chart as shown. The ground speed of the plane is 100 miles an hour due north.

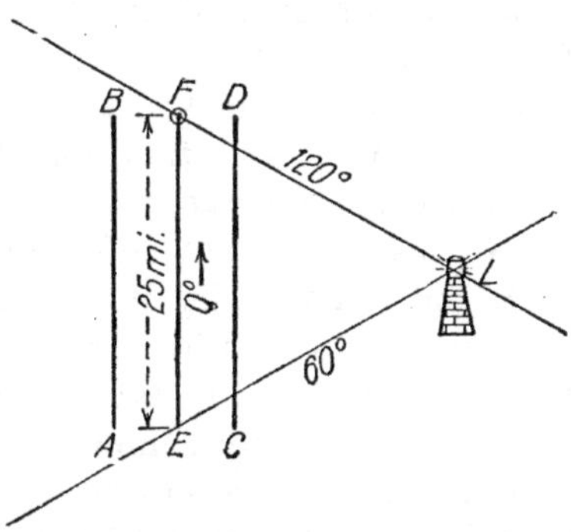

Fig. 51.—Fixing position by two bearings of same object and the run between bearings.

In 15 min. the plane will make good $\frac{15}{60} \times 100 = 25$ miles due north or 0°. From the scale of distance given on the chart a length of 25 miles is marked on a strip of paper. This paper is moved in and out parallel to the course made good of 0° until the marks on the paper just touch the lines of bearing. The position AB is too far out, the position CD is too far in, but EF, which just fits, is the actual track of the plane and F is the position of the plane at the time the second bearing is taken.

Selection of Landmarks.—It is important that the pilot should not confine himself to a few types of especially prominent landmarks such as railroads, highways, rivers, etc., for he may frequently be called to fly new routes where the most direct line does not go near a railroad or a prominent highway. Furthermore, changing weather conditions may obscure some of these landmarks or cause their appearance to vary enough so that they cannot be positively identified.

When a navigator encounters thick weather, it is a very common habit to fly low enough to see the railroad or highway or river as the case may be. This may be very dangerous because of the chance of obstructions high enough to foul the plane. This practice is especially inadvisable in hilly country. In following the coast line this low flying is less objectionable because sudden obstructions can generally be cleared by a sharp turn to seaward.

To follow a direct track to the desired destination often requires the use of minor landmarks. The skillful pilot must be able to locate on the chart these minor features visible on the ground. He must also develop his ability to see at night and to estimate distance accurately under varying atmospheric conditions. The eye can be trained to a remarkable degree of skill for this purpose, but like most other worthwhile accomplishments it requires concentration and much careful and painstaking practice.

Aids to Navigation.—Aids to navigation are needed over the entire country for general air traffic. In the same way that the Federal government has constructed lighthouses, operated radio stations, and supplied free information to the mariner, it has undertaken to develop and maintain lighted and marked airways, to operate a weather service, to broadcast general information, and to furnish free or at cost the charts and other useful information needed by the airman.

Navigation Lights.—For night flying, navigation lights are the principal aid. To make it possible to distinguish any particular light, definite characteristics are used for each one. It is therefore possible for a pilot to identify each definite light and determine his approximate position, simply by comparing the characteristics of the observed light with the light data shown on the chart. The aviator must know, and keep in mind or tabulated in a convenient place, the lights likely to be sighted, and the order in which they will be sighted. Unless the navigation lights are studied carefully, they appear to the pilot a confusing, meaningless jumble.

Marine lights include lighthouses, lightships, and light buoys and are shown on the air charts. It should always be remembered that marine lights may be screened on the landward side and, if visible, may be seen at a much greater distance than stated on the chart, where the visibility is given for a height of eye of 15 ft.

To identify a timed light, it is necessary to use a watch, preferably a stop watch. In addition to the time features of a light, it is necessary to know also the arc it covers, its color, or colors, and the character of the light. A description is given below of light characteristics.

A *fixed light* (F.) is a continuous, steady light.

A *flashing light* (Fl.) is a light showing a simple flash at regular intervals, the duration of light being always less than that of darkness.

An *alternating light* (Alt.) is a light that changes its color, as for example, a white flash followed by a red flash.

A *group flashing light* (Gp. Fl.) is a light showing at regular intervals a group of two or more flashes.

An *occulting light* is a steady light, with, at regular intervals, a sudden and total eclipse; the duration of the light being always greater than, or equal to, the darkness.

A *combination light* might include two or more of the characteristics described above. For instance, a light marked "F. Fl." means that a fixed light is varied at regular intervals, by a single flash of relatively greater brilliancy.

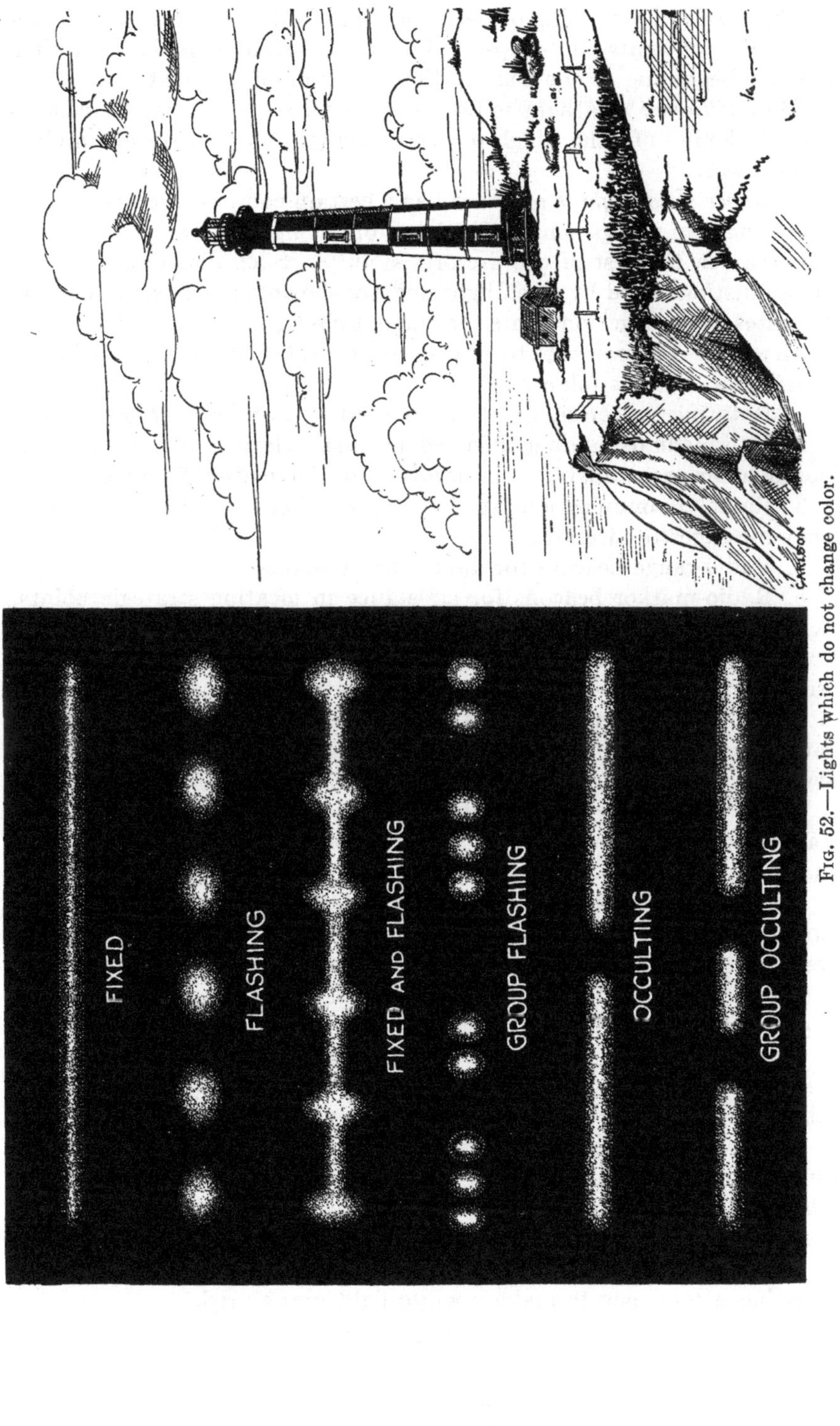

FIG. 52.—Lights which do not change color.

The *period* of a flashing or occulting light is the time occupied by an entire cycle of lights and eclipses; thus for a flashing light it is the time from the beginning of one flash to the beginning of the next.

The *brilliancy* of a light is usually designated by stating its candlepower. Some of the high-powered lights have several million candlepower.

A *danger sector light* is one that has a certain compass sector covering some danger to navigation, in which sector the light is always a different color than in the rest of the arc of visibility. Such danger sectors are usually distinguished by a red light and are shown on charts. Figure 52 illustrates the navigation lights that do not change color.

Airways.—Federal aids to navigation on established airways in the United States are as follows:

1. Rotating beacon lights at approximately 15-mile intervals.
2. Intermediate landing fields so located, relative to airports, that landing areas are available at intervals of approximately 50 miles.
3. Radio-communications stations for weather broadcasts and emergency messages to aircraft.
4. Radio-range beacons for directional guidance.
5. Radio-marker beacons for assistance in locating strategic points, such as intermediate landing fields.
6. Weather-reporting service, involving the use of teletypewriter circuits and point-to-point radio. The teletypewriter circuits are used not only for transmission of weather reports and forecasts but also for transmission of reports on progress of aircraft en route along the airways.

Course lights showing a directional beam along the airway are at or near the revolving beacon, and not only indicate when the pilot is on course but also flash a distinctive signal identifying the field or beacon.

Fixed or *blinker beacon lights* are used under certain conditions to supplement the revolving beacons, and where commercial current is not available. These lights are usually spaced only a few miles apart.

Private beacon lights off the airways that have been certified by the Department of Commerce are also shown.

Characteristics of Lights.—There are only two colors of sufficient contrast to be used with white for distinguishing lights—red and green, although amber lights are used as airport course lights as previously noted. Of these colors, red is the most distinct. If two lights, one red and the other white, appear to be of about equal intensity, the brilliancy of the red light will be greater than that of the white when they are both observed at great distances. This phenomenon is especially marked under certain conditions of the atmosphere and is readily understood, since fog generally has a tendency to make a white light appear red.

The colorations produced by the atmosphere are well known. It has been observed that during foggy weather white lights usually become red, green becomes white, and blue lights disappear or change to so pale a violet tint as to be mistaken for white.

Ranges of Lights.—The distance at which a light may be seen depends upon its intensity and its height. We thus have the luminous range and the geographical range.

a. Luminous Range.—If the lights transmitted their rays through a vacuum, the ranges would be proportional to the square roots of their intensities; but the atmosphere is more or less opaque and weakens the luminous rays. The ranges therefore depend upon the state of the atmosphere and are actually much less than those which would be deduced from the law just referred to.

b. Geographic Range.—The distance at which a light may be seen is limited more definitely by the spheroidal form of the earth than by its intensity. The geographical range depends upon the height of the light above the level of the sea, and the height of the observer.

Operation of the Airways.—At all hours of the day and night aircraft speed from city to city with loads of passengers, mail, and express. Darkness no longer holds any terror for the airman. Bad weather is a less potent enemy than in former years, because pilots are better equipped to cope with adverse conditions.

The Federal airways system, embracing thousands of miles of lighted and radio-equipped air routes, furnishes guidance and assistance to airmen at all times. During daylight hours and in good weather, the aids to air navigation make it easier for the pilot to perform his task efficiently; at night they offer guidance which is even more welcome; and under conditions of poor visibility caused by fog, clouds, rain, or snow, the airway aids are indispensable. Time after time they have enabled aircraft to reach airports or intermediate landing fields safely in circumstances that might have had tragic results if this assistance had been lacking.

Airports.—As airports throughout the country are established, they are listed and rated after inspection by the Civil Aeronautics Authority.

In addition to landing facilities, the larger airports provide for the pilot: aids for fog and night flying, housing, service, and repair facilities, and, most important of all, weather information.

Air-navigation facilities on civil airways require landing fields approximately 50 miles apart to provide a suitable landing field under conditions of stress of weather or in the event of mechanical difficulties.

Intermediate Landing Fields.—Where landing fields and airports are non-existent and where safety demands the establishment of landing facilities, the Federal government establishes and maintains an intermediate field. The landing field is boundary- and obstruction-lighted and day-marked, and is provided with an airway beacon and wind indicator.

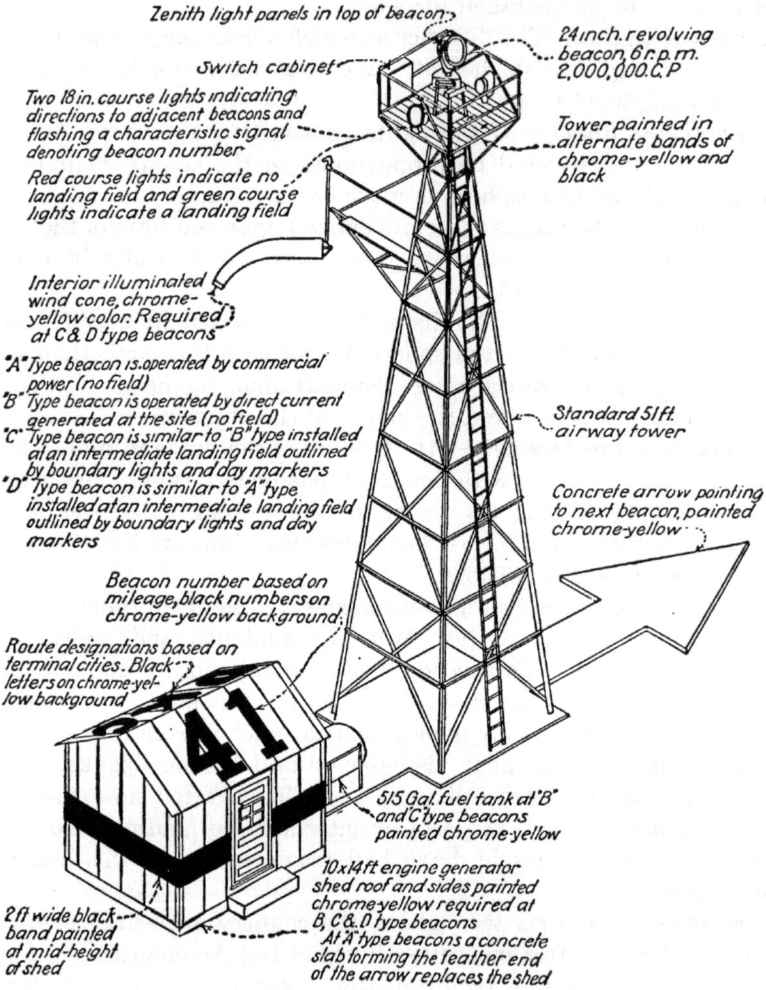

FIG. 53.—Standard installation airways beacon.

Airways are laid out over the safest low-level routes with bad-weather landing fields 50 miles apart, preferably following habitation, roads, communications, and power lines, and adapted for the installation of lighting and radio aids. A high percentage of completed trips on schedule time is essential for the success of air transportation, and the combined use of lights and radio direction is essential to meet this requirement. For safety, the radio-marked airway must coincide with the lighted airways, the pilots making use of all facilities and traveling the same routes in bad weather as on clear nights. When lights are not visible one to another, radio direction takes the pilot over the airway, enabling him to catch a glimpse of lights and fields at close range. This combination of blind-contact flying will greatly increase the flying efficiency and safety.

Airway Lighting.—All airway beacons should have the same light character with distinctive auxiliary lights for identification purposes. Green auxiliary lights are placed at landing fields and red auxiliary lights at airway beacons, where a landing cannot be effected. A typical airway beacon is shown in Fig. 53.

An *airway obstruction light* is a red light indicating an obstruction that is dangerous to air navigation.

Airport Lighting.—The minimum lighting equipment required on an airport is:

1. Airport beacon with green auxiliary light.
2. Landing area flood-light system.
3. Boundary lights.
4. Red obstruction lights.
5. Illuminated wind indicator.
6. Green approach lights on landing runways.
7. Hangar flood lights and roof markings.

1. An *airport location beacon* is a revolving beacon light marking an airport location, and should have an auxiliary light of distinctive character green in color.

2. A *landing area flood light* is a system of flood-light units for illuminating the surface of an airport to sufficient intensity to effect safe landings of aircraft at night.

3. *Boundary lights* are white or amber lights which, from the air, outline the entire usable portion of the landing area.

4. An *airport obstruction light* is a fixed red light indicating the top of an obstruction which is dangerous to aircraft entering or leaving an airport.

5. An *illuminated wind indicator* indicates the true direction of the surface wind.

6. *Landing-field approach lights* are green lights in the boundary lighting circuit at the end of runways and principal landing strips to indicate the favorable points of approach in landing.

Aircraft Navigation Lights.—Aircraft navigation lights are lights carried aboard aircraft flying at night for the purpose of marking aircraft to prevent collision. Between ½ hr. after sunset and ½ hr. before sunrise, airplanes in flight must show the following lights: On the right side, a green light; and on the left side, a red light; each showing unbroken light between two vertical planes whose dihedral angle is 110° when measured to the left and right, respectively, from dead ahead. These lights shall be visible at least 2 miles. At the rear and as far aft as possible, a white light shining rearward, visible in a dihedral angle of 140° bisected by a vertical plane through the line of flight and visible at least 3 miles.

The requirements for airships are the same as for airplanes, except that the side lights shall be doubled horizontally in a fore-and-aft position and the rear light shall be doubled vertically. Lights in a pair shall be at least 7 ft. apart.

A free balloon, between ½ hr. after sunset and ½ hr. before sunrise, shall display one white light not less than 20 ft. below the car, visible for at least 2 miles. A fixed balloon, or airship, shall carry three lights—red, white, and red in a vertical line, one over the other, visible at least 2 miles. The top red light shall not be less than 20 ft. below the car, and the lights shall not be less than 7 nor more than 10 ft. apart.

Airplane headlights are high-intensity projectors mounted on the wing tips for use in landing and taxiing at airports.

A *parachute flare* is a pyrotechnic light attached to a parachute for illuminating a large area at night from an altitude for the purpose of selecting a suitable landing field under conditions of emergency and making a landing.

Marking Structures and Obstructions for Air Navigation.—The Air Commerce Act provides that:

> The persons owning or operating any bridge, causeway, transportation or transmission line, or any structure over navigable waters of the United States shall maintain at their own expense such lights and other signals thereon for the protection of air navigation as the Secretary of Commerce shall prescribe. [Sec. 5 (g).]

Accordingly, the Corps of Engineers, U.S. Army, issues permits for all structures crossing navigable waters subject to the following conditions:

> That if the display of lights and signals on any work hereby authorized by the Corps of Engineers, United States Army, is not otherwise provided for by law, such lights and signals as may be prescribed by the Civil Aeronautics Authority, shall be installed and maintained at the expense of the owner.

The Civil Aeronautics Authority gives detailed instructions for marking obstructions in Form 474.

Radio Aids to Airway Navigation.—Radio is of great and increasing value as an aid to navigation along airways. It is used both for direction finding and for communication. These two uses are treated in detail in the next chapter.

Weather Service Provided by Bureau of Air Commerce and Weather Bureau.—Aeronautical weather information and service for the nation's airways are supplied by the U.S. Weather Bureau and the Civil Aeronautics Authority and disseminated along the Federal airways over teletypewriter circuits.

Airway weather-reporting stations along the airways report the following information hourly in accordance with scheduled sequences: Name of station, ceiling, sky conditions, visibility and general conditions, wind direction and velocity, temperature, dew point, and barometric pressure. Other information such as unusual field conditions and abnormal weather phenomena, if considered sufficiently serious to affect safety of flight, is included in the report.

At 6-hr. intervals the weather service includes also weather maps and forecasts. The more comprehensive reports required for these purposes are transmitted at 2 A.M., 8 A.M., 2 P.M., and 8 P.M. At these times the information regularly available in sequence reports is given a wider distribution and additional information from many reporting stations away from the airway routes is disseminated.

The weather maps, prepared at important Weather Bureau airport stations on the basis of these comprehensive 6-hr. weather collections, cover the entire United States.

The forecasts for airways are drawn up by 10 Weather Bureau forecast centers at Newark, Cleveland, Atlanta, Chicago, Kansas City, Dallas, Salt Lake City, Portland, Oakland, and Burbank, and transmitted over the communications network. Meteorologists at these centers keep a close check on weather conditions within their districts and issue special forecasts between the 6-hr. periods when weather conditions develop so rapidly as to make this necessary.

Navigation along Established Airways.—The airways navigation facilities have simplified navigation along established airways.

The mission of the air-line operating personnel is to complete with maximum efficiency the greatest possible percentage of scheduled flights with due regard to safety, especially where passengers are involved. The principal airway navigation problems to be solved are those connected with poor visibility, ice, and wind drift. Blind-flying and blind-landing methods have been developed to meet poor visibility conditions due principally to fog. Special de-icers have been fitted on the leading edges of the wings and on propellers to permit flying under ice-forming conditions. The effects of variable winds and load conditions in flight are controlled by setting schedules that permit ample factors of safety in speed and range.

Flight schedules are controlled by *dispatchers* who analyze the weather reports, load conditions, and other factors before clearing a plane for flight. Where weather conditions are bad but flight is possible, the pay load is reduced in order to permit an increased amount of gas so that if it is not possible to land at destination, an alternate clear landing field may be reached.

The following procedure is followed by the most successful air lines:

Before taking off, the navigator makes a careful *flight analysis*, which is in the nature of a prophecy as to the compass course and air speed

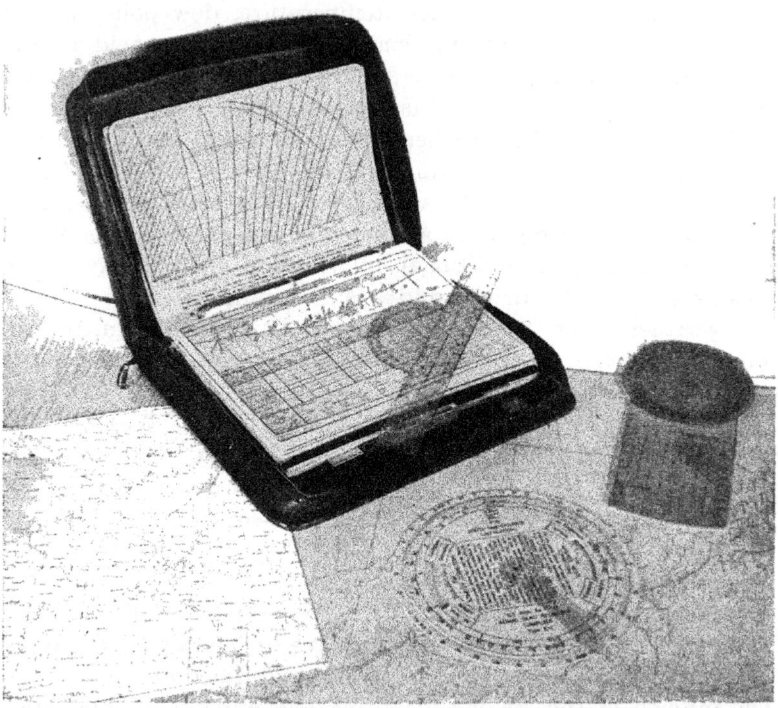

Fig. 54.—Transcontinental and Western Air pilot's navigation kit. (*Courtesy Transcontinental and Western Air.*)

necessary to follow schedule. An important item in making the flight analysis is the determination of the *estimated time of arrival* (E.T.A.). In flight the navigator uses all means at his disposal for the safe navigation of his plane. The best available charts of his route are properly prepared and arranged for convenient use. The charts in common use are the U. S. Department of Commerce sectional and regional charts, and on some special routes, airway strip charts, though the latter are being replaced by the sectional charts. The latest practice is to use two-way radio with a direction finder (D.F.) with rotatable loop in the plane. The D.F. may be used for taking a bearing of any transmitting station

Flight 1	Scheduled time interval	Distance, miles	Scheduled speed, m.p.h.	Cruising speed, m.p.h.	Approx. mid-point position	Distance to go, miles	Time to go
	h m						h m
Newark–Camden 8:00 A.M. E–8:40 A.M. E * 5 min. stop	40	72	108	127	Princeton	36	17
Camden–Pittsburgh 8:45 A.M. E–10:33 A.M. E 10 min. stop	1 48	259	143	152	South leg Harrisburg beam	159	1 03
Pittsburgh–Columbus 10:43 A.M. E – 11:56 A.M. E 10 min. stop	1 13	159	130	142	½ W. Kimbolton Cambridge	69	29
Columbus–Indianapolis 11:06 A.M. C – 12:26 P.M. C 10 min. stop	1 20	182	136	148	Lewisburg	93	38
Indianapolis–St. Louis 12:36 P.M. C – 2:13 P.M. C 10 min. stop	1 37	230	142	152	North leg Effingham beam	102	40
St. Louis–Kansas City 2:23 P.M. C –4:00 P.M. C 20 min. stop	1 37	229	141	151	Columbia Field	121	48
Kansas City–Albuquerque 4:20 P.M. C –9:10 P.M. C 15 min. stop	4 50	759	156	160	Wichita Amarillo	581 285	3 37 1 46
Albuquerque–Los Angeles 7:25 P.M. P–11:50 P.M. P	4 25	673	152	155	Winslow Kingman	446 258	2 51 1 39

* E = Eastern Standard time.
C = Central Standard time.
P = Pacific Standard time.

FIG. 55.—Transcontinental and Western Air navigation schedule sheet.

COLUMBUS-PITTSBURGH-CAMDEN-NEWARK
Mileage Chart Distances Measured Along Radio-range Beams

Check Point	Distance and direction off	Elevation																
Field	....	816	Columbus F.															
Beacon on field	....	880	22½	Newark F.														
Beacon on field	2-N	800	45½	23	Zanesville F.													
Center of field	2-N	800	69	46½	23½	Cambridge F.												
W. edge of town	....	1,319	97	74½	51½	28	Flushing											
Ohio River bank	2-S	1,200	122	99½	76½	53	25	Wellsburg										
Field	....	1,250	159	136½	113½	90	62	37	Pittsburgh F.									
Field	....	1,250	Pittsburgh F.															
W. edge of town	1-S	1,300	16½	Export														
R.R. Junction	2-N	1,300	35½	19	Blairsville													
Beam intersection	....	*	72½	56	37	Intersection of PT-BQ-HX Beams												
Center of town	Over	2,520	83	66½	47½	10½	Altoona											
Field	2½-N	1,000	91½	75	56	19	8½	Williamsburg F.										
Field	1-SSW	605	102½	86	67	30	19½	11	Huntingdon F.									
Center of town	3-NNE	2,340	111	94½	75½	38½	28	19½	8½	Mt. Union								
S. leg of BF range	....	*	112½	96	77	40	29½	21	10	1½	Bellefonte Beam							
Field	2-SSW	750	134	117½	98½	61½	51	42½	31½	23	21½	Blain F.						
Radio-range sta.	....	342	165	148½	129½	92½	82	73½	62½	54	52½	31	Harrisburg					
Center of town	N-edge	640	180	163½	144½	107½	97	88½	77½	69	67½	46	15	Elizabethtown				
Field	2-SSW	390	196½	180	161	124	113½	105	94	85½	84	62½	31½	16½	Lancaster F.			
Field	1½-NNE	600	221½	205	186	149	138½	130	119	110½	109	87½	56½	41½	25	Coatesville F.		
Center of town	2-NNE	500	235½	219	200	163	152½	144	133	124½	123	101½	70½	55½	39	14	West Chester	
Field	....	20	265½	249	230	193	182½	174	163	154½	153	131½	100½	85½	69	44	30	Camden F.
Field	....	20	Camden F.															
Center of field	1-W	200	17	Bristol														
River	....	210	25	8	Trenton													
R.R. Junction	3-SE	350	36	19	11	Princeton												
S.E. leg (MI) range	....	*	39	21	14	3	Martins CR. Beam											
River	1-SE	100	52	35	27	16	13	New Brunswick										
Field	....	10	72	55	47	36	33	20	Newark F.									

* Elevation not available.

FIG. 56.—Mileage chart, Columbus to Newark.

Form 0-331 FLIGHT PLAN

1. Flight #6 Date 5-15-35 Time 2:15 AMC Station KC
2. Cleared to: Columbus Airway KC-SN-ID-CO
3. Total Distance 641 Schedule Elapsed Time 3 hr. 30 min.
4. Assumed True Airspeed: 190

	I	II	III	IV	V	VI	VII	VIII	IX	X	XI
	Leg	From	To	M.C.	Altitude	Temp. °F.	Wind	Drift corr.	M.p.h. gain or loss	Cruising ground speed	Necessary true airspeed
5.	1	KC	SN	90°	10,000	40°	N.W.30	−6°	+23	196	173
6.	2	SN	ID	71°	10,000	35°	WNW23	−5°	+17	196	179
7.	3	ID	CO	85°	10,000	35°	ENE16	−1°	−15	196	211
8.	4										

9. Average — Temp. 37°; Drift corr. +8; Necessary true airspeed 188
10. Estimated Per Cent Power 69
11. Per Cent Power to be Used 71 — Corresponding TAS 190
12. .... Loss (minus) or Gain (plus) +8
13. .... Average Cruising Ground Speed 198
14. Estimated Fuel Consumption...... 101 GPH.
15. Total Fuel on Board.............. 480 Gal.
16. Estimated Fuel Range........... $4^h.43^m$
17. Estimated Climbing Time........ $.27^m$
18. Estimated Cruising Time......... $3^h.01^m$
19. Estimated Elapsed Time......... $3^h.28^m$
20. Estimated Remaining Fuel Range $1^h.15^m$
21. Alternate Airports Akron

Procedure for:

22. Two-way Radio Failure
23. Directional-radio Failure
24. Total Radio Failure
25. Remarks Anticipating good weather conditions
27. Person Consulted Dispatcher
26. I have this date familiarized myself with all recent changes on the airway bulletin board.

(signed) John Doe PILOT

FIG. 57.—Flight plan—a prophecy.

within range, or it may be used as a "homing" loop. Provisions for taking drift observations are at present inadequate and sorely needed on air lines.

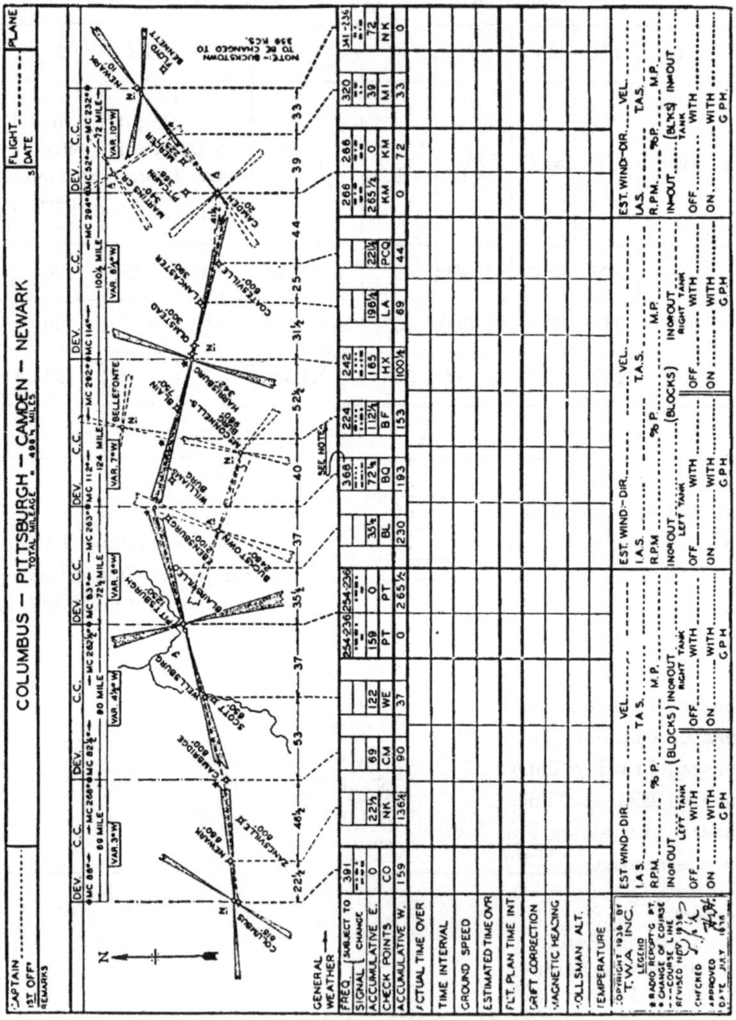

FIG. 58.—Pilot's navigational log sheet.

For practical celestial navigation on established airways, more attention must be paid to suitably placed skylights or windows for taking observations. Provision for this can best be done by the original designer. The latest large air-line planes are designed for accomplishing celestial navigation.

The accompanying forms and explanation of airways navigation are condensed from the navigation data supplied by TWA to each of their pilots. Other air lines use a similar although perhaps a less comprehensive system. Figure 54 shows the TWA navigation kit including the necessary data, equipment, and charts for the flight to be covered. The navigation data and forms have been worked out by Peter H. Redpath, navigation engineer, and his force of experts.

Figure 55 shows the navigation schedule sheet for Flight 1. There are separate sheets for each of 14 or more flights. The schedule times are for elapsed time, block to block. The column "Distance to go" is measured from the approximate mid-point position in the adjoining column.

Figure 56 is the mileage chart for Columbus to Newark. Similar mileage charts are available for all flights. The flight plan (Fig. 57) is filled out by the pilot before taking off. It will be noted that the form is so prepared as to reduce to a minimum the labor of completing it. Detailed instructions for completing the forms, with sample problems, are furnished each pilot. Abbreviations are used where possible.

The actual conditions encountered on the flight are carefully recorded on the pilot's navigational log sheet (Fig. 58).

Both the flight plan and the pilot's navigational log sheet are analyzed and filed in the office of the navigation engineer at Kansas City. Experienced pilots submit flight plans that compare very closely with the log sheet.

CHAPTER V

RADIO

The purpose of this chapter is to give a brief description of the fundamental principles of radio and to describe the apparatus employed in radio aids to navigation. Special attention is devoted to the correct employment of bearings obtained by radio.

Rapid Changes in Radio Aids.—Advance in radio equipment is so rapid that it is impossible to keep the details of the various types of installation up to date in a book such as this, which is not revised annually. For this reason no attempt is made to include here technical details that change from year to year. For further information on aircraft radio, see "Radio Navigational Aids" (*H.O.* 205) and "Radio Weather Aids to Navigation" (*H.O.* 206), recent textbooks on the subject, and also data supplied by the leading manufacturers.

Field of Use.—Radio aids are of great and increasing value to the navigator. These aids are:

1. Radio-range beacons.
2. Two-way radiotelephone.
3. Aircraft direction finders, compasses, and homing devices.
4. Airway broadcasts of meteorological information and notices to airmen.
5. Instrument approach and blind-landing aids, including airway markers and airport-approach markers.

All the preceding aids contribute largely to *safety in flight, the principal problem in aviation.*

Limitations of the use of radio are:

1. The weight and cost of the radio installation.
2. The necessity for wearing earphones except when a visual indicator is used.
3. The necessity for training the operator when the radiotelegraph is used.
4. If fixed stations determine the plane's position, the safety of the plane is placed in the hands of operators on the ground.
5. Radio bearings are great circles on the earth instead of rhumb-line bearings (*i.e.*, compass-course bearings), and therefore need a special plotting chart such as charts of the Department of Commerce on the Lambert projection to plot the bearings; alternatively, the bearings must first be converted to rhumb-line bearings.

Fundamental Principles.—The basic principle of radio, as applied to air navigation, is that Hertzian waves sent from a transmitting station may be detected at a distant receiving station.

A second principle is that the direction from which radio waves are received may be determined to within 1° to 3° by means of the directional characteristics of the loop aerial. This same directional characteristic of a loop (though not yet fully developed) is also applicable, to a limited extent, to transmission as well as to reception, thus making it possible to direct the radio waves in a path. The field patterns of loops, discussed later, are essentially the same whether used for reception or transmission.

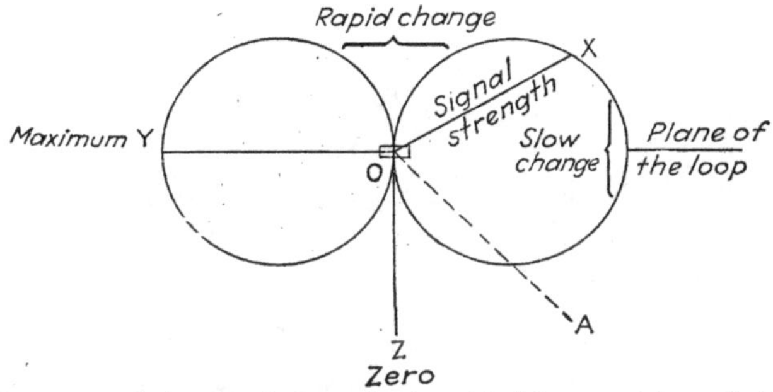

FIG. 59.—Polar curve of the signal strength of an ideal loop aerial unaffected by outside influences.

Radio signals may be sent and received in the form of Morse (or other code) dot-and-dash signals, this method being known as *radiotelegraphy*. By certain additions to the radiotelegraph equipment, the human voice may be transmitted and received. The method of transmitting speech is called *radiotelephony* and has many advantages over telegraphy; for example, trained operators are not necessary, and the speed of communication is higher. Radiotelephone equipment has the disadvantage of being heavier and more complicated in construction, but the operator is more interested in the fact that it is simpler to operate and requires no knowledge of code signals.

Theory.—A radio compass is merely a useful application of the well-known directional characteristics of a loop antenna. When the edge of the loop is pointed toward the transmitting station, the reception is of maximum strength; when the transmitting station is in a direction perpendicular to the plane of the loop, little or no signal will be heard. Therefore, if a dial is mounted on the loop, the direction from which an incoming signal is received may be determined by swinging the loop back and forth until the point of minimum reception strength is found. The point of minimum signal strength is used because that point is much sharper and easier to detect than the point of maximum receptive strength.

The advantage of using the minimum instead of the maximum point is shown by the vector curve of signal strength of a loop antenna (Fig. 59). This curve represents the relative strength of signals received from all points of the compass. The relative strength of the incoming signal received from any direction is measured by drawing a line from the center *O* of the antenna in the direction from which the signal is coming.

DIRECTIONAL RADIO SYSTEMS

Radio-range or Equisignal Beacons.—The equisignal beacon is intended for use on regular air routes, the majority of which are so equipped.

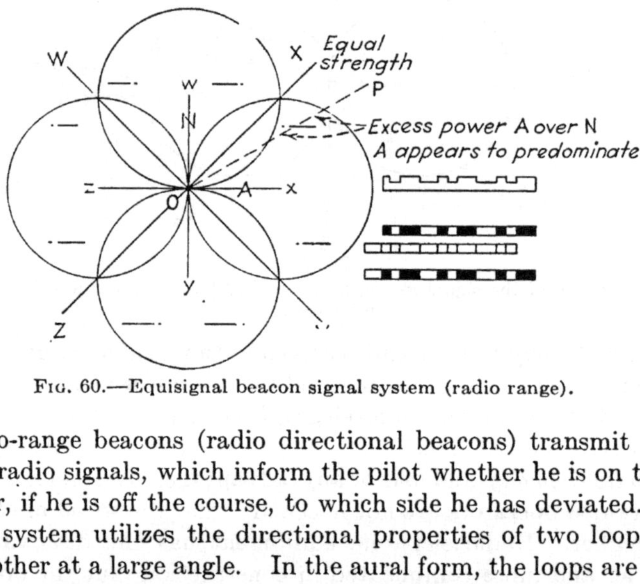

FIG. 60.—Equisignal beacon signal system (radio range).

Radio-range beacons (radio directional beacons) transmit along the airways radio signals, which inform the pilot whether he is on the proper course or, if he is off the course, to which side he has deviated.

This system utilizes the directional properties of two loops inclined to each other at a large angle. In the aural form, the loops are energized alternately with interlocking Morse symbols. For instance, loop *A* (Fig. 60) may be transmitting the letter A (·—), while loop *N* is sending the letter N (—·); the timing is so arranged that the two letters interlock to form a single prolonged dash. Figure 60 shows the polar diagram of radiation of the two loops. In the directions *OW*, *OX*, *OY*, and *OZ* the two signals are exactly equal and interlock. The directions *Ow* and *Oy* receive signals from the *N* loop only, while directions *Ox* and *Oz* receive only the signal A. In all other directions both transmissions are heard, but, owing to the difference in strength, one appears to predominate.

An airplane approaching from the direction *OP* receives a prolonged signal rising in dots and dashes to form the letter A; this is an indication that it is to the left of the defined track. The tracks defined by the beacons need not be at right angles, but may be arranged to suit the convenience of air routes as shown in Fig. 61. This is done by increasing the power in one loop.

Radio-range Orientation.—Following a radio range, once the plane has steadied on the proper course, is quite simple. When, however, the plane's position nearing a station becomes uncertain, one of several

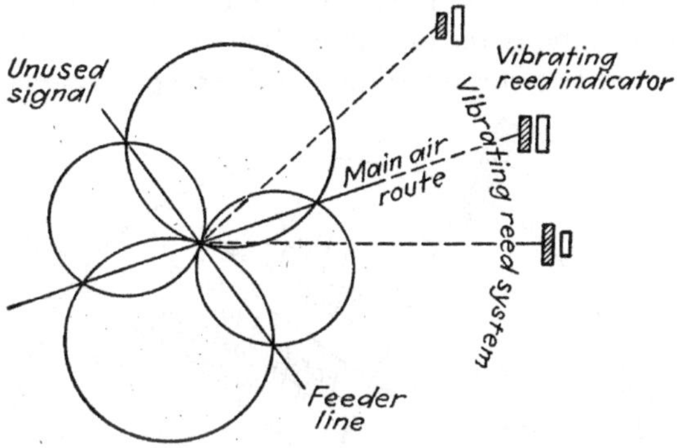

Fig. 61.—Directional control of equisignal beams.

definite plans must be followed in order to approach the beacon on course. Some of these methods are:

1. 90° approach system, where the plane flies at right angles to the "average bisector" of either of two diagonal quadrants until a change of signals identifies his quadrant.

2. Fade-out system where the plane flies parallel to the average bisector.

3. Parallel system where the plane at first flies parallel to the average bisector, then parallel to a range course as shown in Fig. 62.

The parallel system of orientation is used by many air-line pilots. The average bisecting course is flown, and the pilot identifies the quadrant by the fade-out or build-up of the signal level (Fig. 62). If the pilot finds himself in the northern quadrant and wishes to approach the station on the northwest leg, he will fly parallel to the northeast leg in a southwesterly direction. Upon intersecting the northwest leg he will turn to the compass heading of that leg and will stay "on course" by sound. He will then let down to the minimum safe altitude for that leg approaching the station on the final approach. (See latest Civil Aeronautics Authority instructions for details of this and other methods of orientation.)

All aircraft, while progressing away from radio-range stations, are required to keep to the right-hand side and definitely off the "on course" signal at all times.

A pilot should practice flying the radio ranges as much as possible whenever flying cross country, even when it is "clear and unlimited," so as to become familiar with each particular range and the accompanying problems, and should practice intersecting the "cone of silence" at every possible opportunity, until he becomes proficient. He should have a complete understanding of the inherent limitations of radio ranges before attempting to fly them during inclement weather. Range courses, which theoretically should be perfectly straight, may be found to have kinks or bends in them. This is most likely to occur where they pass close to or over hilly or mountainous terrain, large bodies of water, or over mineral deposits. *It should be remembered that radio-range stations are to be used*

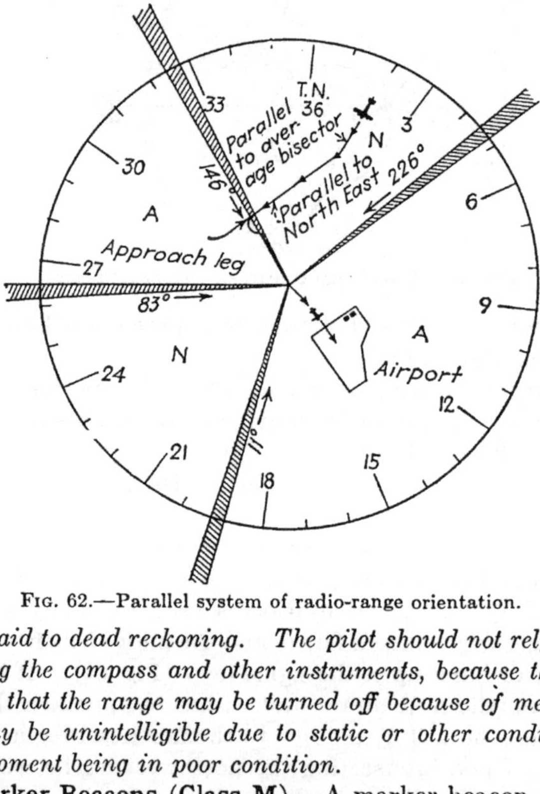

FIG. 62.—Parallel system of radio-range orientation.

strictly as an aid to dead reckoning. The pilot should not rely on the range alone, ignoring the compass and other instruments, because there always is the possibility that the range may be turned off because of mechanical difficulties, or may be unintelligible due to static or other conditions such as receiving equipment being in poor condition.

Radio Marker Beacons (Class M).—A marker beacon (*M*) is a low-powered, omnidirective radio station that transmits a characteristic signal, such as *H* (· · · ·) about every 10 sec. It is also equipped for voice communication with aircraft. Marker beacons have a range of from 3 to 10 miles, depending on the weather conditions and the type and condition of receiving equipment being used. Marker beacons are normally placed at the intersection of two radio ranges, indicating when to tune to the next station. A marker beacon does not operate continuously, but is turned on when the local ceiling is less than "unlimited" and the visibility is less than two miles, or at any time on request.

DESIGNATIONS USED TO DENOTE CLASS OF U.S. AERONAUTIC RADIO STATIONS

B	Broadcast station.	W	Without voice facilities.
RA	Range, Adcock (vertical radiators).	T	Teletype.
RL	Range (loop antenna).	TX	Principal teletype.
M	Marker (nondirective).	P	Point-to-point radio.
ML	Low-powered range (loop antenna).	D	Distantly controlled.
MRL	Medium-powered range (loop antenna).	Z	Zone high-frequency (station location) marker.
MRA	Medium-powered range (vertical radiators).	S	Simultaneous transmission of range signals and voice.
V	Voice communication with aircraft.		

The advantages of the radio-range-beacon system are:

1. Only an ordinary receiver need be carried.
2. No particular skill is required to receive and interpret the signals.
3. A specific track is defined, thus enabling the pilot to correct for drift.
4. Operation is continuous.
5. It reduces danger of collision when used in accordance with regulations, which prescribe that all inbound aircraft fly "on course" and all outbound traffic keep to right of course in the "twilight zone."
6. In addition to affording a means of correcting for drift, the fixed courses afforded by the range beacon direct traffic along prescribed lanes laid out over the most favorable routes equipped with visual and radio markers that warn of unusual hazards.
7. Ranges provide not only a definite fix marked by their cone of silence but in a network of such stations many cross checks are afforded where courses from two stations intersect at any appreciable angle.
8. Radio-range signals, especially those of the now standard "simultaneous" type of facility, may be received on an aircraft direction finder at the same time it is giving course signals for the benefit of those not equipped with loop receivers.
9. The range is adaptable to the use of sharply tuned audio filters in the output of the conventional aircraft receiver, since all such transmitters are modulated at or near 1,000 c.p.s. Filters reduce atmospheric electrical interference to a remarkable degree and increase the distance over which course signals may be used, not to mention reduction of ear fatigue usually resulting from prolonged listening to static crashes.
10. It is a one-way service independent of any transmissions from the aircraft, which needs only a simple receiver. For this reason reliability is increased manyfold.
11. It is capable of giving service to an unlimited number of aircraft simultaneously without creating additional radio interference. It is also quicker.

The disadvantages of the system are:

1. Somewhat elaborate equipment is required on the ground.
2. It is useful only to aircraft approaching from certain directions, except where aircraft are fitted with radio compass or homing device.
3. It jams other radio reception, because it is a modulated note with frequency spread, though the interference is not so great as with voice transmission.
4. It is applicable only to established airways, though the courses may be rotated to serve four or more airways.
5. The limitations considered in this country to be the most important, at least by transport pilots, are:

   a. Multiple courses (due to broken terrain).
   b. Swinging of courses (due to night effect).

Limitation *a* has been largely overcome by the substitution of vertical radiators for loop antennas.

It appears probable that the radio-range system will ultimately be replaced by an ultrahigh-frequency (u.h.f.) operation.

FIG. 63.—Shielded loop aerial. (*Courtesy of Western Electric Company.*)

Rotatable Loop Receiver.—A loop aerial consists of a coil of wire arranged so that the area of the cross section is as large as possible. A standard type of loop for air use is illustrated in Fig. 63. The tube serves three purposes:

1. As a rigid support for the loop in the air stream.
2. As a means of protection against the elements.
3. As an electrostatic shield.

When a loop is coupled to a suitable receiver, bearings of distant radio stations may be obtained by rotating the loop until zero or minimum signal is received. The loop merely indicates the bearing or its reciprocal but does not give the "sense" of the bearing. The latter may be determined by an approximate knowledge of the aircraft's position, or by noting the direction in which the bearing changes as the aircraft moves.

Bearings may be obtained very rapidly with the rotatable loop, and it has the advantage of simplicity. In metal aircraft, the loop must be placed outside the fuselage, thus giving increased drag.

Loop receivers mounted on metal structures such as ships or air frames are subject to an error called "quadrantal error." Currents induced in the metal structure tend to change the apparent direction of incoming signals, the error being zero on fore-and-aft and beam bearings, and a maximum on quadrantal bearings. The aircraft should be swung for a table of errors in a manner similar to that employed for swinging compasses. A table of errors differs from a compass-deviation card in that errors are dependent on the direction of the loop relative to the aircraft's head, and not on the actual direction of the loop relative to north.

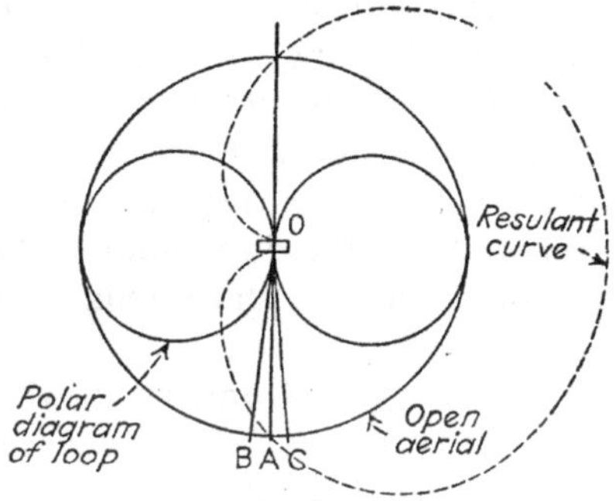

Fig. 64.—Resultant diagram from combined open and loop aerial.

Loop receiving outfits may be arranged for sense finding by the use of an additional aerial, which will be either the trailing aerial or the upper fixed aerial of the aircraft, although use is sometimes made of the antenna effect of the loop itself. When the two are suitably coupled to the same receiver, the resultant polar diagram is the cardioid (heart-shaped) curve shown in Fig. 64, which has only one minimum, thus permitting the determination of sense.

Homing Devices.—Most homing devices are based on the principle of a loop fixed athwartships. A rotatable loop set may be used for this purpose, in which case the procedure is as follows: Lock the loop athwartships and tune in the station toward which it is desired to steer. Alter the course until no signal is heard in the phones, taking care to alter toward the station and not away from it. The aircraft should now be heading directly toward the transmitting station, and the compass reading indicates the course to be steered. When a signal is heard, it indicates that the aircraft has drifted off the correct track. Repeat the procedure given above. The inevitable result of this method is that the aircraft will approach the station in a spiral and will eventually be flying up wind. The cure is to use the loop, not as a homing device, but as a means of obtaining an occasional bearing, the correct use of which is explained later. In some receivers a potentiometer is fitted, by means of which the pilot is enabled to offset the zero to allow for drift. If the loop is rotatable, it may be turned through the amount of the drift angle. The disadvantage of this method is that signals are weakest when the aircraft is heading toward the transmitting station.

Most of the modern outfits make use of the cardioid polar diagram of an open aerial working in conjunction with a loop, as shown in Fig. 64. The open aerial is left connected permanently to the receiver, and the loop is switched in intermittently, either by hand or automatically. If the aircraft is headed toward the station, there is no variation in signal strength; if the station is not straight ahead, the strength fluctuates. The course is altered until rapid operation of the switch produces no effect on signal strength.

When a broadcasting station or airway station of simultaneous type is used for homing, the signal is continuous, and a visual indicator may be used. This is merely some form of meter indicating signal strength. In Fig. 64 a loop is fixed athwartships and the aircraft's head is in the direction *OA*. In this direction the loop has no effect. If signals are received from the direction *OB*, the effect of the loop is in opposition to that of the aerial, and strength drops; signals from *C* give an increased reading.

Only those instruments in which provision is made for correcting for drift can be considered satisfactory. The others, by which the aircraft is merely headed toward the station, may be regarded as a convenient means of obtaining bearings. Homing devices may also be used for going away from a station on a direct track.

Night Effect.—Bearings taken at night, or at dusk or dawn, on the Bellini-Tosi system or loop receivers are liable to be seriously in error. The reason is that part of the signals arrive by reflection from the Heaviside layer in such a manner as to induce e.m.f.'s in the horizontal portions of the loops. During the dark hours, the "sky wave" forms a much greater proportion of the total signal received. In order to obtain a minimum, the goniometer must be turned away from the correct direction. The effect is variable and cannot be predicted. Night effect applies to the radio compass, the direction finder, and the radio beacon; the error from this cause might be 10° to 20° or even more.

The Adcock and "vertical-radiator" aerial systems have been devised to overcome night effect.

Pan American Airways Adcock direction finders have a range of 1,800 miles, giving coverage over 3,600 miles. The accuracy of bearings, even at 1,000 miles, is better than 1½ degrees. Figure 65 shows the actual plot of 32 bearings taken on the run from Alameda to Hawaii.

Blind-landing Radio Equipment.—Instrument approach for making blind landings is an increasingly important phase of air navigation. There are various methods for accomplishing the approach by radio.

As far back as 1907, Schellar in Germany advanced the idea of equisignal beacons, from which the glide-path blind-approach system was developed by Diamond and Dunmore at the U. S. Bureau of Standards. Later, Lorenz in Germany combined the equisignal and glide-path transmitters.

The Lorenz System.—This system consists of a short-wave beacon that provides guidance in the horizontal and vertical planes, and two

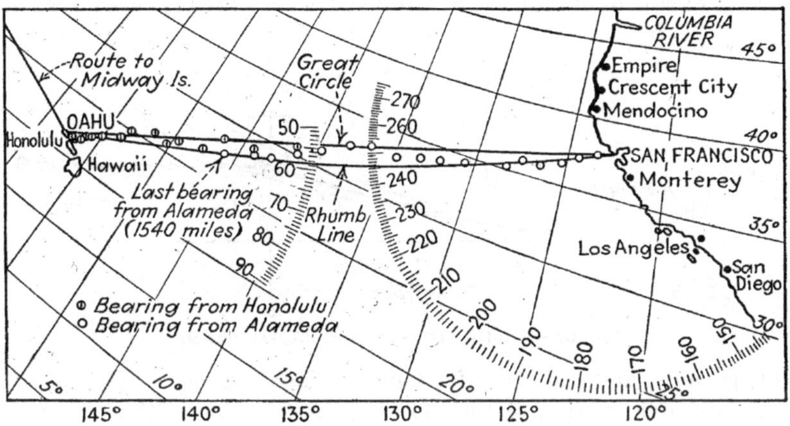

FIG. 65.—Course chart of a flight made by Pan American Airways early in 1935. The crossing between California and Hawaii took slightly over 17 hours, and was checked by 32 radiocompass bearings, which gave positions indicated by the dots. (*Courtesy of Electronics.*)

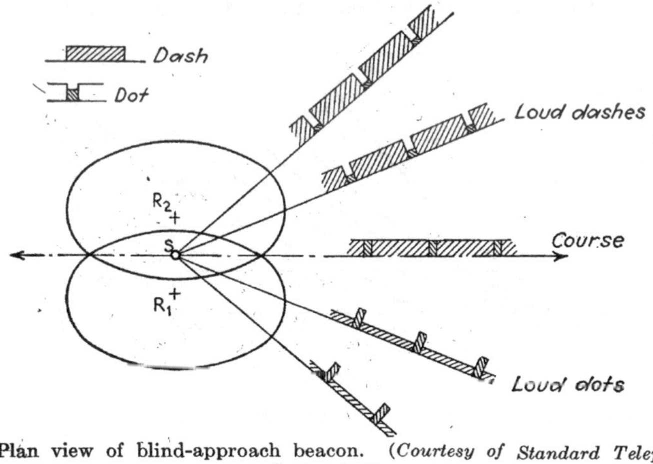

FIG. 66.—Plan view of blind-approach beacon. (*Courtesy of Standard Telephones and Cables, Ltd.*)

marker beacons to indicate distance from the landing ground. The system is considered in two parts—horizontal guidance and glide-path indication.

The main beacon, which is situated at the far end of the landing runway, consists of a small transmitter radiating from a vertical dipole aerial. On either side of it are similar aerials having switches in their centers. When these switches are open, the polar diagram of radiated field strength is a circle. When the switch of one of the aerials is closed, the aerial acts as a reflector, and the diagram of field strength assumes an oval shape. The switches of the two aerials are opened and closed alternately by means of an automatic timing device. One is closed for short,

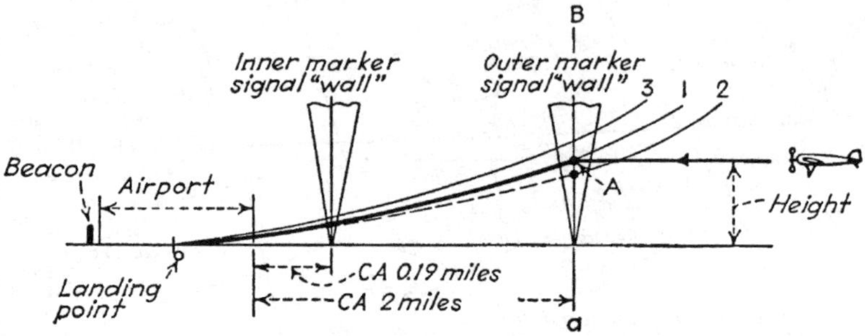

FIG. 67.—Glide-path beam and marker signals.

or dot, periods, and the other for long, or dash, periods. The field thus assumes the oval shapes of Fig. 66 alternately. Along the course, field strength remains the same, whichever aerial is energized.

A pilot approaching from the direction marked "Loud dashes," and listening with headphones, hears a series of dashes. Approaching from the sector marked "Course" he hears a steady note, and from the sector

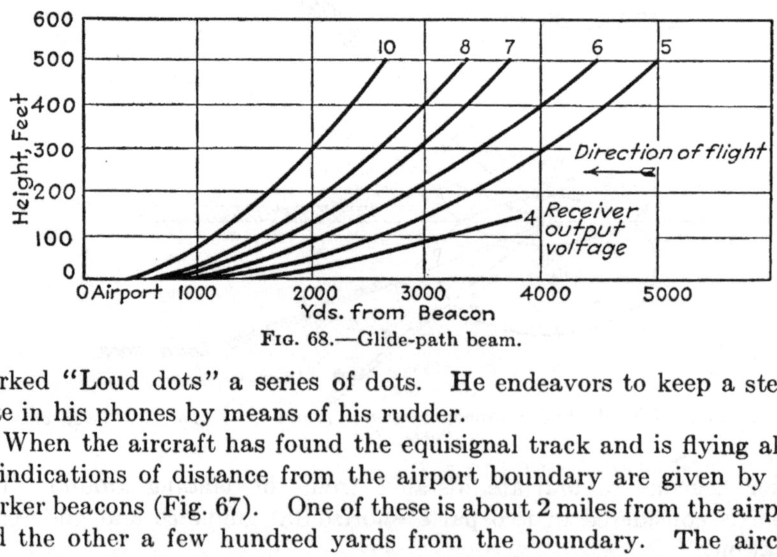

FIG. 68.—Glide-path beam.

marked "Loud dots" a series of dots. He endeavors to keep a steady note in his phones by means of his rudder.

When the aircraft has found the equisignal track and is flying along it, indications of distance from the airport boundary are given by two marker beacons (Fig. 67). One of these is about 2 miles from the airport, and the other a few hundred yards from the boundary. The aircraft approaches the first marker at a standard height, which is determined by means of a sensitive altimeter previously set to the ground pressure as signaled from the airport. On passing this marker, which is indicated by dashes of low-pitched tone, the glide indicator is adjusted to the center of its scale, the throttles are closed, and the aircraft controls adjusted to maintain the glide indicator on its central reading until the machine touches down. Just before landing, the inner marker beacon, which gives a series of dots of high-pitched tone, is heard. This indicates to the pilot that he is just about to pass over the airport boundary.

Figure 68 shows a vertical section through the landing path. The curves represent lines of equal field strength, decreasing in intensity according to their distance from the transmitter. A meter is connected in the output circuit of the receiver and, as the aircraft approaches in level flight, its reading will rise steadily. A certain reading of the meter corresponds to a definite curve of Fig. 68. From this it is seen that the curves of equal field strength converge toward the surface of the airport. By holding the reading constant, the aircraft is guided to a definite point on the airport.

ERRORS IN RADIO DIRECTION FINDING

In using directional radio equipment there are several types of errors that may be encountered. In addition to the night effect, the following errors may be encountered when using the systems shown in parentheses.

Mountain or "Multiple" Effect. (*Radio Compass, Direction Finder, and Radio Beacon*).—This is similar to night effect previously described, but the waves are reflected from surrounding elevated country. The error ranges from 0° to 5° or even more and is extremely troublesome in certain localities.

Mountain effect is generally referred to in this country as "multiple" and "bent" courses as distinct from the swinging caused by night effect. Only the latter phenomenon has been overcome with anything approaching complete success. The remedy appears to be suppression of sky-wave radiation through use of a vertical transmitting antenna which confines emission to direct propagation along the earth's surface (ground wave).

Interstation Interference. (*Radio Compass*).—It may be found sometimes that stations cannot be sufficiently separated by the radio compass on account of lack of receiver selectivity. This may not be obvious to the user, as the interference may not be audible. Sometimes this effect may be detected by wandering of the bearing.

General Failures. (*Radio Compass and Sometimes Direction Finder*). These may arise from a failure of part of the antenna system, or because the receiver has been maladjusted or improperly aligned.

Heavy Static. (*Radio Compass*).—A radio compass will deflect on ordinary heavy static and the static course laid out will be toward the center of the electrical disturbance.

Miscellaneous errors (*Radio Compass*) such as those due to heavy rain static, where the compass ceases to indicate, are obvious, as a change in heading produces no visual change in indication.

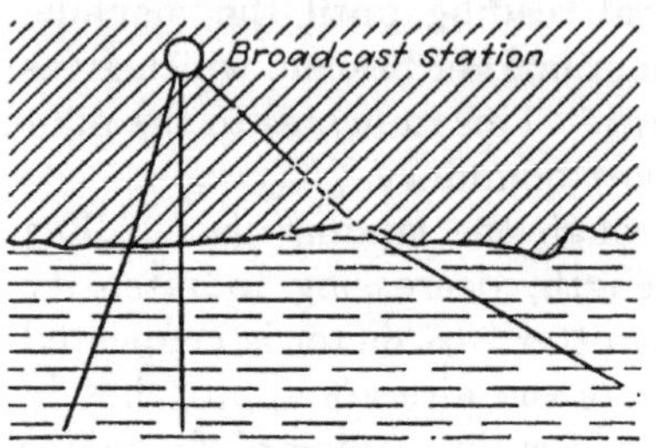

FIG. 69.—Coast refraction.

Pulsing of the indicator needle on keyed signals such as A or N can be corrected by averaging the readings.

Coast Refraction.—When radio waves cross a coast line at an angle, they are bent or refracted, as shown in Fig. 69. This bending is not noticeable at large angles, but at small angles it may be considerable. Bearings that make a small angle with the coast, or that pass alternately over land and water, should be regarded with suspicion.

THE TECHNIQUE OF RADIO NAVIGATION

Convergency.—Radio waves travel along great circles, which cut different meridians at different angles, so that a bearing of B measured at A (Fig. 71) would not be the reciprocal of the bearing of A measured at B. The difference is called *convergency*. Convergency is the angle between the tangents to two meridians and is found from the formula:

$$\text{Convergency in degrees} = \text{DLo.} \sin L_m$$

where DLo. = difference of longitude, and L_m = mid-latitude. The convergency may be found quite simply by graphical means, as shown in Fig. 70, or from tables.

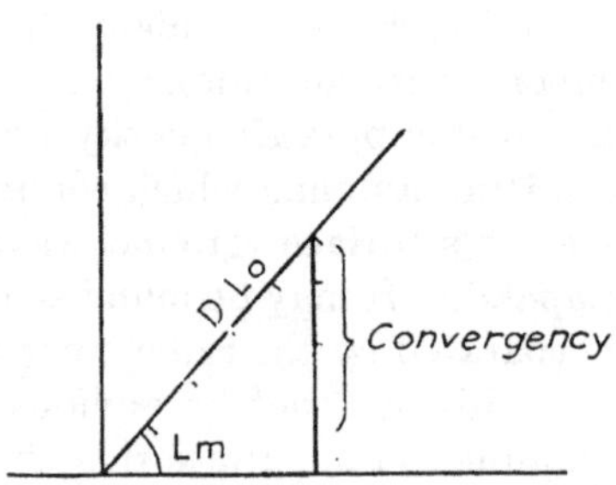

FIG. 70.—Finding convergency of meridians.

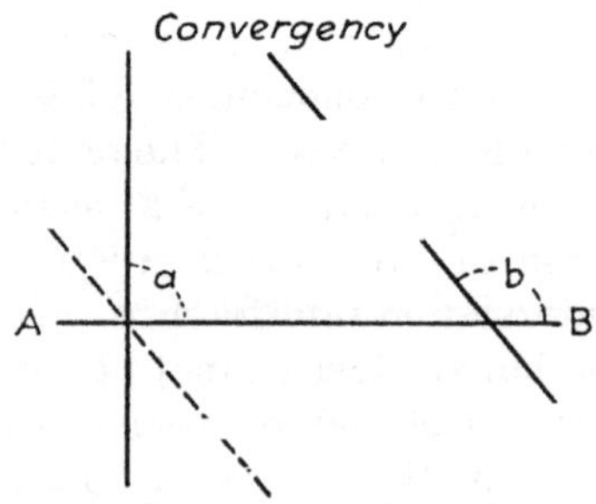

FIG. 71.—Convergency of meridians.

If a ground station at A (Fig. 71) observes the bearing of an aircraft at B, the bearing is plotted directly with reference to the nearest meridian. If the aircraft takes the bearing, convergency must be applied before plotting from A. The bearing must make the angle b at the aircraft and is plotted as the angle a from the station. In this case, convergency is subtracted from the observed bearing. In applying the convergency angle, remember that the great-circle course passes into higher latitudes than the rhumb course.

Example.—An aircraft observes the bearing of a radio station to be 250°; the difference of longitude is 3°, and the mid-latitude 45°S., so that the convergency is 2° (to the nearest degree). A rough sketch (Fig. 72) shows that the bearing should be plotted as the reciprocal of 252°, *i.e.*, 72°, from the station.

Convergency must also be applied when comparing bearings taken at the ground station with bearings taken at the airplane, as is sometimes done when calibrating the airplane for D.F. It should be remembered that radio bearings are relative to the head of the plane and should be converted to true bearings. This may be done by adding starboard or subtracting port radio bearings from the true heading of the plane.

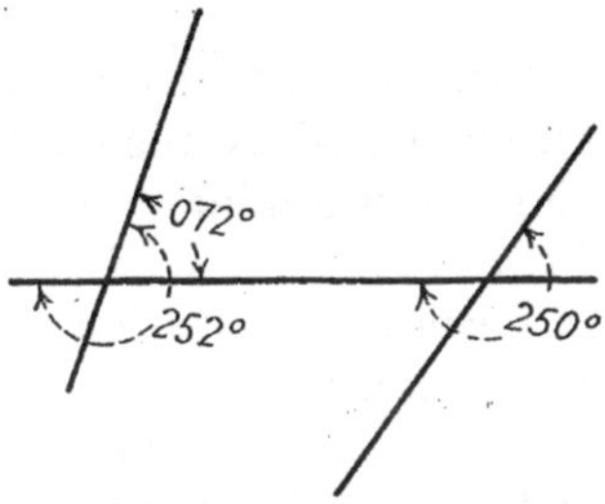

FIG. 72.—Plotting radio bearings.

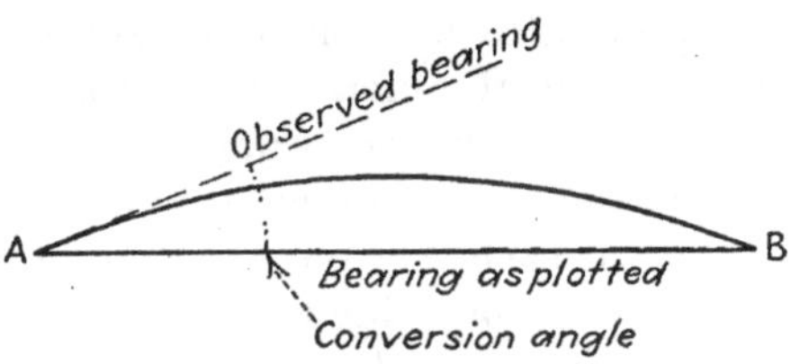

FIG. 73.—Converting great circle to Mercator bearings.

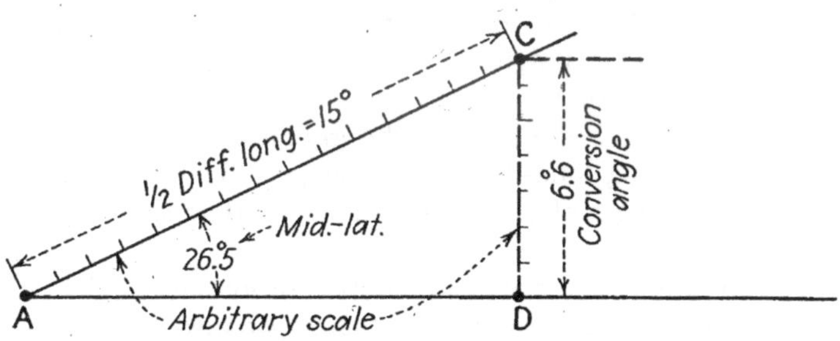

FIG. 74.—Finding conversion angle by plotting.

Conversion Angle.—In navigation we often wish to convert the radio or great-circle bearing to the rhumb-line bearing for plotting on a Mercator chart. This conversion is not necessary on the United States sectional and regional charts on Lambert projection. The angle between these bearings is half the convergency discussed above and is known as the *conversion angle*. It must always be applied, whether the bearing is taken by the radio station or by the aircraft.

In Fig. 73, if a radio station at *A* observes the bearing of an aircraft at *B*, the conversion angle must be added; if the aircraft observes the bearing, the correction is subtracted. Since the *conversion angle* is half the convergency, the same rules apply and may be found graphically as shown in Fig. 74. It is so easy to find the conversion angle graphically, that tables for doing the same thing should be ignored.

Example.—A plane in Lat. 38°03'N., Long. 55°00'W., by dead reckoning receives a radio bearing of 118° from Bar Harbor Station (Lat. 44°19'N., Long., 68°11'W.).

To find the Mercator bearing:

Bar Harbor: Lat. 44°19'N., Long. 68°11'W.
Dead-reckoning position: Lat. 38°03'N., Long. 55°00'W.
L_m, 41°11'N., DLo. 13°11'W.

Find graphically as explained above or enter table with middle latitude 41° and difference of longitude 13.2°. The correction is found to be +4.4° (since the vessel is in north latitude and eastward of station).

Mercator bearing is the radio bearing plus correction, or 118° + 4.4° = 122.4°.

Electrical "Deviation."—Bearings taken with a loop antenna or radio compass mounted in a plane are subject to errors caused by the electric conductivity in the metal of the plane affecting the incoming waves, so that the plane must be swung until signals have been received from all directions and a table of corrections similar to a compass deviation table made out for each relative direction of incoming signals.

The "deviation" due to local metal can be almost entirely eliminated in most installations by the use of compensators, which are grounded wires over or under the loop. By varying their number and distance from the loop the effect of local metal can be balanced out. This is exactly equivalent, in a magnetic compass, to compensating it so that it points to magnetic north.

Special Projections.—Charts on the gnomonic projection are sometimes used for plotting long-distance bearings. The compass roses centered on the observing stations are distorted, so that the bearings will be correct when plotted.

The best practice is to use Lambert projection charts on which both the great-circle courses and the distances may be measured with required accuracy.

Curves of Equal Bearing.—Where it is intended to do long-distance navigation by means of short-wave D.F., it may be of advantage to prepare a chart showing curves of equal bearing to the radio stations that it is proposed to use. If the bearings are observed at the stations, these curves are great circles. If the aircraft observes the bearings, the curves are lines passing through all places from which the bearing of the station is the same when taken at the aircraft. A Mercator chart prepared for the north Atlantic is shown in Fig. 75. At long distances, it is almost essential to use such charts, owing to the large difference in angle between the true position line (the curve of equal bearing) and that plotted as a rhumb line.

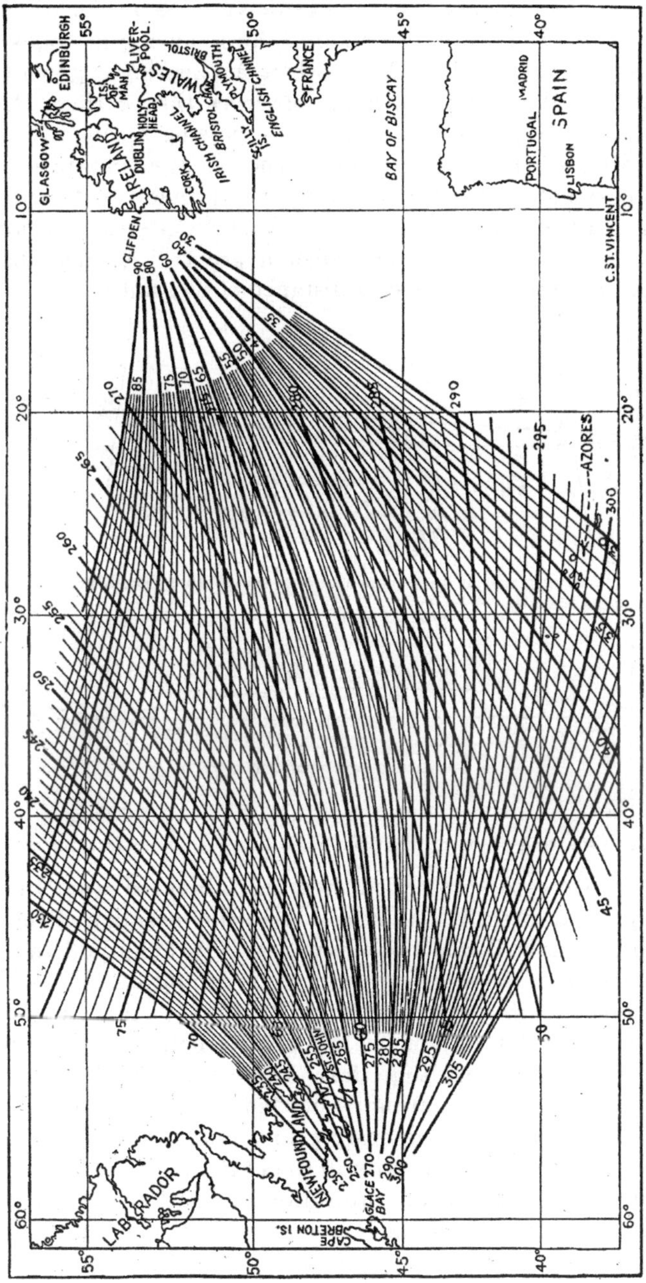

FIG. 75.—Curves of equal radio bearings to Glace Bay and Clifden from points in the north Atlantic, plotted on a Mercator map.

Plotting Position Lines from Long-distance Radio Bearings.—If a chart of equal bearings is not available, the following methods may be used. Two cases are considered: (1) when the bearing is given from the ground station, (2) when the aircraft observes the bearing. In case 1 the position line is a part of the great circle passing through the position of the aircraft and the radio station. The position line in case 2 is a part of the curve of equal great-circle bearing previously described.

Plotting may be done on a map or chart on Mercator's projection, or on some other projection for which a great circle does not differ sensibly from a straight line over a distance of 600 miles.

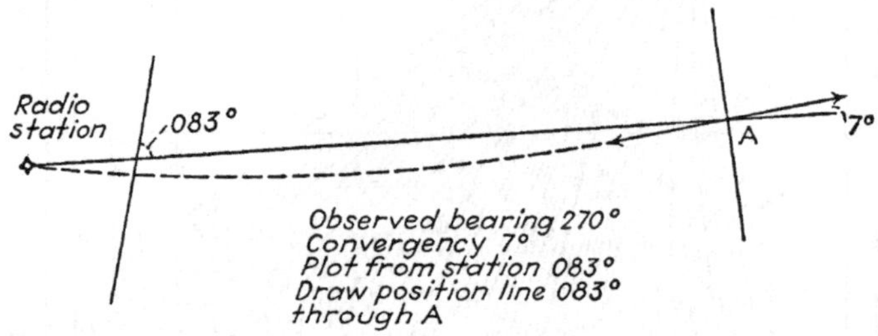

FIG. 76.—Plotting position lines from radio bearings on a non-Mercator chart.

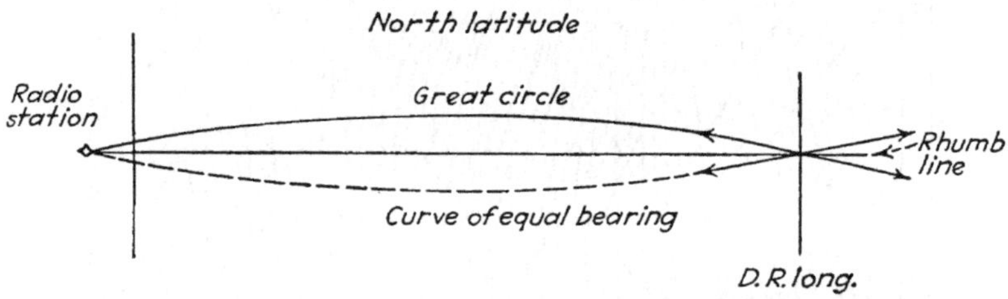

FIG. 77.—Plotting position lines from radio bearings on a Mercator chart.

For a non-Mercator projection, bearings observed from the ground station are plotted directly with reference to the meridian of the station. For a bearing observed from the aircraft, a D.R. longitude and latitude are chosen, and from these the convergency is computed. This is applied to the observed bearing and the reciprocal of the corrected bearing is plotted from the radio station. Through the point *A* (Fig. 76) where this line cuts the D.R. longitude draw a line making an angle equal to the convergency with the bearing as plotted. The line so obtained is a tangent to the curve of equal bearing and does not differ appreciably from the true curve, even if the error in the D.R. longitude is considerable.

For plotting on a Mercator projection, a D.R. position is assumed as before and the conversion angle computed. For a bearing given by the radio station, the conversion angle is applied, and the rhumb-line bearing from the station plotted. When the aircraft observes the bearing, the conversion angle is applied, and the reciprocal of the rhumb line so obtained plotted. Through the point where the rhumb line cuts the assumed longitude, draw a line making an angle equal to the conversion angle with the rhumb line, as shown in Fig. 77. If the bearing is observed from the radio station, the line is drawn to represent the tangent to the

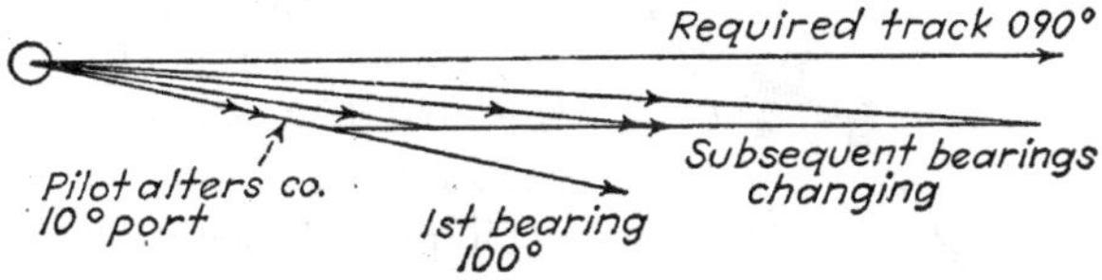

Fig. 78.—Use of radio tail bearings.

great circle. If the aircraft observes the bearing, the position line is plotted as a tangent to the curve of equal bearing. The correct direction for plotting the position line is determined by the rule that the great circle is convex to the nearer pole, whereas the curve of equal bearing is concave.

Use of Tail Bearings.—When attempting to maintain a track by radio tail bearings of the point of departure, it is desirable that the actual track line should be followed. In Fig. 78, the radio bearing discloses an error of 10° in the track. If the pilot merely turns parallel (*i.e.*, alters course

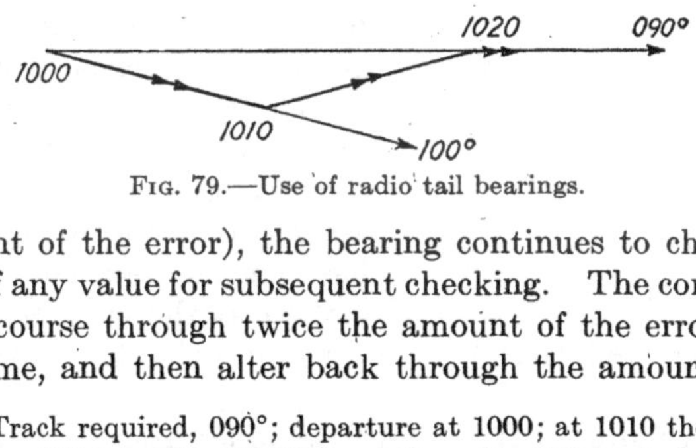

Fig. 79.—Use of radio tail bearings.

by the amount of the error), the bearing continues to change and thus ceases to be of any value for subsequent checking. The correct procedure is to change course through twice the amount of the error for an equal interval of time, and then alter back through the amount of the error.

Example.—Track required, 090°; departure at 1000; at 1010 the bearing is 100°. Change the course 20° to port until 1020 and then alter 10° to starboard, as shown in Fig. 79. Any subsequent change of bearing then indicates an error in the track. The careful navigator, when estimating the actual alteration of course and the time to fly on it, will consider the general direction of the wind and its effect on drift angle and ground speed.

Use of Homing Bearings.—When radio bearings of the destination are obtainable, either by means of a homing device or from the ground, a definite procedure should be followed if it is desired to fly on a line as straight as possible. With a heavy beam wind, this would necessitate traveling a slightly greater distance but it does not make much difference in fuel. This method, however, is nearly foolproof, provided the sending and receiving sets do not get out of order. Also, it has been demonstrated that practically no time is lost by using the method. Figure 80 illustrates the effect of merely setting the successive bearings on the compass and heading toward the station. It will be seen that the course is changed through about 60°, the change being more rapid as the station is approached. If bearings are obtained from the ground, a point is reached where the bearing changes too rapidly to be followed, and, if the visibility is poor, the pilot overshoots the station.

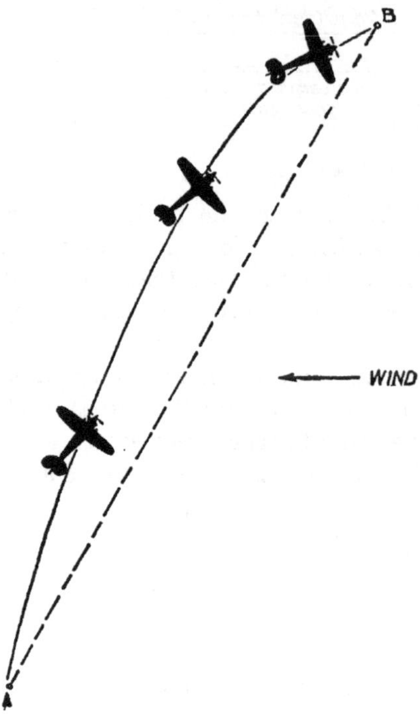

FIG. 80.—Allowance for wind drift.

The first case to be considered is that of approaching by D.F. in conditions where the surface is visible. Before a bearing is obtained, a drift reading should be taken. Suppose that the course steered is 240° and that the drift is 10° to port, *i.e.*, the track is 230°. If the observed bearing of the destination is 235°, drift allowance will be about the same for 235° as for 230°, so the pilot should steer 245°. The important point is to use radio bearings and drifts together whenever possible.

The second case deals with approach by successive bearings when drift cannot be observed. Let us assume that the navigator or pilot has some idea of his distance from the station at the moment of obtaining the first bearing, and that he has an approximate knowledge of his ground speed.

In Fig. 81, the aircraft *A* is about 100 miles from the station *B* when the bearing is found to be 090°. After flying about 25 miles, the bearing

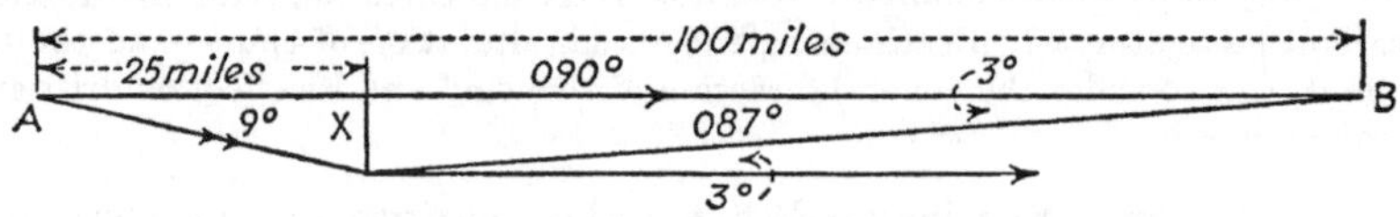

FIG. 81.—Correction for homing course.

changes to 087°; what alteration of course should be made? The aircraft has drifted laterally a distance X, which subtends an angle of 3° at the station. By simple proportion, the angle at *A* should be about $3 \times 3 = 9°$. If the pilot alters 9°, he merely makes good a track of 090°. A further alteration of 3° must be made to make good a track of 087°. The total alteration is 12° or four times the change of bearing;

the distance flown is one quarter of the initial distance. Hence the rule is to multiply the change of bearing by the reciprocal of the fraction of the distance flown, and alter the course in the same direction that the bearing changes.

Example.—If the bearing changes 4° clockwise while flying one-fifth the initial distance, the course must be altered $4 \times 5 = 20°$ to starboard.

It is better to work on time than on distance. The slide rule may be used to deal with awkward figures.

Example.—The pilot expects to reach his destination at 1500. At 1408 the bearing is 44°; at 1422 it is 41°. The initial time is 52^m, the time flown 14^m, and the change of bearing is 3° left. Since $52/14 \times 3 = 11$, the course is altered 11° to port.

If the pilot has no idea of his distance from the destination, he must rely on overcorrecting. Each time the bearing changes he should alter course by three times the amount of the change. A couple of corrections of this description will nearly always ensure a straight approach.

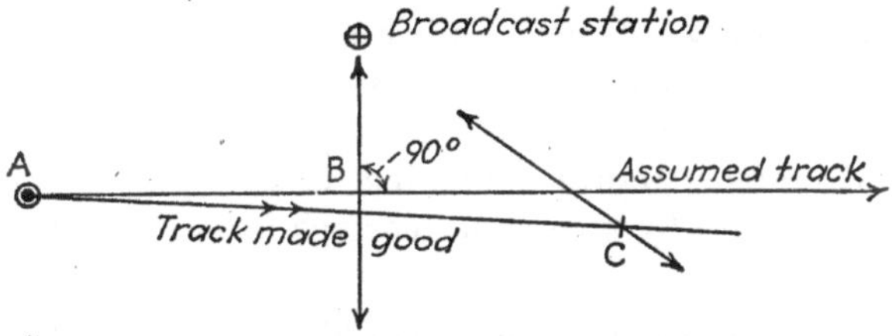

FIG. 82.—Use of radio station not on the route.

Use of Stations Not on the Route.—An aircraft equipped with a direction finder is to fly between two airports, neither of which has a radio station, but there is a broadcasting station some distance to one side of the track. How can best use be made of the direction finder?

Set the course by using a forecasted or assumed wind. Obtain a bearing of the station when it is as nearly as possible at right angles to the track, and note the time. Plot the bearings and, using the distance AB (Fig. 82), calculate the ground speed. Take and plot a second bearing, noting the time. Multiply the ground speed by the interval elapsed since departure, and cut the second bearing at this distance (AC). This gives the track made good, and any error may now be corrected. If it is not possible to get a right-angle bearing, the problem must be attacked in a different manner.

The point of departure (Fig. 83) is not treated as an actual position but merely as a point on the position line joining it to the station. The track is found by using a scale of time as in the well-known method of finding track from three bearings. The intersections of the track with the position lines represent fixes, within the limits of accuracy of the bearings.

Theoretically a plane flying 100 miles on the fixed-loop homing device in a 50-mile cross wind at 180 m.p.h. air speed would lose 1.5 min. over the time required for straight flight. The smallness of this loss in time is due to the fact that for straight-line flight the plane must be continually headed into the wind by the amount of the drift angle. This reduces the speed over the entire course. On the curved-homing course the speed is not reduced at the start, although the reduction at the end of the flight is greater.

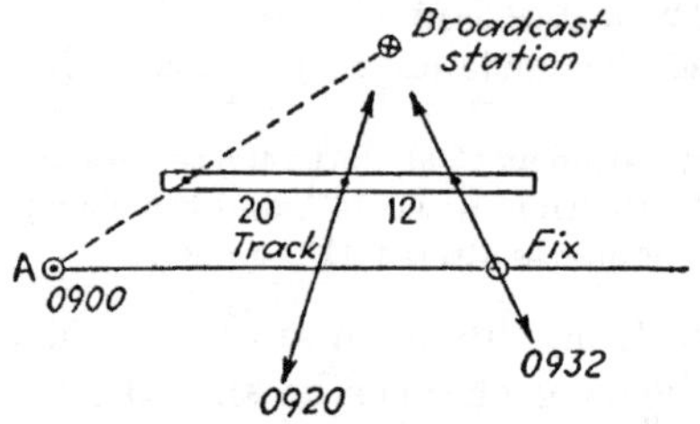

FIG. 83.—Homing by course and run between bearings.

Although the time lost in the homing method is not serious, there are other objectionable features as noted, so that a means for homing on a direct course on any suitable station is desirable. This may be accomplished by means of a D.F. loop that can be rotated to allow for wind drift. With allowance for drift feasible, the homing method has the advantage of simplicity, longer range, and more transmitting stations available.

Indirect Homing.—Indirect homing is the process of navigating to a given destination by means of bearings from a radio station off the direct route.

Example.—It is desired to fly to an airport 300 miles away. Fifty miles on the near side and 20 miles off the track, there is a radio station. Navigation in thick weather should be via the radio station, as this ensures accuracy, while hardly affecting the total distance. The station is approached on a steady track by means of the homing procedure. The last 50 miles are flown by holding the appropriate tail bearing on the radio station.

Example.—The destination is 250 miles due north, and there is a radio station 30 miles northeast from it. Fly well to the west of north until the radio station bears northeast; then turn on to that bearing.

Flights navigated by radio should be carefully prepared in advance by noting the transmitting times and frequencies of the various stations. Bear in mind that "the longest way round is often the shortest way home" in this class of work.

Radio-compass Station.—Radio-compass or RC stations are operated by the Navy Department and are usually located along the seacoast. These stations are operated principally for the use of the mariner, but are also available for the use of the airman. Since some of these stations are not calibrated for arcs over the land, *i.e.*, the bearings corrected for local conditions, the navigator is cautioned to get complete data on the station used.

Aircraft Radiotelephone.—The use of aircraft radio for communication from plane to plane, from plane to ground, and for summoning aid in an emergency has been greatly facilitated by the development of the radiophone. Radiophone, for aeronautical use, has been a goal sought since commercial air lines began scheduled operations calling for flying irrespective of weather conditions and at night.

FIG. 84.—Practical use of the radiophone. (*Courtesy of Boeing System.*)

The ground-plane and plane-to-plane telephone service calls for complicated engineering in installation, but its use by the pilot is simplicity itself. The entire equipment is nearly automatic in operation, requiring no adjusting on the part of the pilot, whose full attention can be directed to the operation of his engine and to keeping the airplane on its course.

As soon as the pilot takes off, the ground station checks on the pilot and adjusts its mechanism to tune in with the pilot, who does not adjust his set in flight. The pilot turns the switch to *receive* and waits for the periodic reports that give him the latest reports on the weather, on wind velocity at various altitudes, dispatching company orders, and other helpful information. If he wants to talk to the ground, he switches to *send* on the panel board.

Distance range of transmitters in planes and at stations is about 200 miles and, when half that distance away from the ground station last passed, the pilot picks up the next ground station ahead, and so on during the flight. The device is set to *receive* at all times the pilot does not want to call, and at *neutral* only when the plane is on the ground. Transport planes have a pilot and a copilot, who can also fly the plane. Figure 84 shows the practical use of the radiophone for navigation.

Conclusion.—The navigator who proposes to make effective use of radio bearings would do well to investigate the principles underlying the examples given in this chapter. Let every bearing be chosen for a definite purpose and see that the maximum value is obtained from it. Most pilots obtain far more bearings than they actually need, thus going a long way toward overloading the ground organization. Bear in mind that an isolated radio bearing has only three possible uses:

1. A check on track.
2. To check ground speed.
3. As a leading line.

Make sure that the bearing chosen serves one of these purposes.

A navigator's capabilities may be judged by the inverse ratio of the number of bearings he requires on any given flight. Radio direction-finding facilities should be used as guides, not as a substitute for railroads.

Government weather information is discussed under aids to navigation in Chap. VI.

CHAPTER VI

DEAD RECKONING—THEORY

Dead reckoning is the method of determining a position by keeping an account, or reckoning, of the direction and distance run from a previously known position called the point of departure. The navigator's duty is to keep a careful record of his dead-reckoning position at all times and to check this position by piloting, by celestial navigation, and by radio.

Dead reckoning requires a knowledge of the direction and the speed of flight. In still air direction is obtained from the compass, the speed is read from the air-speed indicator, and the navigator's task is simple. Generally, however, the position of an aircraft on the surface of the earth is the resultant of two distinct motions: the motion of the plane in the direction in which it is headed and the motion of the air itself relative to the earth, which is commonly referred to as the wind.

The accuracy with which the course and distance flown are determined depends upon the accuracy of the instruments used, the skill of the operator, and the conditions in the air. At best, dead reckoning may not be expected to produce extreme accuracy.

Value of Dead Reckoning.—The advantage of dead reckoning as compared with other methods is that it is available when lowered visibility obscures landmarks and makes piloting difficult; when the sky is overcast, making celestial navigation impossible; when flying over unmapped country or over water; and when for any reason navigation by wireless is not available. It is the one method that is at all times available, and the finished navigator will always make use of it, whether or not other methods are employed.

The disadvantage of dead reckoning lies in the fact that the farther the plane travels, the greater is the probable error in the dead-reckoning position, and the further fact that it does not determine a position definitely after that position has once become uncertain. Once the dead-reckoning sequence of a plane is broken, dead reckoning is of little value until the position has been reestablished by other methods of navigation. A pilot at such a critical time may find that simple piloting will not help him, or that the direction finder will not work, or that celestial navigation cannot be used. But the skillful and resourceful navigator who has mastered all four methods will almost always be able to find one method that will work. If no method can be made to work, the pilot should make the best landing possible.

Unless an account or reckoning is kept of an aircraft's course and speed, it would be "lost" continuously except for instants when definite positions are determined by the other three methods. By keeping a simple record of course, distance, and time, and estimating the wind, the navigator may estimate the plane's position at any instant with an accuracy dependent on the reliability of his determination of these quantities.

Of the various methods of navigation, dead reckoning is unquestionably the most universally used. It is always required in some form, however simple, even when reliance is placed on one or more of the other methods. Dead reckoning can never be dispensed with entirely.

The Importance of Wind.—In actual practice, finding one's way by dead reckoning hinges largely upon the success with which the wind drift may be determined. A good compass gives the navigator his heading with reasonable accuracy. The air-speed meter gives a close approximation to the speed of the plane through the air. However, the essence of the problem is not the course and speed through the air but the course and speed made good over the ground, and these depend upon the accuracy of the determination of wind direction and velocity.

The most practical means for obtaining wind data, and the method ordinarily used, is for the navigator, before taking off, to obtain directly from the Weather Information Center all the data needed. Expert observers working together collect and digest the data and prognosticate the wind and weather. Since wind effect is inseparably associated with dead reckoning, we shall deal with wind before proceeding any further.

WIND AND ALLIED PROBLEMS

When an airplane leaves the ground and flies in any direction, the motion through the air alone is not sufficient to enable its motion relative to the ground to be stated; it is necessary to know also the motion of the air itself relative to the ground, *i.e.*, the wind.

The problem is easily visualized by thinking of a man in a boat rowing over a river and trying to reach a pole driven into the river bottom and projecting above the surface. The man and the boat move relatively to the water and the water also moves relatively to the pole. The man soon finds that from positions straight up the river or straight down the river as measured from the pole he can row straight for the pole, but that his rates of approach are different; he also finds that from a position abreast of the pole he must direct his boat to some extent upstream as well as toward the pole. In the air the conditions are very similar, with the wind taking the place of the river current, and it is just as necessary to set a course to some extent upwind as it was to head to some extent upstream.

Definitions.—When flying through moving air, it is necessary to use definite names for certain directions and speeds. The following names have been standardized:

The *heading* is the direction in which the airplane is pointing. This may be relative to the true north, the magnetic north, or the compass indication of north, and so may be *true heading, magnetic heading,* or *compass heading.* See also the official definitions in Chap. I. Some duplication is given here to make the picture more complete.

The *course* is the direction in which it is desired that the plane should travel. This may be relative to the true north, the magnetic north, or the compass indication of north, and so may be *true course, magnetic course,* or *compass course.* It is the course laid out on the chart and may be defined as "intended track."

The *track* is the direction of the actual path over the ground covered by the airplane. Except when there is a dead calm or when the wind is blowing exactly along the *heading* or the *heading reversed,* the track will differ from the *heading.* It will also be *true track, magnetic track,* or *compass track* as before.

The distinction between these three terms is very fine. Often they are carelessly used to mean the same thing, but their proper use ensures an easier and more exact understanding of the problem.

The *drift angle* is the angular difference between the heading and the track. The drift is said to be *right* when the airplane is driven by the wind to the right of the direction in which it is pointing; if the airplane is driven to the left, the drift is called left.

The *air speed* is the true speed of the airplane through the air and is therefore always measured along the *heading.*

Indicated air speed (I.A.S.) is the reading of the air speed indicator.

Calibrated air speed is the reading of the air-speed indicator, corrected for instrumental and installation errors.

The *ground speed* is the speed of the airplane over the ground and is therefore measured along the track.

The *no-wind position* of an aircraft at any moment is the position where the aircraft would have been at that moment if no wind had been blowing.

The *wind direction* is the direction *from* which the wind is blowing. Surface wind directions are reported by the Weather Bureau as one of 16 compass points, such as E.N.E., whereas winds aloft are reported as one of 36 ten-degree directions, such as 260°.

The *wind speed* is the speed at which the wind blows. Wind direction and wind speed may also be considered together as just *wind;* for example, wind east, 30 m.p.h. means that the wind is blowing from the east at 30 m.p.h. The wind is always given as a true direction (*i.e.,* with reference to true north); consequently, when working wind problems the *heading*, *course*, and *track* will also necessarily be the true heading, true course, and true track and this will be assumed in all subsequent work.

The above elements are clearly shown in Fig. 85.

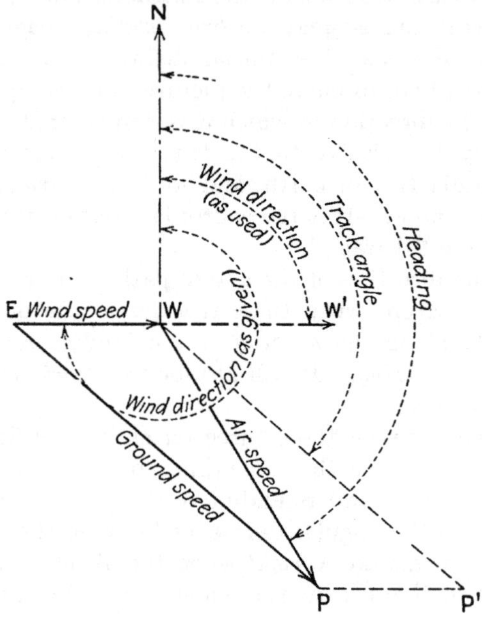

FIG. 85.—Dead-reckoning terms.

To save himself from making careless slips, the student is advised to judge every triangle of velocities he draws, by the following four common-sense considerations:

1. The three sides of the triangle must be heading and air speed (AS), track and ground speed (GS), and wind speed and direction, coupled like that.
2. The airplane is blown by the wind *off* its heading *onto* its track.
3. The heading will always be a little into wind from the track.
4. If the wind is blowing from appreciably ahead of the beam, the GS must be less than the AS; if it is blowing from appreciably abaft the beam, it must be more than the AS.

Triangle of Velocities.—The speed and direction of an airplane flying through the air, which itself is moving bodily in some direction, are given by the triangle of velocities.

Any velocity may be represented by a straight line with an arrow, by making the direction as shown by the arrow represent the direction of the motion, and by making the length of the line proportional to the speed. The resultant of two simultaneous velocities can then be found by starting from any point and drawing a line so that its length and direction represent one of the velocities, and from the end of this line drawing another line to represent the second velocity. The resultant velocity is found by closing the triangle in the reverse direction.

In Fig. 86 let the line $E \to W$ represent a wind of 30 m.p.h. from the east; and let the line $W \to P$ represent the heading and air speed of an airplane flying northeast through the air at 80 m.p.h. These two velocities form two sides of the triangle EWP, starting from E to W for the wind direction and speed and going from W to P for the heading and air speed. The resulting motion due to the combination of these two velocities can be represented by the line $E \to P$, which closes the triangle in the reverse direction; in other words, $E \to P$ represents the track and ground speed. This triangle of velocities is of the utmost importance in all problems of dead reckoning, and it must be thoroughly understood.

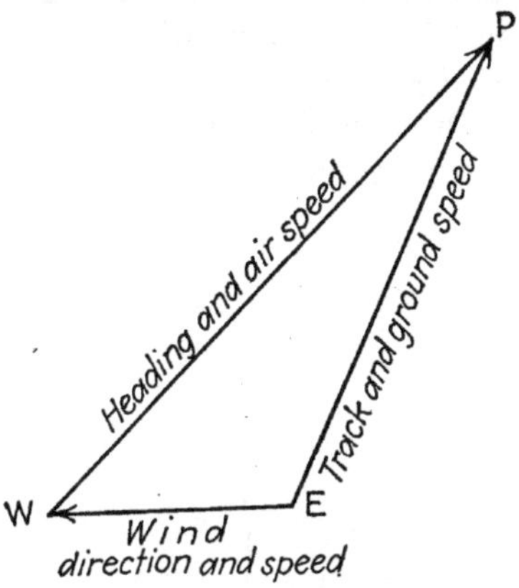

Fig. 86.—Triangle of velocities.

In the chosen lettering:

E = earth.

W = wind.

P = plane.

S = ship, while two letters connected by an arrow indicate direction and distance (or velocity) between the two points (Fig. 90).

For example, $E \to P$ indicates the speed and direction of the plane relative to the earth—in other words, the plane's track and ground speed; $W \to P$ represents the speed and direction of the plane relative to the wind, which is the heading and air speed.

The triangle of velocities will now be used to solve a series of typical problems:

Problem 1.—To find the heading to make good any given track when the wind and air speed are given.

Draw $E \to W$ to represent the wind in speed and direction, and from E draw $E \to P$ to represent the track (Fig. 87). With center W and radius equal to the air speed, draw a circle cutting $E \to P$ at P and join $W \to P$. Then $W \to P$ represents the heading, and the length of $E \to P$ gives the ground speed.

Problem 2.—To find the heading and air speed to make good any given track and ground speed, given the wind.

Referring to Fig. 86, it will be clear that a little variation of the construction is all that is necessary. Draw $E \rightarrow W$ to represent the wind, and draw $E \rightarrow P$ to represent the track and ground speed; the difference from Prob. 1 is that the length $E \rightarrow P$ is now known before completing the triangle. Join $W \rightarrow P$; then $W \rightarrow P$ represents the heading and air speed required.

Problem 3.—To find the wind, given the heading and air speed and also the track and ground speed.

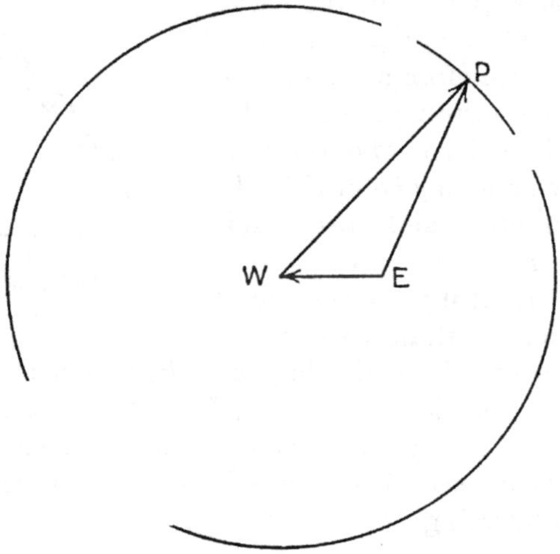

FIG. 87.—Determination of heading and ground speed.

The order of drawing the sides of the triangle is different (see Fig. 86). From W draw $W \rightarrow P$ to represent the heading and air speed, and from P draw $E \leftarrow P$ (with arrow reversed) to represent the ground speed and track; join EW. Then $E \rightarrow W$ represents the wind.

Problem 4.—To find the wind from two or more drift observations on known headings for a given air speed.

Referring to Fig. 88, from a center W draw $W \rightarrow P$, $W \rightarrow P_1$, and $W \rightarrow P_2$ to represent the known headings, with a radius equivalent to the given air speed, and from P, P_1, and P_2 draw PE, P_1E, and P_2E so that the angles WPE, WP_1E, and WP_2E are equal to the observed drifts. It will be found (if the observations are all absolutely accurate) that the three lines PE, P_1E, and P_2E cut in E; in actual practice the three or more lines will form a small triangle or other figure instead of intersecting at one point, and the center of the small figure can be taken as the point E. Join EW; then $E \rightarrow W$ represents the wind.

Problem 5.—The out-and-home problem.

In Fig. 88, P_1 has been so chosen that the tracks $E \rightarrow P$ and $E \rightarrow P_1$ are exactly opposite in direction and therefore represent out-and-home tracks respectively. Thus if $E \rightarrow P$ is the outward track, $E \rightarrow P_1$ will be the track home, and vice versa. The length $E \rightarrow P$ also represents the outward ground speed and $E \rightarrow P_1$ the homeward ground speed; these are not equal unless there is no wind or the wind is at right angles to the track.

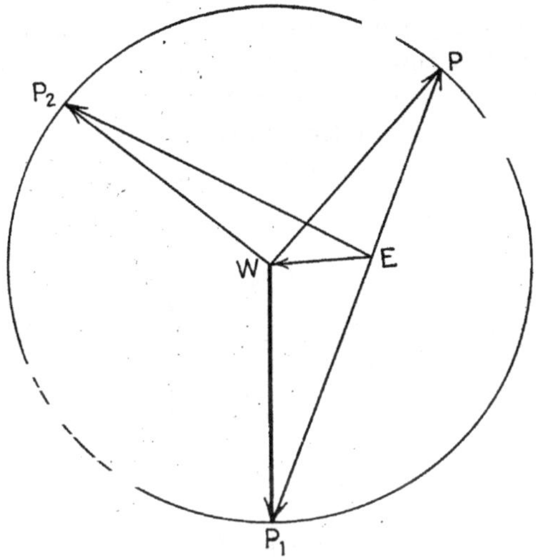

FIG. 88.—Determination of wind from drift observations.

Now, by geometry, because $W \rightarrow P = W \rightarrow P_1$ (being radii of the same circle), it is known that the angles WPE and WP_1E are equal,[1] or that the drift on the outward journey is equal and opposite to that on the homeward journey. For example: given wind 30 m.p.h. from east, air speed 80 m.p.h., and outward track 25°, the outward heading was 45°, giving 20° left drift. Now because the two drifts are equal and opposite, the navigator just notes that 45° less 25° is 20°, and 20° from 25° is 5°, the opposite of which is 185°; thus 185° is the return heading. If the outward drift had been right drift, say 10°, and track 25° as before, the outward heading would have been 15° and the calculation would have been 15° to 25° is 10°, and another 10° is 35°, the reverse of which is 215°, the homeward heading.

Put into words, the procedure is to go from the outward heading to the outward track and then by an equal step to the homeward heading reversed, when adding or subtracting 180° gives the homeward heading. This property is most important and is to be utilized on all suitable occasions; for instance, when a solo pilot flies into mist he can at once turn around or retrace his *track* if he so desires.

[1] By geometry, angles opposite equal sides of an isosceles triangle are equal. By construction (see Prob. 4) triangle PWP_1 is isosceles.

Problem 6.—The out-and-home problem continued.

Time to Turn and Radius of Action.—Redrawing Fig. 88 as Fig. 89, and omitting $W \rightarrow P_2$ and $E \rightarrow P_2$ as not of interest at the present moment, we have $E \rightarrow P$ as the outward ground speed and $E \rightarrow P_1$ as the return ground speed. These are unequal on an out-and-home flight, while the distance out and home is the same, whence it can be shown that the length $E \rightarrow P_1$ can represent the time taken on the outward flight and $E \rightarrow P$ the time taken to return, both measured on some unknown scale of time; consequently the ratio of $(E \rightarrow P_1)$ to $(E \rightarrow P)$ gives the ratio of the out flying time to the home flying time.

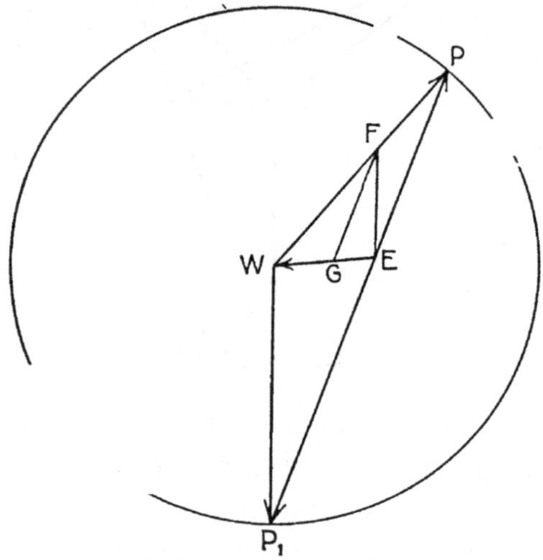

Fig. 89.—Determination of radius of action.

If preferred, this may be stated as follows: the ratio EP_1 to PP_1, the sum of the two lengths, is the proportion of the total time after which it is necessary to turn back. From this it would not be difficult to calculate the radius of action or distance at which to turn back, as given below, but this can be more easily arrived at as follows:

Draw EF parallel to $W \rightarrow P_1$ to cut $W \rightarrow P$ in F, and from F draw FG parallel to PEP_1 to cut $E \rightarrow W$ in G. Then FG is the radius of action per fuel hour, and, multiplied by the number of hours for which fuel is available (less safety margin), gives the total radius of action. The proof of this may be left to the student as an exercise in plane geometry.

Formulas for Radius-of-action Problems

The problem here considered is that of determining how far a plane may fly from its base under different wind conditions.

R = distance from base at time of turning back, *i.e.*, radius of action.

t_1 = time of outward flight.

t_2 = time of return flight.
T = total time of flight = $t_1 + t_2$.
S_1 = rate of departure (ground speed) from base.
S_2 = rate of returning (ground speed) to base.

Since

$$t_1 = \frac{R}{S_1} \quad \text{and} \quad t_2 = \frac{R}{S_2}$$

we have

$$T = \frac{R}{S_1} + \frac{R}{S_2} = \frac{R(S_1 + S_2)}{S_1 S_2}$$

from which

$$R = \frac{T S_1 S_2}{S_1 + S_2}$$

But $R = t_1 S_1$, so

$$t_1 = \frac{T S_2}{S_1 + S_2}$$

$$t_2 = \frac{T S_1}{S_1 + S_2} = T - t_1$$

Moving Base and Relative Wind (Interception).—In flying over the sea it is quite normal practice to fly from the aircraft carrier and to return to it, so that the aircraft carrier is really a moving base, of which the course and speed are known. The same type of problem is now presenting itself in civil aviation when airplanes fly out to transfer passengers and mails to ships that have already put to sea and are out of sight of land, and conversely when airplanes fly from ships approaching land in order to land passengers and mails before the arrival of the ship.

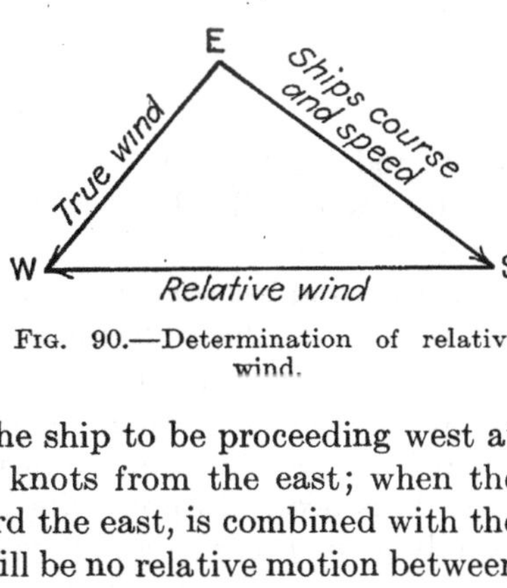

FIG. 90.—Determination of relative wind.

All such problems can be greatly simplified by considering the ship or moving base at rest and combining the ship's motion reversed with the actual wind to constitute what is called the *relative wind*. The truth of this can be easily seen by supposing the ship to be proceeding west at 30 knots and the actual wind to be 30 knots from the east; when the ship's speed reversed, *i.e.*, 30 knots toward the east, is combined with the wind, *i.e.*, 30 knots from the east, there will be no relative motion between ship and air and the relative wind is zero. All the problems to which the triangle of velocities has been applied can be solved in exactly the same way by first combining the actual wind with the course and speed of the ship or moving base to obtain the relative wind, and then proceeding as before.

To illustrate this, suppose a carrier is proceeding at 24 knots on course 127° while the wind is 18 knots from 37° (see Fig. 90). From E draw $E \rightarrow S$ to represent the ship's speed and course, and from E draw $E \rightarrow W$ to represent the wind, drawing the line from E toward 217° (180° + 37°), the direction toward which it is blowing; join $S \rightarrow W$.

The relative wind furnishes a very easy method of extending the problems solved to cover other problems.

Problem 7.—The out-and-home problem (moving base).

If the wind is determined by drift observations (see Prob. 4) of the aircraft carrier, the wind found, represented by $S \rightarrow W$, is the relative

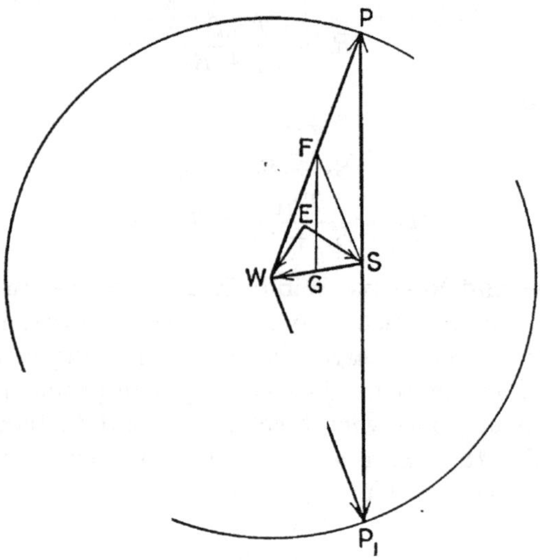

FIG. 91.—Problem of moving base.

wind, and all the information required (relative to the ship) can be obtained in exactly the same way as for the fixed-base problem given in Probs. 5 and 6.

If the wind has been ascertained by some other method and is given as the true wind, this only means that two extra lines must be drawn as shown in Fig. 91. From E draw $E \rightarrow S$ to represent the ship's course and speed and from E draw $E \rightarrow W$ to represent the wind, drawing the line from E in the direction toward which the wind is blowing, and join $S \rightarrow W$ as described above for finding the relative wind. Now proceed exactly as in Probs. 5 and 6, remembering that all the quantities found are relative to the aircraft carrier.

It will now be preferable to call the quantities $S \rightarrow P$ and $S \rightarrow P_1$ by other names. In the case of the fixed base they were the outward and homeward tracks and ground speeds, but these terms can no longer be used as they are no longer relative to the ground but to the carrier or moving base; they are now called the *bearing* and *rates of departure and approach*, respectively.

Problem 8.—The problem of intercepting a ship.

This problem is now extremely important and can be handled in the same way as the moving base. From the ship's known position or bearing, course, and speed, at the time the airplane takes off, the direction $S \rightarrow P$ is determined (Fig. 91). Using the ship's course and speed and the true wind, which is known (or if not known can be found as described previously), draw $E \rightarrow S$ and $E \rightarrow W$ and find $S \rightarrow W$, the relative wind. From S set off the ship's bearing $S \rightarrow P$ and join $W \rightarrow P$, which gives the heading for air speed $W \rightarrow P$.

Problem 9.—The problem of flying from the ship to land.

That is similar to Prob. 1. The ship's position being fixed for the time of the plane's departure makes it a fixed-based problem. $E \rightarrow W$ is true wind and $E \rightarrow P$ track or bearing. Adding $E \rightarrow S$ and obtaining $W \rightarrow S$ (relative wind) is superfluous and makes a complicated problem out of a simple one. True wind can always be found aboard ship before taking off.

Problem 10.—The problem of flying out from one fixed base and returning to another fixed base.

This problem becomes exactly the same as Prob. 7 by considering the first base to move uniformly toward the second base in such a manner as to cover the distance in the predetermined time of flight. By this means the outward and homeward headings, and the times out and home, can be found; the radius of action will now be measured from some point between the two bases; if the track and ground speed are drawn, using the true wind, the knowledge of the outward and homeward times gained from the use of the relative wind can be used to determine the exact position of the turning point and the outward and homeward tracks.

NOTE: The scope of this book does not allow for an exhaustive treatment of relative movement problems. For a complete presentation of the subject the reader is referred to Tornich's "Radius of Action of Aircraft," by Weems System of Navigation, Annapolis, Md., 1940.

Wind Diagrams.—Figures 86 to 91 are diagrams in which each line represents direction and speed and must not be confused with other diagrams in which the lines represent direction and distance. The attentive reader will at once refer to line FG of Fig. 89, which is stated to represent the radius of action per fuel hour; this is a rate of so many miles per hour and, although not usually regarded as a speed, should be so regarded. In all the wind diagrams the same letter has been used for the same purpose.

An attempt has been made to treat the subject systematically by making W, the center of the only circle used, the intersection of the air speed and wind; this was done to avoid the confusion that may easily arise when arcs or circles are drawn with their centers at different corners of the diagram.

Dead-reckoning Methods.—Having discussed the important subject of wind and having previously discussed the means of determining direction (under Compass), we are now ready to take up the dead-reckoning methods of tracking a plane's position in the air. The means for determining air speed (air-speed meter) and for measuring ground speed and wind drift will be discussed in the next chapter.

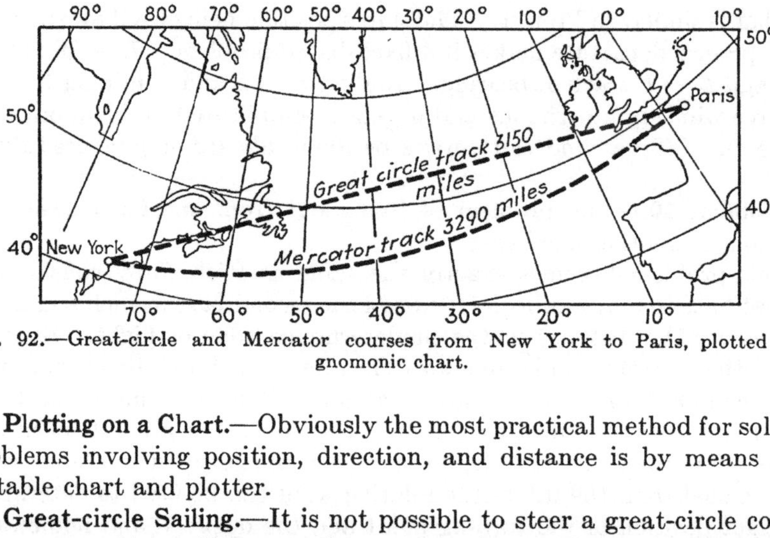

FIG. 92.—Great-circle and Mercator courses from New York to Paris, plotted on a gnomonic chart.

Plotting on a Chart.—Obviously the most practical method for solving problems involving position, direction, and distance is by means of a suitable chart and plotter.

Great-circle Sailing.—It is not possible to steer a great-circle course by compass, since this course by compass is continually changing. In practice, however, suitable chords of the great-circle course may be steered, and when so steered give a close approximation to the great-circle course.

The most practical way of laying down a great-circle track is by working directly on a navigation chart drawn on the Lambert projection. It will be recalled that this is a conical projection so drawn that it will give the minimum distortion over the greatest percentage of area. A straight line drawn on this chart, although it does not fall exactly along the great circle, is within the limits of required accuracy.

For those who wish to compare the Lambert track with the great-circle track, the method of laying down the true great-circle track is given. It may be used where there is no sectional, regional, or skeleton Lambert chart available.

First the track is laid down on a great-circle chart by connecting any two desired points by a straight line. Unfortunately, a great-circle chart is not suitable for navigation; therefore, the latitudes and longitudes of convenient points along the great-circle track are transferred to the navigation chart. Compass courses and distances are then measured on the navigation chart for the chords or legs transferred from the great-circle chart (see Figs. 92 and 93).

In this connection small-scale skeleton great-circle and Mercator *control* charts will be found convenient. The great-circle control chart may be a blank chart showing fixed latitude and unnumbered longitude lines, which may be numbered as required.

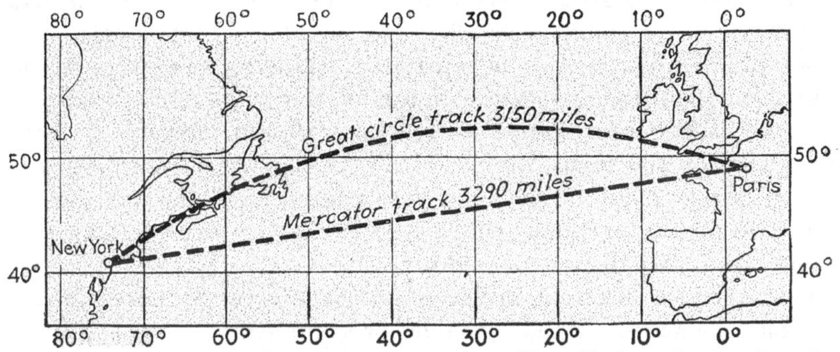

FIG. 93.—Great-circle and Mercator courses from New York to Paris, plotted on a Mercator chart.

Trigonometrical Calculation of Great-circle Course.—For more accurate work, such as computing the shortest distance between two points for record flights, the great-circle distance is calculated by spherical trigonometry as described in *Hydrographic Publication* 9 (Bowditch).

Great-circle Versus Mercator Flying.—In practice, we steer Mercator courses, *i.e.*, compass courses. When a long flight is made, these courses should be altered at suitable intervals in order to approximate the great-circle track. For distances greater than about 400 miles, and especially when in high latitudes, the saving in distance by approximating the great-circle track is considerable.

On his epoch-making flight from New York to Paris, Colonel Lindbergh followed the great-circle track, which he had previously computed carefully and plotted on a Mercator chart of the north Atlantic Ocean. The saving in distance by following the great circle on that flight was 140 nautical miles or 160 statute miles. The true compass course was changed to conform to the great-circle track at intervals of 100 miles.

Figure 92 shows the appearance of the great-circle course and the rhumb-line or Mercator course when laid down on a chart constructed on the gnomonic or great-circle projection. Figure 93 shows these courses as they appear when plotted on a chart constructed on the Mercator projection. It will be noted that the initial great-circle course is north of east (54°), while the final great-circle course is south of east (112°), with a continuous change along the course. While on the great-circle track, Lindbergh was always headed directly for Paris; if he had followed a Mercator track his plane would have been headed to the right of Paris until he got close to it. The great-circle track appears longer than the Mercator track in Fig. 93, because the Mercator projection distorts the areas in high latitude.

Conversion of Great-circle and Mercator Courses.—On long flights it is desirable to follow the great-circle track. Owing to the convergence of the meridians, the great circle cuts each meridian at a different angle, while the Mercator course cuts all meridians at the same angle. Since it is not practicable to steer a constantly changing course, it becomes necessary on long flights to steer chords of the great circle, each chord or leg being a Mercator course steered for a chosen interval of distance or time, say for 100 miles, or for 1 hr. At the end of each leg a new Mercator course must be set, the new course including not only the convergence of the meridians, but also the changes in variation, deviation, and wind drift discussed elsewhere.

The conversion of great-circle courses to Mercator courses and the reverse may be accomplished by tables or graphically.

We give below an example solved graphically. See also Chap. V for converting radio bearings to Mercator bearings. The formula is

$$\text{Conversion angle} \doteq \tfrac{1}{2}\text{DLo. } \sin L_m$$

Figure 74 shows graphically how the conversion angle is obtained in practice. A right triangle is constructed by using the mid-latitude as an angle between the base and hypotenuse; the hypotenuse is drawn to some convenient scale to equal ½DLo. The side opposite the mid-latitude is then the conversion angle in units equal to those of the ½DLo.

Figure 94 shows the application of the modifying angle to the conversion angle to convert to a series of chords approximating the great circle. This modifying angle is merely the initial conversion angle (at point of departure) divided by the total number of legs to be flown.

Applying the Conversion Angle.—Since the great-circle track, except when it coincides with a meridian or the equator, passes into higher latitudes than the Mercator track, it is necessary to keep only this fact in mind when making conversions. For example, in north latitude on a westerly course the conversion angle would be added to a Mercator course or subtracted from a great-circle course.

Practical Use of the Conversion Angle.—In practice it is not possible to keep continuously on the great-circle track. In fact, it might become necessary to make radical changes in course due to storm areas or other causes. Under such conditions new great-circle courses will be required from time to time. A method whereby these course conversions may be done simply and quickly is therefore most desirable, and the simple formula above, when worked graphically, seems to meet the need.

The great-circle distance is also needed by the air navigator. It may, of course, be computed at leisure before the take-off but, once in the air and committed to the flight, the important factor is the course. The speed is affected by wind and other factors, but eventually, if the *correct course* is followed, the plane will pass over the destination and, since the ground speed may be controlled within wide limits by means of the

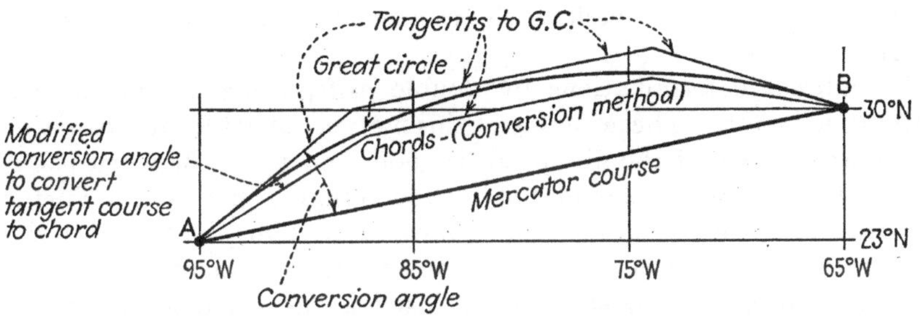

FIG. 94.—Determination chord courses from great-circle course on Mercator chart.

engine throttle, the desired schedule may usually be followed. On the other hand, if off the course, the plane might miss its objective entirely; hence, the importance of maintaining the proper course.

The Determination of Ground Speed in Flight.—There are three methods available:

1. Direct determination by noting the time elapsed between passing over two points defined on the chart and dividing the distance by the time. The accuracy of the result is good if the points observed are sufficiently far apart.

2. Indirect determination by noting the time an object takes to cross the interval between two cross wires in a drift frame or other similar arrangement. In this case the actual height of the aircraft above the object observed is required, and this height is subject to considerable error, which is transferred to the ground speed; thus the resulting error in the ground speed may be relatively large.

The errors in the ground speed so found may be classified as follows:

a. Error in determining the time of passing between the wires. The time varies according to the type of instrument used, but is generally about 5 to 15 sec., so that an error of $\frac{1}{5}$ sec. in the timing may give a 4 per cent error in the ground speed.

b. Error in the assumed value of the height of the aircraft above the object observed. Here there are two sources of error, namely, the error of the altimeter due to change of barometric pressure, and the usual lack of exact knowledge of the height of the observed object above the reference plane to which the altimeter reading is referred. Over wild, unmapped country the height of the observed object can only be guessed.

As the percentage error in the ground-speed determination is approximately equal to the percentage error in the altitude, it will be appreciated that this error when flying at 3,000 ft. can easily amount to 10 per cent when flying over well-charted territory, and even 20 per cent over unknown country. Thus the ground speed determined by timing over a drift frame and making use of the altitude of the aircraft above the object timed can be as much as 25 per cent in error when circumstances are unfavorable.

3. Indirect determinations by double drift, utilizing the air speed as the scale factor. The accuracy of this method, apart from the question of the exact value of the air speed, is good, and the error when using a well-made instrument should not exceed about 1 per cent. Further, as three drifts can be worked just as easily as two, the use of a third drift observation gives a measure of the accuracy obtained.

The value of the air speed to be used is the corrected or true air speed, and this can be readily obtained from the air-speed meter reading and a correction table, or air-speed computer; the true air speed so obtained is not absolutely accurate, but the error is not likely to exceed about 3 or 4 per cent, and is usually less at normal flying altitudes of, say, 4,000 ft. Thus by this method the ground speed can certainly be determined within 5 per cent.

In order to use the double- (or triple-) drift method, the course has to be altered twice, involving an increase in the distance flown, but this disadvantage is more apparent than real, as will be seen later. The increase of distance flown is not large if the courses are arranged as follows:

Course 1. Normal course.
Course 2. 45° to right for 3 min.
Course 3. 90° to left for 3 min.
Course 4. Revert to normal course.

Neglecting the second-order effect of wind on the distance flown on the above courses, we shall have made good on the normal course, during the 6 min. that the aircraft has been off that course, a distance equivalent to 4.2 min. flying, so that our real loss of distance has been equal to 1.8, or less than 2 min. flying. In return for this expenditure there has been obtained an accurate knowledge of the ground speed and the wind. This knowledge will enable the navigator to set improved courses which will save all that 2 min., and probably more, so that the actual air mileage flown is not really increased by the intentional small excursion to obtain accurate knowledge of the ground speed and wind.

In the case of aircraft operating over wild uninhabited areas, the reduction of the uncertainty in the dead-reckoning position when using the above method of determining ground speed and wind is of the greatest value in those rare but inevitable cases where forced landings have to be made.

Navigation Errors Introduced by Incorrectly Determined Ground Speeds.—Of the three methods of ground-speed determination just considered, the first direct method introduces no errors into the navigation, as in every case a new point of departure is gained.

In the third, or double-drift, method practically the whole error arises from a small uncertainty in the true air speed. As this error will not vary greatly during any one flight, the general effect on dead reckoning is merely a small-scale error.

In the second or timing method, the errors arise principally from uncertainty of the true altitude, and also uncertainty of the height of the observed object; the latter quantity varies independently of everything else, while the altimeter uncertainty usually becomes greater as the distance and time interval from the beginning of the flight increase. Thus there will be an increasing divergence of the ascertained ground speed from the true value, and the effect on the dead reckoning is, of course, similar.

If the courses to be made good during the flight are point-to-point courses to avoid natural obstacles, to keep within reasonable distance of emergency landing grounds, or for other reasons, the change of course to be calculated by the navigator (owing to wind) will not be accurate when method 2 is used, but will be more accurate when method 3 is used.

CHAPTER VII

DEAD-RECKONING EQUIPMENT

Navigation has necessitated the development of several new types of instruments to assist the flier in piloting his craft and to aid the navigator in determining his position. Several of these, such as the inclinometer and the turn indicator, have no direct bearing on determining the position of the craft but are required for navigation under conditions of limited visibility.

Correct Arrangement of Instruments.—Proper instrument flying can be done only by using the instruments in combination with one another; therefore it is of the utmost importance that they be properly grouped so that the eye can travel from one to the other with maximum ease. The arrangement of the instruments on the panel should be carefully planned with this fact in mind, and the student should then be trained to take advantage of this arrangement and learn to read the instruments in their correct sequence.

It is the purpose here, however, not to discuss instrument flight (the reader is referred to "Instrument Flying," by Weems and Zweng, for detailed study of this) but rather to point out the existence of instruments of two classes—those for navigation and those for plane control—and to describe those navigational instruments necessary for dead reckoning. These latter may be grouped as follows:

a. Compass (magnetic).

b. Drift sight.

c. Altimeter.

d. Air-speed indicator.

e. Watch, or elapsed-time clock.

f. Charts, chart board, and aircraft plotter.

Other instruments not strictly necessary, but desirable for dead reckoning in certain circumstances, are:

g. Turn-and-bank indicator.

h. Inclinometer or bank indicator of gravity type and fore-and-aft level.

i. Bearing plate and pelorus, or observer-type compass.

j. Rate-of-climb indicator.

k. Directional gyro.

l. Gyro horizon.

m. Dead-reckoning computers.

The Compass.—The compass, as discussed in Chap. III, indicates the direction in which the plane is flying and also enables the navigator to take bearings of visible objects.

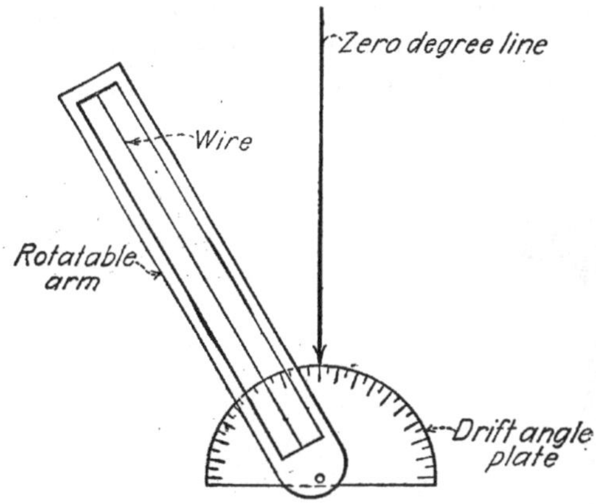

FIG. 95.—Diagram of a drift indicator.

Drift Sights.—The fundamental components of any drift sight are means for setting a line or wire parallel to the apparent direction of movement of the ground and some means of measuring the drift angle, *i.e.*, the angle between the line or wire and the longitudinal axis of the airplane.

FIG. 96.—Longines speed timer, permitting ground-speed and other calculations to 0.2 sec. and up to 60 sec. on outer dial, or up to 60 min. on small top dial, or up to 12 hr. on small lower dial. The plunger at the left permits time to be "taken out," leaving desired flying-time record. (*Courtesy of Longines-Wittnauer Company.*)

The plain drift sight (Fig. 95) is installed where the navigator can look through it at objects on the ground. It is installed so that the zero degree line is parallel to the fore-and-aft line of the plane. To use it the navigator, with eye held in one position, looks through the slot in the rotatable arm at objects on the ground and turns the arm until the objects appear to travel down the slot, parallel to the wire or to the sides of the slot. He then reads the angle of drift from the scale on the drift-angle plate.

To find the drift angle in this way, it is *not* necessary to know the altitude; it is only necessary to be able to see stationary objects on the ground. It is therefore easy to get the drift when flying over land. When flying over water, objects to sight on are not easy to find, but the break of waves or ripples can generally be seen and a reasonably accurate drift reading obtained. By adding two sliders and an eyepiece to the plain drift sight shown in Fig. 95, it becomes also a ground-speed meter. The eye is held at a fixed distance above the drift arm, and the sliders are set to correspond to the altitude. The time required for the observed object to pass from the first to the second slider determines the ground speed. Special timers, such as the one shown in Fig. 96, are used for timing the observations.

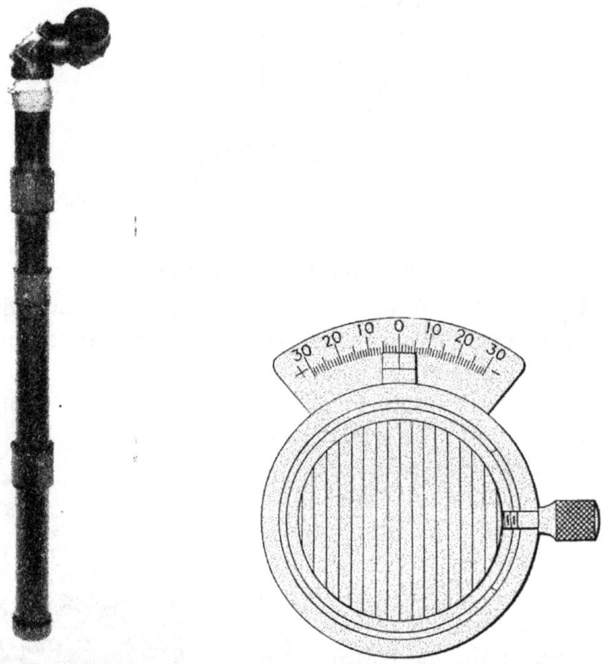

FIG. 97.—Gatty drift indicator. Grid, index, and scale are shown at the right.

Gatty Drift Indicator.—Figure 97 shows the drift indicator designed by Harold Gatty to provide a means for measuring drift from inside a plane by a pilot at the controls with the eye in normal position. The drift grid and ground appear in the same plane through a periscope. Drift is determined by observing an object on the ground or on the water. The drift grid is placed parallel to the apparent motion of the observed body by means of the knurled screw, and the drift is indicated on the scale. The Gatty drift indicator may be mounted in any plane and is easy to operate.

The Altimeter.—The altimeter is required, not only for use in the direct method of determining ground speed, but also for correcting the readings of the air-speed indicator, as well as for general dead-reckoning purposes; for example, a change of altitude may be regarded as a warning to check the wind, as this generally varies with the altitude.

The altimeter (see Fig. 98) is an adaptation of the aneroid barometer to indicate altitude. Its face is graduated in altitude in feet, instead of in barometric pressure. The scale is uniformly graduated, *i.e.*, the angle between the zero mark and the 1,000-ft. mark is the same as that between any two adjacent 1,000-ft. marks. This, together with the fact that the face can be rotated by means of a knob provided for that purpose, makes various adjustments possible. For instance, suppose the plane takes off from an airport at an altitude of 2,000 ft. above sea level, and that the pilot is interested, not in his altitude above sea level, but only in that above this airport or the surrounding country. He can then set his altimeter at zero before taking off, instead of at 2,000; it will then register directly the altitudes in which he is interested.

Fig. 98.—Barometric-setting altimeter.

It must be remembered that the altimeter, being a barometric instrument, is graduated for an altitude scale that is purely arbitrary. The atmospheric pressures for which the instrument registers various altitudes are pressures which on the average do correspond to those altitudes; but on any particular day the pressure at any given altitude may vary from this arbitrary standard by an amount that will register several hundred feet on the altimeter. This may be readily understood when one remembers that clear weather is generally accompanied by a high barometer, and stormy weather by a low one.

If the face of the altimeter were fixed and the zero mark corresponded only to the arbitrary standard pressure for zero altitude (29.92 in. of mercury), an altimeter at a sea-level airport might register −500 ft. one day in clear weather and +400 ft. the next day if stormy. The fact that the face of the altimeter can be rotated permits the pilot, before taking off, to set it to the true altitude of the airport, or to zero for sea level.

The Air-speed Indicator.—The air-speed indicator (Fig. 99), as its name implies, shows the speed of the plane through the air. Its use is necessary to the navigator (*a*) if he is flying a straight point-to-point course and desires to predict his time of arrival; (*b*) if he is flying a broken course and must turn at a point not indicated by a prominent landmark; (c) if he wishes to find the speed and direction of the wind.

As explained before, the navigator can obtain the ground speed from the drift indicator, but he cannot fly at a steady ground speed unless he flies at a steady air speed. If he knows the ground speed, drift, compass course, and air speed, he knows that the factor causing the difference between the air speed and compass course on one hand and the ground speed and course made good on the other is the speed and direction of the wind, and this can be obtained by easy plotting.

FIG. 99.—Air-speed meter.

The Pitot tube is placed where it will receive the pressure of the relative wind caused by the passage of the plane through the air at the air speed. It is readily seen that this pressure depends upon the air speed. It has been determined that it is directly proportional to the square of the air speed and to the density of the air; the latter depends chiefly on the barometric pressure, but also to some extent on temperature.

The barometric pressure falls with increasing altitude. Therefore, although the air speed of a plane may be the same at 10,000 ft. as at sea level, the pressure produced by the Pitot tube will be less at 10,000 ft. than at sea level; hence, the air-speed indicator will register less at the higher altitude than at the lower. The correction for this altitude effect is roughly 2 per cent per 1,000 ft. of altitude above sea level. For example, at 5,000 ft. above sea level the true air speed is 10 per cent greater than that shown by the indicator; thus if the indicator registers 100 m.p.h., the true air speed is 110 m.p.h. These corrections may also be made by means of various mechanical devices such as the height and air-speed computer and the Dalton Mark VII computer described later in this chapter.

In using the air-speed indicator for accurate navigation, it is essential to make this altitude correction. Although air-speed indicators are accurately built instruments, they are connected by long leads of tubing to the Pitot tube. This tubing may leak, in which case the indicator will not register accurately. If he needs accurate air speeds, the pilot must ensure that there are no leaks in the tubing and, as a final check, determine the other errors by flying the plane up and down wind over a measured course. A good air-speed indicator properly installed should not have errors exceeding 2 or 3 m.p.h.

Most air-speed indicators are graduated in land miles per hour, but those used by the Navy are marked in knots or nautical miles per hour. Since the nautical mile equals, for all practical purposes, the minute of latitude, the use of a knot indicator helps to simplify aerial navigation and is recommended particularly for flying over seas. If, however, a miles-per-hour (m.p.h.) indicator is used, its readings may be converted to knots by diminishing them by $\frac{1}{8}$, or, more accurately, by multiplying by 0.868. Various computers also make this conversion.

Air-speed indicators that show true air speed regardless of the altitude have been built, but their complexity, cost, and weight have restricted their use to highly scientific work.

The Watch.—The use of a watch in dead-reckoning navigation is merely to multiply speed by time to give distance. Any reliable timepiece—even a wrist watch—may be used for dead reckoning, although special watches with stop-watch features will usually be found more desirable.

A timepiece now installed as standard equipment in many military aircraft is known as the *elapsed-time clock.* This instrument operates on the stop-watch principle and provides a convenient method of keeping an account of the elapsed time between check points. Its other uses in the dead-reckoning problem are readily apparent.

The Gyro Turn Indicator.—The construction and theory of this instrument will be described in Chap. IX, but its application will be dealt with here. It has by custom been saddled with a wrong name, as it is really a rate-of-turn indicator. It indicates very small rates of turn (rates of change of direction) and hence gives a definite warning that the magnetic-compass reading may be affected by the accelerations causing the turn. It is usually combined with a transverse level which indicates the correctness of the banking during turns, or that the plane is level when flying straight, and thus serves as an indicator of whether side slip is taking place and, if so, to which side.

Turn indicators are of two types, air driven and electrical, according to the arrangements made for driving the gyroscope. The air-driven type (Fig. 100) is usually driven by suction from a Venturi tube or a suction box and is extremely reliable, but care must be taken that the suction used is of the correct amount for the particular instrument. Excess of suction causes the rotor to revolve at too high a speed, with consequent rapid wear and also excess of sensitivity of the instrument. The electrical type necessitates a suitable supply of electric current.

FIG. 100.—Combined turn-and-bank indicator.

Bearing Plate and Pelorus.—Formerly, and in accordance with marine practice, circles with sighting vanes were sometimes mounted on a compass and used for taking bearings; such a fitting was called an azimuth circle or *bearing plate*, and the bearing was read against the compass card. When the compass was fitted in a position from which it was difficult to take bearings, another position was chosen, and the sighting gear was mounted on a dummy compass card which could be set to correspond to the real compass card some distance away. Such an arrangement is called a *pelorus* (Fig. 101).

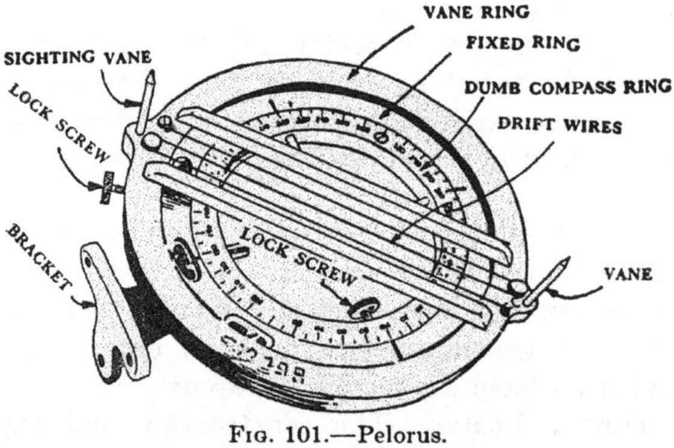

FIG. 101.—Pelorus.

The introduction of the aperiodic compass and the transfer of the compass card graduation to the grid ring, with the accompanying lowering of the compass position, made the use of the bearing plate on the steering compass impracticable; the pelorus required to be set correctly, which often presented considerable difficulties. As a result, an azimuth circle of prismatic type was fitted to another compass in which a light card was secured to the magnetic element, thus forming an observer compass.

The Rate-of-climb Indicator.—The rate-of-climb indicator shows in hundreds of feet per minute the rate of climb or rate of descent. Figure 102 shows the appearance of the instrument, which is self-contained without any outside connections. It is used indirectly for navigation, since the rate of climb or descent affects the speed of the plane. It acts on the principle of the altimeter with the addition of a capillary tube for the slow escape or intake of air as the pressure on the diaphragm changes.

FIG. 102.—Rate-of-climb indicator.

Artificial Horizon or Gyro Horizon. The use of this important instrument is described in Chap. IX.

Directional Gyro.—The directional gyro is a gyroscopic instrument that has the property of remaining in any direction in which it is set over a period of time. It is not affected by accelerations, as is the magnetic compass, but has no azimuth-seeking properties; consequently, it will wander from the set direction after a short time and do so in an irregular manner. Its purpose is to indicate direction when the compass is not available owing to turns. In use it is set to the compass indication before making turns, and then used instead of the compass while carrying out the turns; it is also used instead of the compass in blind flying. Owing to the absence of any azimuth-seeking quality, it requires frequent setting.

Relative-bearing or Drift Lines.—An extremely simple and useful scheme for the navigator is to paint on the fuselage, tail, and wing surfaces of his aircraft plainly marked lines showing relative bearings. It is then possible for the navigator to take approximate relative bearings or wind-drift observations by sighting along these lines. Since these lines radiate from one center the observer's eye must be at or near this point when observations are taken and this fact should be kept in mind when deciding upon the point from which these lines radiate and which governs their location. This scheme will be found particularly useful in single-seater planes where the pilot must also be the navigator.

DEAD-RECKONING COMPUTATIONS

The greater part of navigation, other than piloting, as done by the average aviator, consists of:

1. Computing the ground speed and compass heading, given the track, air speed, and wind force and direction.

2. Computing the ground speed and the track, given the compass heading, air speed, and wind force and direction.

3. Computing the time of flight, or the distance covered in a given time at a given ground speed.

4. Converting compass courses to magnetic or true, and vice versa, and plotting courses, positions, and distances on a chart.

With the wind force and direction, the air speed, and the compass course known, there still remains a considerable amount of computation to be done in order to lay the proper heading and to keep to a definite schedule of flight. The wind drift and ground speed must first be determined, and then the distance traveled in a given time must be computed frequently.

The computations for ground speed, drift, distances, and time may be made by:

1. Separate calculations for each problem.
2. Special tables giving quick solutions.
3. Diagrams.
4. Plotting on a chart.
5. Various types of navigational computers.

Separate calculations for each problem require the solution of a right triangle and are little used in practice.

Special tables for convenient solution of dead-reckoning problems were included at some length in the earlier editions of this book but, with the development of efficient computers, the tables have been largely replaced by such computers as the Dalton Mark VII. Some of these tables are shown in Appendix A.

Dead-reckoning Data from Diagrams.—The tables discussed above and other data may be put into graphical form by means of diagrams. In using a diagram the interpolation required is done by eye and is usually easier and faster. Since the various navigational computers (described later) have for the time being largely supplanted tabular and graphical methods, no further mention will be made of diagrams.

Dead-reckoning Plotting on Charts.—This is the usual method of accomplishing dead reckoning and, because of its importance, is discussed at length in Chap. II on Charts.

Dead-reckoning Computers.—It is not possible to describe all the various types of dead-reckoning computers that have been produced, but the following are given as representative of their class.

The Jenson Aircraft Computer.—This computer (Fig. 103) was designed by Captain H. M. Jensen, U. S. Navy, to solve problems involving air speed, ground speed, wind angle, wind velocity, and drift angle—any two being easily determined when the other three are known. Other problems, such as compass correction to parallel track when miles flown and miles off course are known, may also be solved. The device consists of a base plate on which the diagram is printed and to which is pivoted a transparent spinner with a scale and arrow.

To give one example of the facility with which the device may be used to solve any of several problems, suppose that we are given a course of 150°, a wind of 15 knots from 30°, and a true air speed of 120 knots, and that we require the ground speed and course correction. Set the arrow of the rotating (top) disk at the zero of the outer (fixed) circle. Pencil-mark the point on the top disk corresponding to the intersection

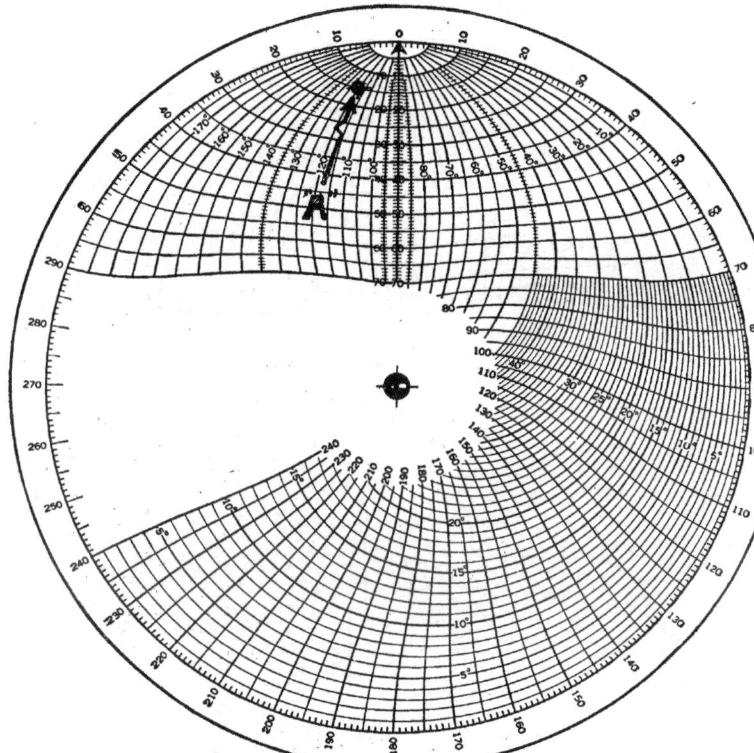

FIG. 103.—Jensen aircraft computer.

of the 120°-wind-angle curve (150 − 30) and the force 15-knot curve as shown at *A* in Fig. 103. Then rotate the top disk until the point *A* is over the 120-knot air-speed curve as shown in Fig. 104. The course correction is equal to the number of curves between the point *A* and the outer circle, or 6½° as shown under bracket *B*. Opposite the arrow read the ground speed 127 knots as indicated at *C*.

The Aircraft Navigational Computer Mark VII.—Aerial dead reckoning requires that the course be corrected for wind drift, magnetic variation, and the compass deviation, and that the air speed be corrected for the air-speed-meter calibration, the pressure altitude and air temperature, and the increase or decrease of speed due to the wind. None of these corrections is in itself particularly difficult, but it takes time and considerable mental effort for the navigator to be sure of his results. A series of computers, starting with the Dead Reckoner by the author, has been developed since 1932 to solve these various problems; the Mark VII is one of the best of these. It was developed originally by the late Lieut. Philip Dalton, U.S.N.R., for military purposes. Soon afterward, it was widely adopted by air-line pilots.

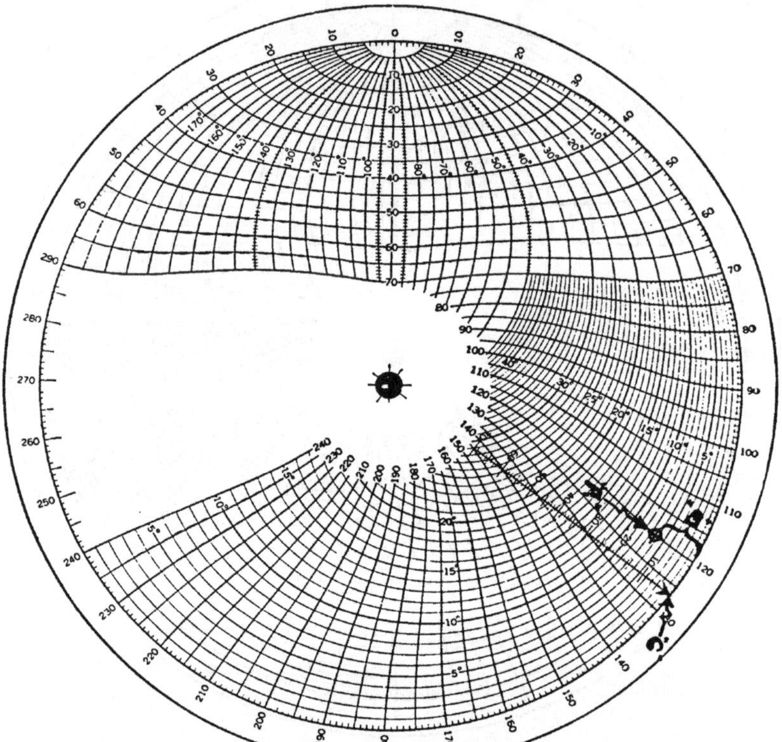

FIG. 104.—Jensen aircraft computer aligned to show ground speed.

Figure 105 illustrates the side of the instrument used for solving wind problems. *A* is a transparent rotatable celluloid plotting disk on which a wind arrow, such as *OW*, is pencil-marked in proper orientation with respect to the disk compass rose and to the scale of the transparent grid piece *B*. *C* is the base piece, the center line of which has an air-speed scale on which are pencil-marked the true air speed, as at *H*, and the magnetic variation, as at *M*. With these data pencil-marked on the computer, a drift and ground-speed problem may be solved as follows: The plotting disk *A* is rotated to set the track or course desired to be made good at the track index *Tr* and held there while the grid piece *B* is rotated to adjust the points W and H along the same vertical grid line, as shown in the figure. The ground speed HW can then be read from the vertical scales of the grid piece, and the magnetic heading can be read directly at the pencil-marked magnet-heading index M, the true heading at T being disregarded.

The computer can be used to solve all types of wind problems. For example, if the course and air speed and the track and ground speed had been known, and the wind speed and direction required, the procedure described above could be reversed to plot the unknown wind arrow OW.

Figure 106 illustrates the circular dead-reckoning slide rule on the back of the computer. Like all circular slide rules, it has continuous scales that can be used to solve any problem in multiplication, division, or proportion. As shown at the left-hand side of the figure, the main outer scales are labeled "Miles" and "Minutes," and adjacent to the "minutes" scale is an extra scale labeled "Hours." The 60-min. and 1-hr. divisions of the time scales are marked with an arrow labeled "Ground-speed Index." With this index set to read a given speed on the "miles" scale, the time required to make good any distance at that speed, or vice versa, can be read directly from the time and distance scales. Or, by setting an elapsed time opposite the distance made good, the ground speed can be read from the distance scale opposite the speed index.

The 33- and 38-min. marks are labeled "Naut." and "Stat.," respectively, for conversions from nautical to statute miles, or vice versa. By setting the index "Naut." opposite a distance given in nautical miles, the "Stat." index is opposite the equivalent number of statute miles, or vice versa. Obviously these indexes also give the relation between knots and miles per hour.

In addition to the scales for speed-time-distance computations and nautical-statute conversions, this slide rule has scales for applying air temperature and pressure altitude (see below) to correct the air-speed meter and altimeter readings for variations from standard atmospheric conditions.

Air-speed meters should be calibrated for instrumental and installation errors. Then the true air speed can be accurately obtained by correcting the calibrated air speed for variations from standard sea-level pressure and temperature. At high altitudes and in hot weather an air-speed meter reads low because the air that operates it is less dense. Installation errors may amount to 2 or 3 per cent, but at an altitude of 15,000 ft. an air-speed meter may read as much as 30 per cent low because of the less dense air.

To make the density correction to air speed, the Mark VII computer is provided with scales labeled "For Air-speed Meter Corrections."

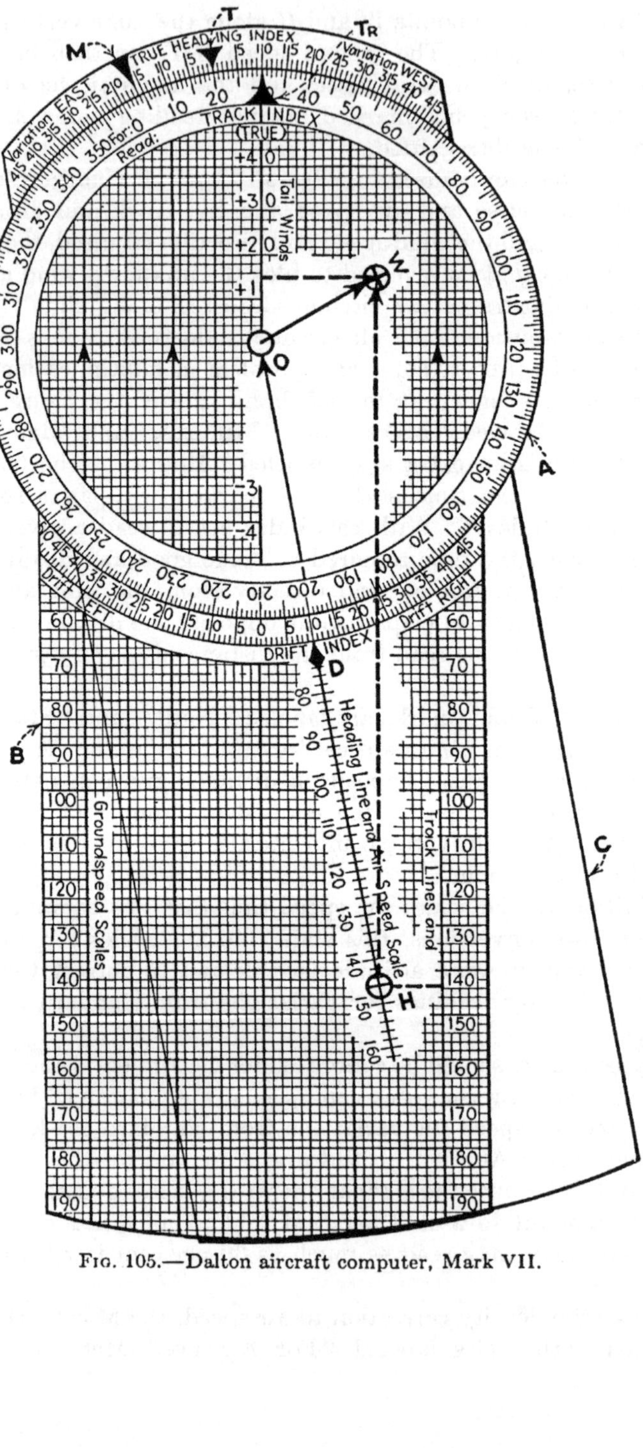

Fig. 105.—Dalton aircraft computer, Mark VII.

As shown in Fig. 106, set the air temperature A opposite the pressure (barometric) altitude B, pressure altitude being the altimeter reading when it is set to read zero for standard sea-level pressure. Having thus set the slide rule for temperature and pressure, find the calibrated (or indicated) air speed C on the "minutes" scale, and opposite it the corrected or true air speed D.

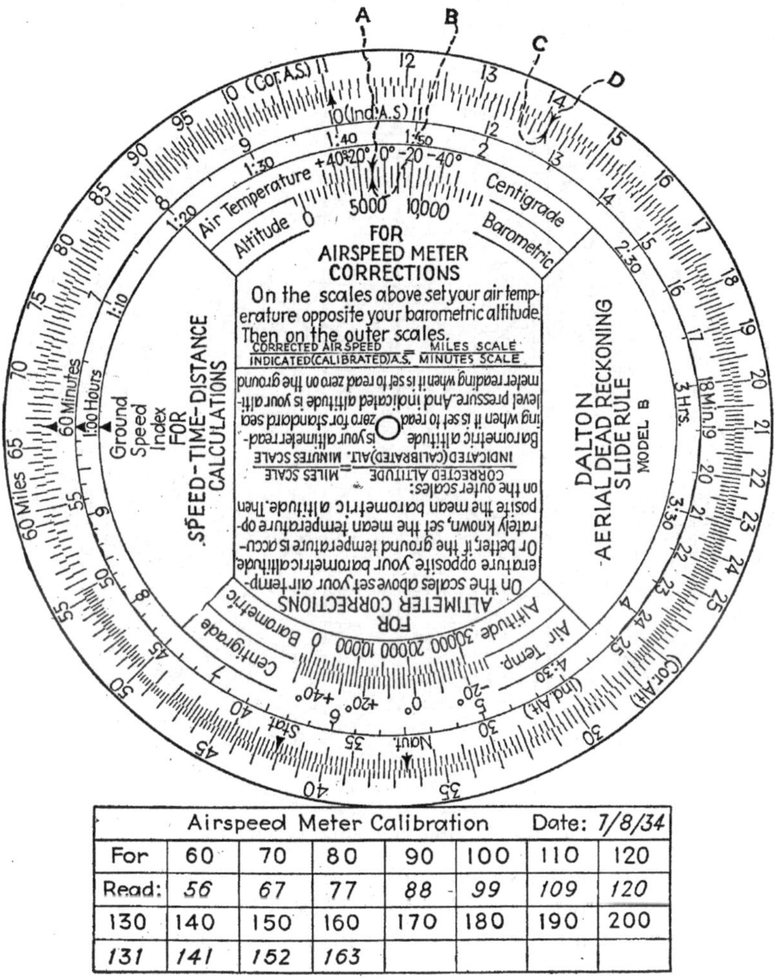

Airspeed Meter Calibration					Date: 7/8/34		
For	60	70	80	90	100	110	120
Read:	*56*	*67*	*77*	*88*	*99*	*109*	*120*
130	140	150	160	170	180	190	200
131	*141*	*152*	*163*				

FIG. 106.—Slide-rule feature of the Dalton Mark VII computer.

Altimeter corrections are made by a similar procedure. If the ground-level temperature and pressure altitude are not known, set the temperature aloft on the scales labeled "For Altimeter Corrections," and read the corrected altitude from the "miles" scales opposite the calibrated altitude on the "minutes" scale. This correction is based on the assumption that the standard temperature lapse rate of 2°C. per 1,000 ft. of altitude exists. If the ground-level temperature and pressure altitude are known, set the mean temperature opposite the mean pressure altitude and read the corrected altitude as before; this assumes only that the temperature lapse rate is uniform.

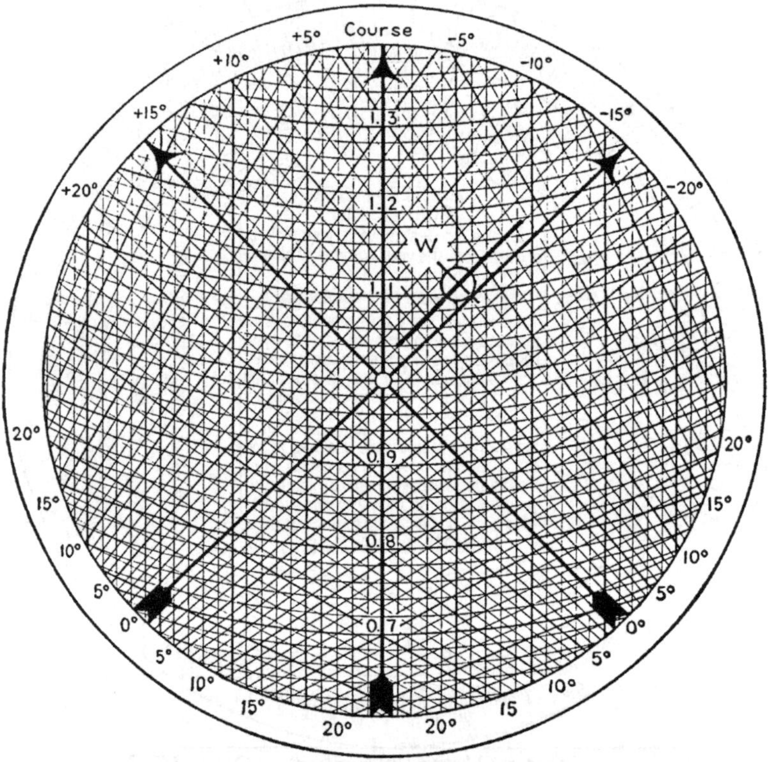

Fig. 107.—Dalton computer, type E–1A.

In addition to the uses mentioned above, an alert R.A.F. officer has listed more than thirty uses for the Mark VII computer. Complete instructions with sample problems are furnished with the Dalton Mark VII computer.

Dalton Computer, Type E–1A.—This is an enlarged edition of the circular slide rule found on the back of the Dalton Aircraft Navigational Computer, Mark VII, but having the back printed with a three-color double-drift diagram (Fig. 107).

To use the three-color double-drift diagram, first obtain two drift measurements on courses 45° either side of the track to be made good. The diverging green lines and the green scale of the diagram represent drift measurements on the course 45° to the right of the track, and the red lines and scale represent drift measurements on the course 45° to the left of the track. The intersection of any green line with any red line represents the wind point for those two drift measurements. The position of this wind point with respect to the black lines parallel to the black track arrow, as read from the black scale at the top of the diagram, gives the track correction necessary to allow for the wind. The position of the wind point with respect to the black circles and scale along the track arrow gives the factor to be applied to the true air speed to find the ground speed. An illustration follows.

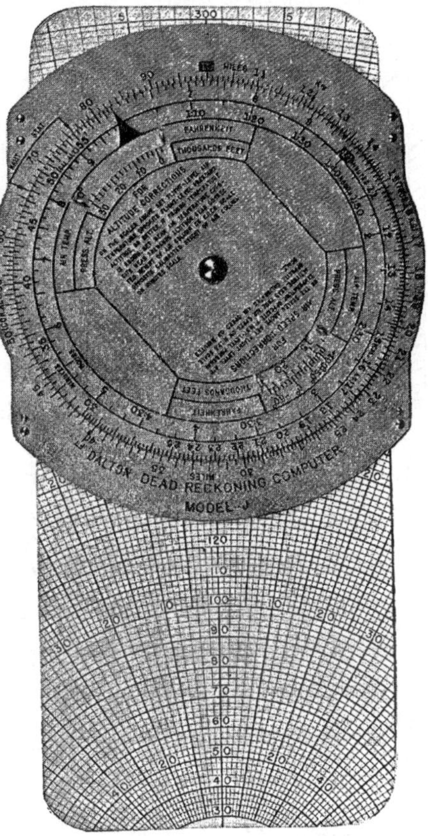

Fig. 108.—Dalton model J computer.

Example.

Given: Track to be made good, 60°.
Drift, 1° left on course 105° (45° to right of track).
Drift, 8° right on course 15° (45° to left of track).
True air speed, 150 m.p.h.

Required: Heading to make good the track, and corresponding ground speed.

As shown in Fig. 107, mark the wind point W at the intersection of the green line representing drift 1° left and the red line for drift 8° right. Then read from the black scale at the top of the vertical black lines the correction −5°, which, when applied to the track 60°, gives the heading of 55° necessary to make good the track. From the black circles and the scale along the track arrow read the ground-speed factor 1.11, which, when multiplied by the air speed 150 m.p.h., gives the ground speed 166 m.p.h.

Dalton Dead-reckoning Computers.—The *Model G computer* is a device for quickly obtaining the solution of drift and interception triangles and similar graphical problems encountered in aerial dead reckoning, without the necessity of plotting the complete triangles. For example, two or three small pencil marks made on the face of the computer and the manipulation of two operating knobs give the solution of any wind problem. The computer is also designed to eliminate all the mental arithmetic ordinarily involved in applying variations, drift angles, wind angles, ground-speed factors, etc., all given data and answers being read directly on the scales of the instrument. The computer includes a circular slide rule for speed-time-distance computations with additional scales for air-speed meter and altimeter corrections.

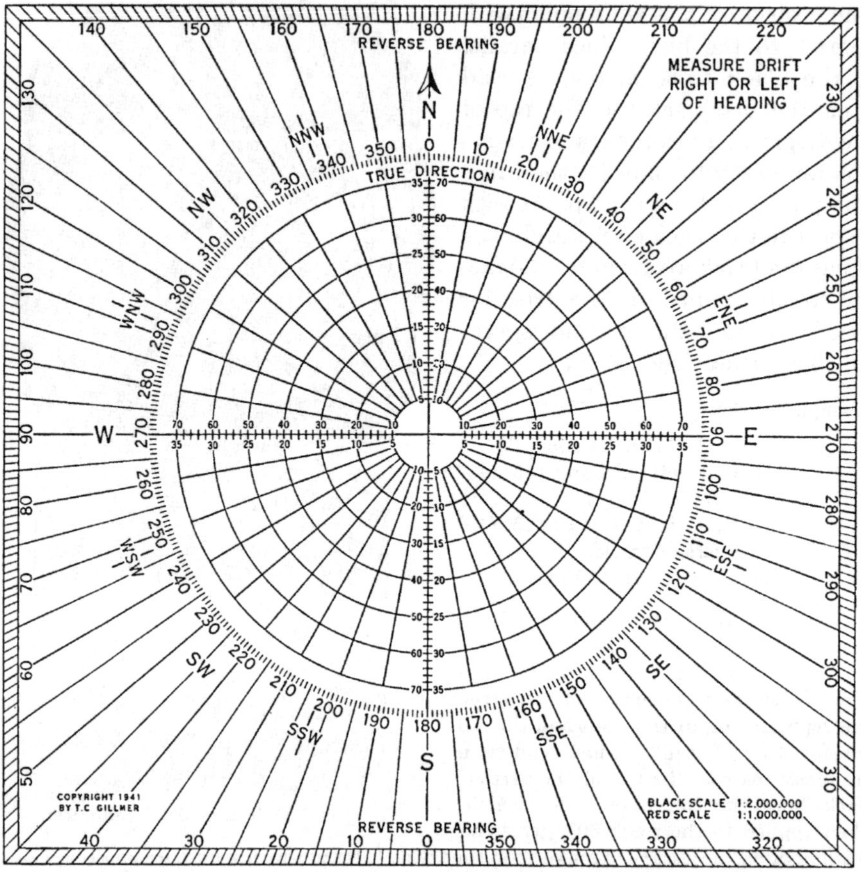

FIG. 109*a*.—Gillmer computer.

The instrument consists of a flat box 3 in. wide, 6¼ in. long, and 1 in. thick with a hinged cover on top and two operating knobs on the right-hand side. It is provided with leather straps for fastening it to the operator's knee. The instrument cover carries the circular slide rule on its upper side and, when it is hinged back, a note pad fastened to its underside is made available and the face of the wind computer is exposed. The wind computer consists of a transparent plotting disk mounted in a compass ring, rotated by an operating knob. A section of a polar coordinate chart, in the form of an endless belt mounted on rollers, is visible directly beneath the transparent plotting disk.

The *Dalton Model J computer* differs from the Model G only in construction. Instead of the drift grid being mounted "belt style" on

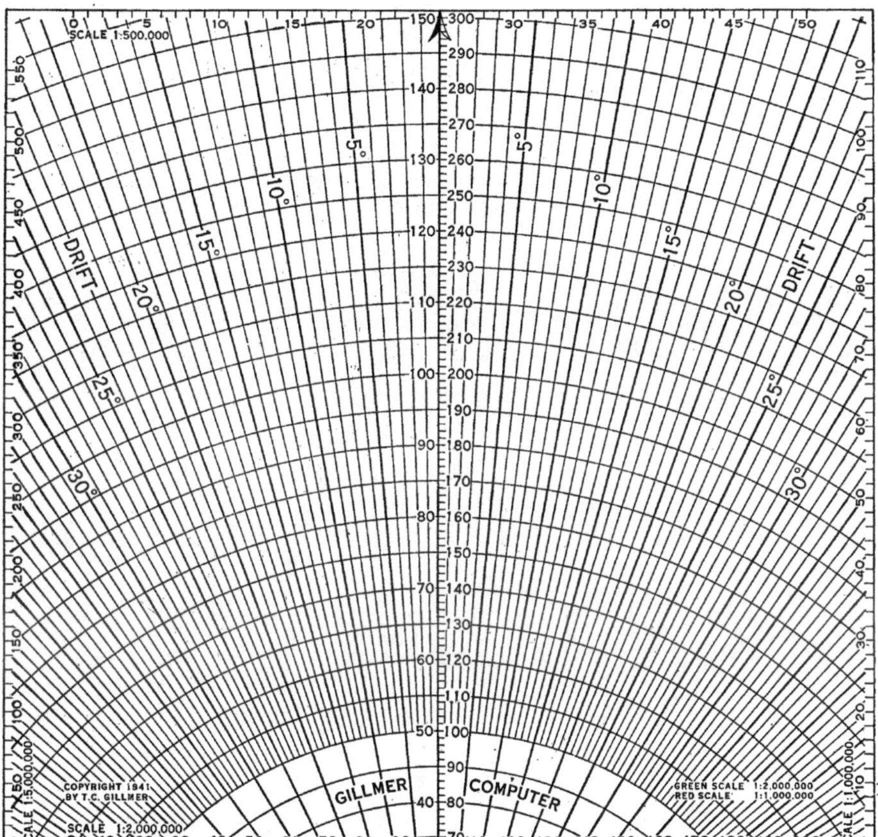

FIG. 109*b*.—Gillmer computer.

rollers, it consists of a sliding card inserted under the compass rose. The circular slide rule is mounted on the other side. The operation is identical to that of the Model G. The popularity of both of these computers has grown considerably since they were first developed because they both have individual features that increase their desirability. Model G is easy to operate in the air or under flying conditions. Model J is compact and easily stowed or carried in the pocket. It has become a standard item to Army air navigators and is shown in Fig. 108.

Gillmer Flight Computer.—Similar in principle to the preceding computers, the Gillmer computer may be used as a successful substitute.

The drift grid is in the form of a transparent plastic plate to be oriented freely on top of the compass rose. By marking the wind direction and velocity with a pencil mark on the compass rose and wind grid plate, the velocity triangle (wind side) is located. The triangle is completed

FIG. 110.—Gillmer computer.

by placing the drift grid plate on top of the compass rose plate, the center drift line on the course of the compass rose, and the grid's speed circle passing over the penciled wind mark. The readings are then taken off directly.

This type of computer is capable of handling any wind drift, radius of action, or interception problems possible on other types. Its most advantageous features are low cost, accuracy, simplicity in operation, and extensive speed range. It makes no provision for magnetic correction, all directions being true, and there is no time, speed, or distance scale. These latter omissions should not be considered defects or lack of refinement, since they cannot rightfully be considered part of the graphical problem.

Figure 109 shows separately the two parts of the Gillmer flight computer. Figure 110 shows the two parts together.

Operation: To find the heading and ground speed for a given track and air speed.
Given: Wind, 20 m.p.h. from N.W.
Track to be made good, 75° (T).
Air speed, 150 m.p.h.
(NOTE: Use black and green scales.)

Procedure: (See Fig. 110.) Locate the wind by a pencil mark in the quadrant it is blowing from on 20-mile wind circle on N.W. bearing. Place plate B on top of plate A with the course arrow (center line of plate B) through the center of the compass rose on plate A and through 75° (track). Keeping the plates thus aligned, move in or out until the 150 (AS) speed circle falls over the penciled wind mark of plate A. Now read 6½° drift from the drift lines through the same wind mark and read the ground speed at the center of the compass rose, 159 m.p.h. The drift (6½°) is between the heading (drift line through wind) and the track. Drift is always named right or left of heading. In this case it is 6½° right drift and is subtracted from the track to obtain the heading. Thus,

$$\text{Heading} = 75° - 6\tfrac{1}{2}° = 68\tfrac{1}{2}° \text{ (T)}$$

Remarks: Remember, *air speed* is always measured along *heading*. *Ground speed* is always measured along *track*.

Lyon Computer.—This is a circular computer designed by T. C. Lyon and described in *Civil Aeronautics Bulletin* 24. On one side is a slide rule similar to that on the Dalton Mark VII computer; on the other side is a wind-drift computer.

Several air lines, Cox and Stevens Aircraft, and others have produced special aircraft computers, but space does not permit of detailed descriptions of them here.

Dalton Plotting Boards.—These plotting boards may be used to solve many navigation problems quickly and simply. To do this, however, the procedure must be thoroughly understood by the operator. By orienting the transparent celluloid top disk, the dead-reckoning track of the plane may be recorded. The bearings, courses, etc., may be drawn in by hand and accuracy maintained if only the critical points are definitely marked. The principles used with this board are those described at length in Chap. VI. Also, the same features are included in this board as on the grid side of the Mark VII computer. These boards are at present widely used in the air arm of the U. S. Navy. Many of its applications are of a restricted nature.

CHAPTER VIII

DEAD RECKONING—PRACTICE

For daylight flying in fair weather, having plotted the desired course on the chart, it is customary and quite practicable merely to note on the chart any characteristic landmarks along the route to be followed and to shape the course in the air with reference to them. This is air pilotage and has been discussed in Chap. IV.

However, many accidents have resulted because pilots, running unexpectedly into fog or bad weather, have been forced to rely wholly upon instruments with which they were not thoroughly familiar. It is strongly urged, therefore, that pilots practice instrument flying, in order that they may be prepared for such emergencies.

Accuracy of Dead Reckoning.—Reasonably efficient dead reckoning should produce an accuracy well within 5 per cent of the distance flown, or within 5 miles in 100, 10 miles in 200, or 15 miles in 300. Efficient dead reckoning requires frequent observations for the wind force and direction, and also requires that an accurate record be kept of all data obtained. It profits the navigator little if, while working dead reckoning, he notes the craft's position by pilotage but fails to make the necessary record of the time and the new point of departure.

A good compass provides the heading with reasonable accuracy, though in bumpy air it is sometimes difficult to steer an accurate course. The air-speed meter gives a close approximation to the speed of the plane through the air. The *track* and *ground speed*, however, are the data required for keeping a continuous reckoning of the plane's position, and these values are affected by the inaccuracy in determining the wind force and direction.

It is a good point to remember that if a plane flies 60 miles on a course 1° in error, it gets 1 mile off its proper course. A 3° error in course over a distance of 120 miles causes an error to the right or left of the course of $3 \times {}^{120}\!/_{60} = 6$ miles. Had Colonel Lindbergh steered a course 3° in error on his flight from New York to Paris, he would have arrived no closer to Paris than $3 \times 3{,}600/60 = 180$ miles. An error in speed alone is no more complicated than an error in the speedometer of an automobile —the plane will arrive ahead or behind schedule by the amount of time gained or lost. In case of excessive errors in estimating the speed, fuel might become exhausted or a night landing might become necessary.

Dead reckoning alone affords no means of determining definitely a plane's position, once the position becomes uncertain. When in sight of known objects on the earth's surface, the position may be determined by methods of pilotage; otherwise, recourse must be had to radio or celestial navigation, discussed later.

Suitable Charts.—One of the most important items of navigational equipment is an accurate chart. In fact, some of the other instruments are of fundamental value only as related to the chart. For example, the pilot may be able to follow any given course by means of the compass, but there are no practical means of determining the course to be followed except by measuring it on a chart. Again, the altimeter registers the height of the plane above sea level, while the pilot is chiefly concerned with his height above the ground. At present the height above the ground may be had only as the difference between the altimeter reading and the ground elevation indicated on the chart. The distance to be traveled, intervening landmarks, location of aids to navigation, airports, airways, magnetic variation, etc., may be determined readily only by reference to charts. Thus it is evident that a chart is essential to all navigational control, regardless of the method of navigation used.

Importance of Keeping Records of Position.—If flying over a fog bank, or at other times when we cannot see the earth, it is necessary to keep an accurate account of the courses steered, the air speed, and the time on each course. For instance, if we fly 40 min. on course 77° by compass and then steer 59° for 58 more min., we must step off on the chart the distance on each leg in order to find the position after the second leg has been completed.

LOG OF PLANE

FROM SMITHTOWN TO JONESBORO DATE July 1, 1942

TIME	REMARKS
1010	Took off. Wind force 20 from 45°. AS 90.
1015	Reached altitude 800. Set course 87°PC, 70°T.
1028	Directly over Millville.
1052	CC to 83° to allow for increased wind.
1120	Speeded up to 95.
1150	1 mile north of Nixon. Altitude 2,000.

PC is abbreviation for per compass.
T is abbreviation for true.
CC is abbreviation for changed course.

The importance of keeping an accurate record of the dead reckoning cannot be stressed too much. Each time the course or speed is changed, this fact together with the time should be noted. Also, in actual practice, piloting is combined with dead reckoning. Therefore, there should be a careful running record of all possible data pertaining to both methods. A good way to arrange the notes is to put the time in hours and minutes in the left column, then to make the notes in as abbreviated a form as possible to the right. The reason the notes should be abbreviated is because of the difficulty of writing clearly owing to the vibration of the plane, and further, to save time. A suggested form is something like the one shown on page 151.

A running record, or log, like this will be found helpful in many ways. When all goes well, these notes might not be necessary, but in order to keep them correctly, one must practice, and when the data are needed, they will be needed urgently.

Preliminary Work.—Before taking off, the navigator should arrange the equipment, collect all required data, and in fact do everything that he possibly can in advance. The course should be plotted, and suitable intervals of either time or distance or both should be stepped off along the course. Twenty-mile intervals will often be a convenient scale to measure along the course.

From the latest wind data, air speed, and true course (track to be made good), all of which are usually known in advance, the true heading and ground speed are computed. Then the variation, taken from the chart, and the deviation, taken from the compass deviation table, are applied to the true heading to obtain the compass heading (or course to steer). This and other pertinent data are tabulated for convenient reference and noted on the chart as indicated in the preceding paragraph.

Operations in the Air.—On taking off, the pilot gains the proper altitude and sets the compass heading previously computed. For short flights under 200 miles where piloting alone is required, the navigator, who is also the pilot in a single-place plane, frequently checks the plane's position by comparing recognized objects on the ground with those shown on the chart. Ranges are followed where possible.

If it is found that the wind drift does not agree with the drift allowed for, a new check is made on the wind, or an estimated correction is applied.

If the visibility is low, or if on flights longer than about 200 miles, the navigator should keep a careful written record, or log, of navigational data, plotting changes in the original course laid down, or new courses, as necessary. It is particularly necessary to keep an accurate account of the time, ground speed, and compass course steered.

Where the pilot is also navigating the plane, particular care should be taken to have the equipment reduced to a minimum and arranged for the most convenient use.

When piloting is possible, advantage of this method should be taken to keep the plane's position *determined continuously* and *recorded*. There will be times when keeping a running record of the plane's position will appear unnecessary and useless work. However, when low visibility from any cause is encountered, the latest recorded data of the plane's position, the course steered, and the average air speed are vitally necessary.

For (magnetic).............	N.	330°	300°	W.	240°	210°
Steer (compass)............	N.	328°	297°	W.	243°	212°
For (magnetic).............	S.	150°	120°	E.	60°	30°
Steer (compass)............	S.	148°	117°	E.	63°	32°

FIG. 111.—Deviation card.

Since taking drift observations is somewhat difficult and also inaccurate, many navigators shun this operation. Where a plane has two-way radiotelephone equipment, the wind data will usually be obtained in flight from ground stations with more accuracy and with less trouble than it may be determined in the air by drift observations. Notwithstanding the inaccuracies and the difficulties in making the observations, it is advisable for navigators to make frequent observations of wind drift.

To head plane (magnetic)	Steer with compass reading
0°	0°
30°	32°
60°	63°
90°	90°
120°	117°
150°	148°
180°	180°
210°	212°
240°	243°
270°	270°
300°	297°
330°	328°

FIG. 112.—Deviation card.

Where conditions permit, the plane's position should be determined continuously, that is every 2 or 3 min., by piloting, and ample data recorded. When piloting methods cannot be applied, especial attention should be given to dead reckoning.

Correction for Compass Deviation.—Owing to the magnetic disturbances in the plane itself, the compass does not register the correct magnetic course on all headings, but deviates a few degrees to the right or to the left on some headings. The amount and direction of deviation are recorded on a *deviation card*, which should be tacked up in the plane for ready reference. It is recommended that deviation be checked frequently (see Chap. III).

Figures 111 and 112 are typical forms recording compass deviation. Both cards record the same deviation for the same compass but in different terms and arrangement. It will be noted that this compass is so compensated that true magnetic readings are obtained on north, south, east, and west headings, and that deviations of 2° and 3° occur on the other headings, *i.e.*, to head plane on a course of 60°, the compass should read 63°.

For deviation cards having wording and arrangement different from these, the principles are the same, and by a comparison with these, the elements involved may be identified.

NOTE: To correct a magnetic course falling between those listed on the deviation card, note the correction for the nearest reading, whether it is increased or decreased and the number of degrees of difference. Apply this correction to the magnetic course to get the correct compass heading to steer. Examples:

Desired magnetic course	Nearest heading	Corresponding compass reading	Correction indicated	Steer with compass reading
65°	60°	63°	+3°	65° + 3° = 68°
145°	150°	148°	−2°	145° − 2° = 143°
275°	270°	270°	0°	275° + 0° = 275°
310°	300°	297°	−3°	310° − 3° = 307°

Correction to Course in Flight.—Since the direction and velocity of the wind vary with altitude, the initial wind corrections, made before leaving the ground, should be checked in flight when possible, as these will affect both the ground speed and the compass course.

To check ground speed in flight, transfer the 10- and 50-mile intervals from the border scale of the chart to the course line by means of pencil marks along the straight edge of a piece of paper. Use short cross lines for the 10-mile marks and diamonds for the 50-mile intervals. By noting the time of passing landmarks near the scale markings, the distance covered in 1 hr. of flight may be read directly; this gives the ground speed of the plane; *i.e.*, if 120 miles of the course scale has been covered in 1 hr. then the ground speed is 120 miles per hour, etc.

In practice there is no such thing as a constant compass course. The initial course must be corrected in flight for every change in direction or velocity of wind. Wind corrections are always made toward the wind.

To Find Compass Course by Flight.—Having drawn the course line on the chart and divided it for the 10- and 50-mile intervals as in Fig. 113, the pilot may begin his flight *in clear weather,* without previously determining his compass course. Follow the course by landmarks for a reasonable distance and until the compass card has come to rest, then read the course directly from the compass. This is the *correct "initial" compass course,* including correction for magnetic variation, wind force, and compass deviation.

It must be noted, however, that this is *only* the *initial* compass course and that *further correction must be made in flight* for change in magnetic variation, convergence of the meridians, and changes of the wind. Correction for changes in compass deviation are to be disregarded.

The correction for change in *magnetic variation* may be made by applying *one-half of the difference* between the magnetic lines at the beginning and end of the course.

Correction for convergence of the meridians may be made by noting the *number of degrees of longitude* crossed by the course line and *on an easterly course by adding* 1° to the compass course for every 2° of longitude crossed. *On a westerly course subtract* 1° from the compass bearing for every 2° of longitude.

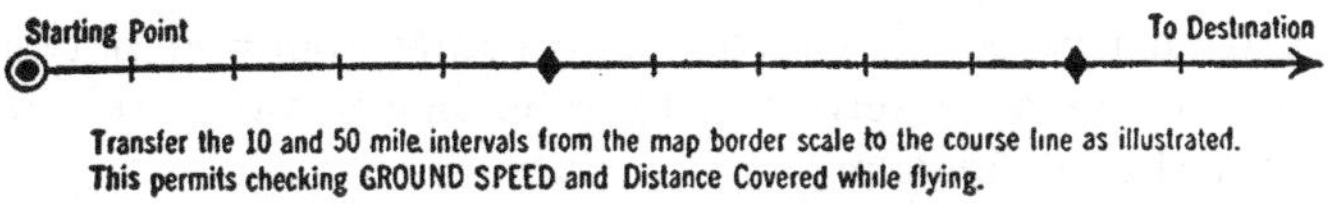

FIG. 113.—Checking ground speed in flight.

NOTE: This correction, for convergence of meridians, also applies when the true course has been measured from the initial meridian instead of the mid-course meridian. The amount of this correction varies with the latitude but the value given will serve all practical purposes for flying in this country.

Correction in flight for change in direction and velocity of wind can be made only by estimate, after a thorough understanding of the foregoing has been gained through experience.

To measure the initial compass course in flight, as described above, is so simple and practical that many pilots are tempted to use this method exclusively. The result is that when they are called upon to use the previously described methods for thick-weather flying, they are either unable to do so, or become so confused with the elements involved that the results will more than likely be wrong.

Though the pilot uses the simpler method normally, it is recommended that he also thoroughly acquaint himself with the use of the first methods described and employ them at least once a month, or preferably fortnightly, to keep in practice and protect himself and the good name of aviation in time of need.

Identifying the Ground Position.—In fog flying it is an easy matter for those not thoroughly familiar with the magnetic compass and turn indicator to lose all sense of direction; the compass may even oscillate until it begins to turn slowly, like a top, as explained in Chap. III.

Even if the sense of direction is not lost, it is not always easy to identify one's position on coming out of a fog. At such times a properly air-marked city is most welcome. But, in the absence of such air marking, nearly every city or town has its own distinctive marking.

A city may be located where two main railroads cross; this is a mark that will not be duplicated within a radius of many miles. Even the angle of their crossing is faithfully pictured on the chart. Another city lies in the V where two rivers meet; another is just beyond the intersection of one of those rivers with a railroad.

County seats are marked on the chart with a circle and dot symbol. In flight, the distinctive architecture of the courthouse—usually a dome—is an easily recognized landmark. Even the county lines and township lines may prove of value, as minor highways and fences generally trend with these lines, taking their direction from them.

If one has identified one's position and is still uncertain of the direction, a straight course may be steered in any direction until a second ground position can be identified. The course followed between these two points can now be determined by reference to the chart, and a new course set for the destination.

Highways have been advisedly omitted from some aviation charts. It is realized that in some sections highways do constitute important landmarks, and for this reason a number of experimental charts were prepared including highways. It was found, however, that to include even major highways introduced an element of confusion that only obscured more important and reliable details—a conclusion that has been confirmed by many leading pilots.

For those desiring a strip chart over a regular route it is suggested that after plotting the course and dividing it into 10- and 50-mile intervals, a strip 10 in. wide, 5 in. on each side of the course, be cut and mounted on cloth and folded accordion style. Some prefer mounting on thin plyboard, and continuing the map with some overlap on the reverse side of the board.

When the route passes from one chart to another, the charts should be joined so that the plotted course forms a continuous line; the transition from one chart to another may be made more easily if the adjoining edges of the charts are divided into 10-mile intervals perpendicular to the course.

Source of Wind and Weather Information.—The Weather Bureau forecasts *wind and weather conditions for aviators* at frequent intervals. This information should be obtained from your local airport, or the nearest principal airport, and those along your route, by phone before taking off, or by radiophone in flight. This is the only safe way (see Chap. XII on Meteorology).

Conclusions drawn from charts, graphs, diagrams, and general statements of average or prevailing wind and weather conditions for years or seasons are likely to be erroneous for any particular flight.

The pilot's interest and safety lie in definite knowledge of conditions to be expected along his immediate route, and this can be obtained only from current weather reports and forecasts.

Wind direction is that direction *from which* the wind blows. The U. S. Weather Bureau uses the eight principal *true* directions given in Fig. 114.

Before working out the wind correction for the magnetic course, true directions must be converted into magnetic directions by adding west variation or subtracting east variation, as the case may be (*i.e.*, for

Maine a northwest wind equals 315° + 20° west variation = 335° magnetic. In Washington state the same wind equals 315° − 23° east variation = 292° magnetic).

Wind velocity is the rate, in statute miles per hour, at which the wind is blowing. The U. S. Weather Bureau uses the *terms* given in the Beaufort scale, Fig. 144, for reporting wind velocities.

Direction (used by U. S. Weather Bureau)	Corresponding true bearing	45° sector included
1 N	0° (360°)	337½° to 22½°
2 NE	45°	22½° to 67½°
3 E	90°	67½° to 112½°
4 SE	135°	112½° to 157½°
5 S	180°	157½° to 202½°
6 SW	225°	202½° to 247½°
7 W	270°	247½° to 292½°
8 NW	315°	292½° to 337½°

Fig. 114.—Table for converting wind directions into degrees.

SOLUTION OF PRACTICAL PROBLEMS

1. *Given:* An aircraft leaves a fixed base on true track 270°, air speed, 80 m.p.h. Wind is from 136°, 37 m.p.h.

Required:

a. Ground speed and flying time out for 1 hr. flying.
b. Ground speed and flying time in for 1 hr. flying.
c. Same as (*a*) and (*b*) except total flying time is 2 hr.
d. Same as (*a*) and (*b*) except total flying time is 3 hr.

Procedure: (Solution from "Radius of Action of Aircraft," by Mary Tornich.) Figure 115 is the unit triangle for 1 hr.'s flying.

$E \rightarrow W$ = wind speed and direction = 136°, 37 m.p.h.
$W \rightarrow P$ = heading and air speed flying out = 251°, 80 m.p.h.
$W \rightarrow P_1$ = heading and air speed flying in = 109°, 80 m.p.h.
s_1 = rate of departure = 100 m.p.h.
s_2 = rate of return = 50 m.p.h.

Apply the values found to the formula

$$R = \frac{T(s_1 \times s_2)}{s_1 + s_2}$$

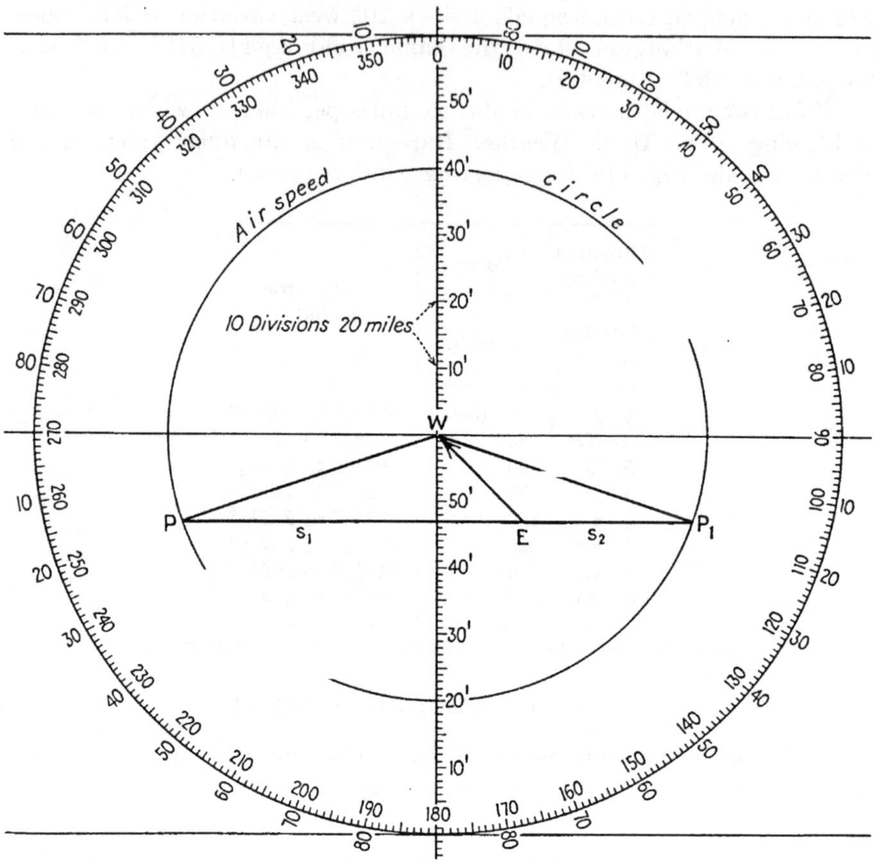

FIG. 115.—Graphical solution of radius of action from fixed base for 1 hr.

a. Radius of action for 1 hr.:

$$1 \times \frac{100 \times 50}{100 + 50} = 33.3 \text{ miles}$$

b. Radius of action for 2 hr.:

$$2 \times \frac{100 \times 50}{100 + 50} = 66.6 \text{ miles}$$

c. Radius of action for 3 hr.:

$$3 \times \frac{100 \times 50}{100 + 50} = 100 \text{ miles}$$

Or, solving for t, to find the flying time out to the point of turning, in the formula,

$$t_1 = \frac{T \times s_2}{s_1 + s_2}$$

$$(a)\ \frac{1 \times 50}{100 + 50} = 0.333 \text{ hr. or } 20 \text{ min.}$$

$$(b)\ \frac{2 \times 50}{100 + 50} = 0.667 \text{ hr. or } 40 \text{ min.}$$

$$(c)\ \frac{3 \times 50}{100 + 50} = 1 \text{ hr.}$$

It is seen in the following that $R = t \times$ ground speed out:

$$(a)\ 0.333 \times 100 = 33.3 \text{ miles}$$
$$(b)\ 0.667 \times 100 = 66.7 \text{ miles}$$
$$(c)\ 1 \times 100 = 100 \text{ miles}$$

If $R = t_1 \times$ ground speed in, then

$$\frac{R}{\text{Ground speed in}} = \text{flying time in}$$

Applying the values found for R to the above formula,

$$(a)\ \frac{33.3 \text{ miles}}{50 \text{ m.p.h.}} = 0.666 \text{ hr., or } 40 \text{ min.}$$

$$(b)\ \frac{66.7 \text{ miles}}{50 \text{ m.p.h.}} = 1.334 \text{ hr., or } 1 \text{ hr., } 20 \text{ min.}$$

$$(c)\ \frac{100 \text{ miles}}{50 \text{ m.p.h.}} = 2 \text{ hr.}$$

RECAPITULATION

Track	Ground speed	Heading	Air speed	Flying time	Radius of action
(a) Out 270°	100	251°	80	20 min.	33.3 miles
(a) In 90°	50	109°	80	40 min.	
				60 min.	
(b) Out 270°	100	251°	80	40 min.	66.6 miles
(b) In 90°	50	109°	80	1 hr. 20 min.	
				2 hr. 00 min.	
(c) Out 270°	100	251°	80	1 hr. 00 min.	100 miles
(c) In 90°	50	109°	80	2 hr. 00 min.	
				3 hr. 00 min.	

2. *Given:* True course 300°, variation 5°W., deviation 3°E. Distance 350 miles, indicated air speed 108 m.p.h. After flying at 6,000 ft., temperature 52°F., for 2 hr. the navigator determines his position to be 200 miles from his point of departure and 60 miles to the right of his course.

Required:

1. The compass heading to destination.
2. Estimated flying time to destination.

Solution (From "Instrument Flying," by Weems and Zweng): Refer to Fig. 116. Use Mark II plotter and Mark VII computer.

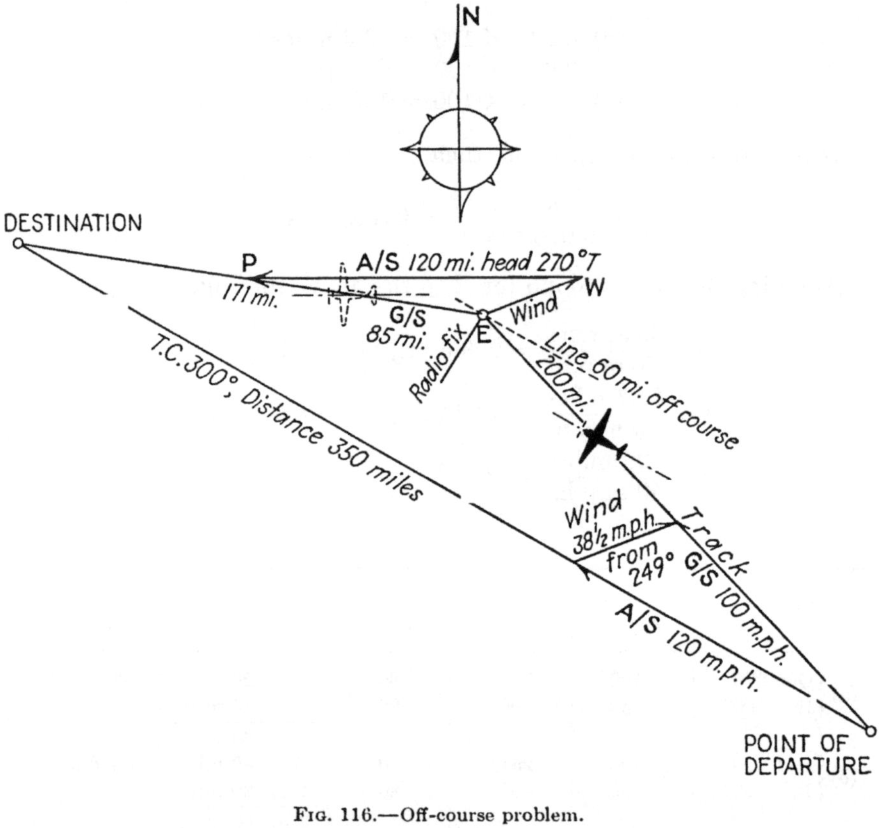

FIG. 116.—Off-course problem.

1. Ground speed out: 200/2 = 100 m.p.h.

2. True air speed for indicated AS 108 m.p.h., temperature 52°F., at 6,000 ft., is 120. The indicated AS may be converted to true AS direct by the Mark VII computer, which gives true AS 111 per cent of indicated AS.

3. Plot the wind for 1 hr. by connecting a point 120 miles along the course line with a point 100' miles along the track line, giving a wind of 38½ m.p.h. from 249° true.

4. Transpose this wind to the position determined at the end of 2 hr. and plot for heading and GS in to the destination, giving heading = 270° true, GS = 85 m.p.h.

5. Compass heading for 270° true, variation 5°W., deviation 3°E., is 270° + 5° − 3° = 272°.

6. Distance in = 171 from the plot; therefore 171/85 = 2 hr. 1 min., flying time to the destination.

NOTE: A practical method for computing true from indicated AS is to assume 2 miles increase in indicated air speed for each 1,000-ft. increase in altitude. In this problem 6,000 ft. increases the indicated air speed 12 m.p.h., giving a true air speed of 108 plus 12, or 120 m.p.h.

Explanation: The first step is to lay off your course from the point of departure or last known position. This is the heading that has been maintained. Along it, with a suitable scale, lay out the air speed or the distance at the end of 1 hour's travel.

Now from the point of departure by radio fix you determine that you have been set off this course a definite distance and a certain number of miles from the point of departure. From the account of the time and the distance between this new fix and the point of departure the ground speed may be determined. It is laid out on the track line determined above. The triangle is then completed by connecting the ends of the air-speed and ground-speed lines with the wind force and direction line. As soon as this wind is determined, lay out a new course to the destination from the present position. This now becomes your track, or the ground you must actually cover. From your position then place your wind force and direction arrow and connect the two lines by striking an arc from the head of the wind line equal to the air speed. This line, then, is the new heading. (See Fig. 116, *EWP*.)

LETDOWN PROCEDURE

In order that a pilot flying on instruments may make allowance for descending from normal cruising altitude to break out over or near the location where he desires to approach his field for landing and determine the time for this descent, the following procedure should be adhered to.

The indicated air speed being controlled at constant value, the pilot may determine the true air speed for the letdown by averaging the true air speed at his cruising altitude and the true air speed at the altitude he desires to level off near his landing. With this as the letdown air speed, we consider the letdown as a separate drift problem, solving for our ground speed and heading thus: With the new air speed, construct the air-speed side of the triangle from the head of the wind line to determine the new ground speed. The wind used here is the same wind used throughout the course or a corrected wind determined during flight as being the most accurate estimate. With the new ground speed determined, the time for beginning the letdown may be set and the letdown begun at the proper time to break out at the desired spot. A typical problem is here given for illustration.

3. *Given:* $C_n(T)$ = 270°(track to be made). Wind = 225°(S.W.), 50 m.p.h. Distance = 200 miles. (Indicated) AS = 100 m.p.h. at flying altitude 10,000 ft. Letdown AS = 100 m.p.h., rate 500 ft. per minute to break out at 1,500 ft.

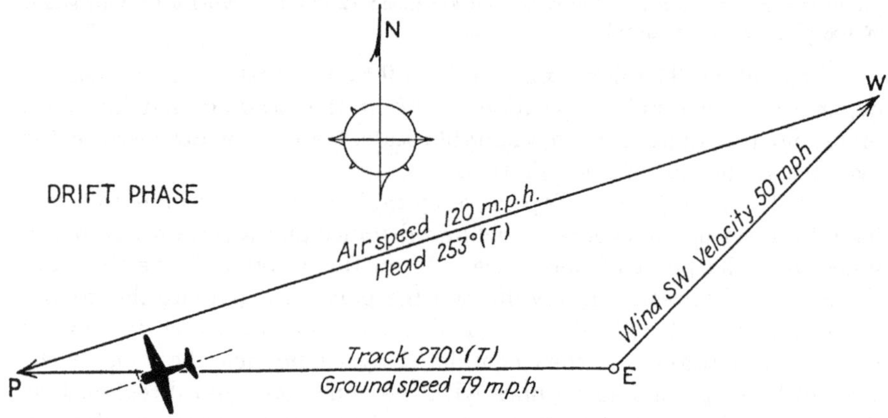

FIG. 117.—Letdown problem—drift phase.

Required: What course to fly [heading (*T*)]; the total flying time.

Solution: True AS = 120 m.p.h. at 10,000 ft. (2 m.p.h. for each 1,000 ft.).

1. Drift phase: See Fig. 117.
2. Letdown phase: See Fig. 118.

10,000 ft. to 1,500 ft. = 8,500 ft. to drop at 500 ft. per minute = 17 min. to letdown.
True as at 10,000 ft. = 120 m.p.h.
True as at 1,500 ft. = 103 m.p.h.
Average true as for letting down = 111.5 m.p.h.
17 min. at 70 m.p.h. GS = 19.8 miles
19.8 miles from total distance of 200 miles leaves 180.2 miles.

180.2 miles at 79 m.p.h. =	2 hr. 17 min.
19.8 miles at 70 m.p.h. =	17 min.
Total time to fly 200 miles	2 hr. 34 min.

At 2 hr. 17 min. out from the point of departure, change the heading from 253° *T* to 251°.5 *T*, and start letting down at indicated air speed of 100 m.p.h. and rate 500 ft. per minute.

NOTE: Another component might be considered in that the sloping track during the letdown would actually be longer than the horizontal track. However, on long sloping descents this difference may be considered negligible along with the component of the horizontal wind.

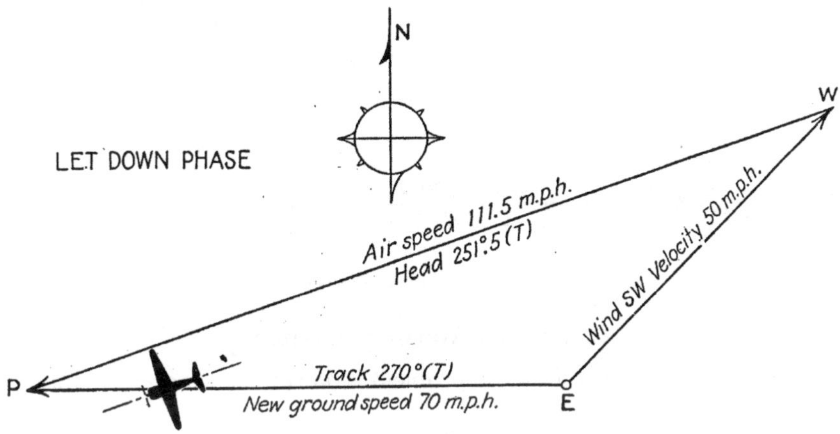

FIG. 118.—Letdown problem—letdown phase.

CLIMB PROCEDURE

The problems involved in climb with regard to the navigational factors are very similar to those of letdown. Because of varying load and its effect on safe rate of climb, however, a wider range of factors results. The indicated air speed on most air-line planes is normally maintained at 120 m.p.h. throughout the climb. The rate of climb at sea level is, however, far higher than that upon leveling off around 10,000 ft. So in reaching a single value for computing the time for climb, an average must be struck between the two.

4. *Given:* True course = 270°(track). Wind = S.W., 50 m.p.h. Rate during climb = 400 ft. per minute. Indicated air speed throughout climb = 120 m.p.h.

Required: Heading and distance covered during climb.

Solution: AS (true) at 10,000 ft. = 140 m.p.h. (2 miles per 1,000 ft.).

Average true AS = 130 m.p.h.

At 400 ft. per minute, it requires 25 min. to reach 10,000 ft. (See Fig. 119.)

For 25 min. at 90 m.p.h. GS = 37.5 miles.

Level off after 25 min., 37.5 miles from the point of departure.

At 37.5 miles from the point of departure the plane is at its cruising altitude (10,000 ft.) and on track to the destination, and the problem is computed for simple drift, as in Fig. 117. The heading should be adjusted, and allowance made for a new ground speed.

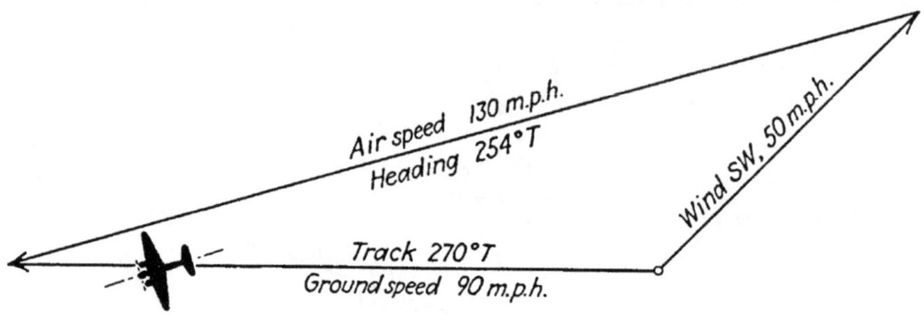

FIG. 119.—Climb procedure.

ALTERNATE AIRPORT PROBLEM

The following radius-of-action problem is one given on recent examinations for instrument rating:

5. *Problem:* A pilot desires to fly from airport A to airport B under instrument conditions and gives as the alternate airport C. The true course from A to B is 50°. The distance is 350 miles. The true course from B to C is 180° and the distance 140 miles. The air speed of the plane is 90 miles per hour. Wind is from 270°, 30 miles per hour. Aside from the required reserve, the plane has fuel supply for just 4 hr. How far may he proceed toward B and still have fuel enough to reach C if advised by radio that the weather at B has closed in altogether?

The solution and explanation of this problem are given in Fig. 120.

Solution: Plot B 50°, 350 miles to scale from A and plot C 180°, 140 miles from B. Draw lines connecting AB, AC, and BC. Draw the wind line AW from 270°, 30 miles to scale. From W, the end of the wind vector, with a radius equal to the air speed of the plane, 90 miles to scale, strike an arc. It will intersect the course line AB at P_1. The direction of the line WP_1, 38°, is the heading necessary to make good the course of 50° true. Measurement of the line AP_1 to scale gives the ground speed, 111 miles per hour.

To determine how far toward B the pilot may fly and still arrive at the alternate airport C within 4 hr. we proceed as follows:

Assume airport A moves along AC at such a rate that it will reach C in 4 hr. Measurement of the length of AC to scale shows it to be 281 miles. In one hour A must move at the rate of $281/4 = 70.25$ miles per hour, to A_1. From P_1 through A_1, draw a line of indefinite length. From W, with a radius equal to the air speed of the plane, 90 miles to scale, strike an arc. It will intersect the line P_1A_1 extended, at P_2.

P_1A_1 is the rate of departure (S_1); P_2A_2 is the rate of return (S_2). WP_2 is the heading. Now, if we connect AP_2 with a line, its direction gives us the track after turning, 163° true, and its length to scale, the ground speed of 94 miles per hour. Since this is the track to C from the point

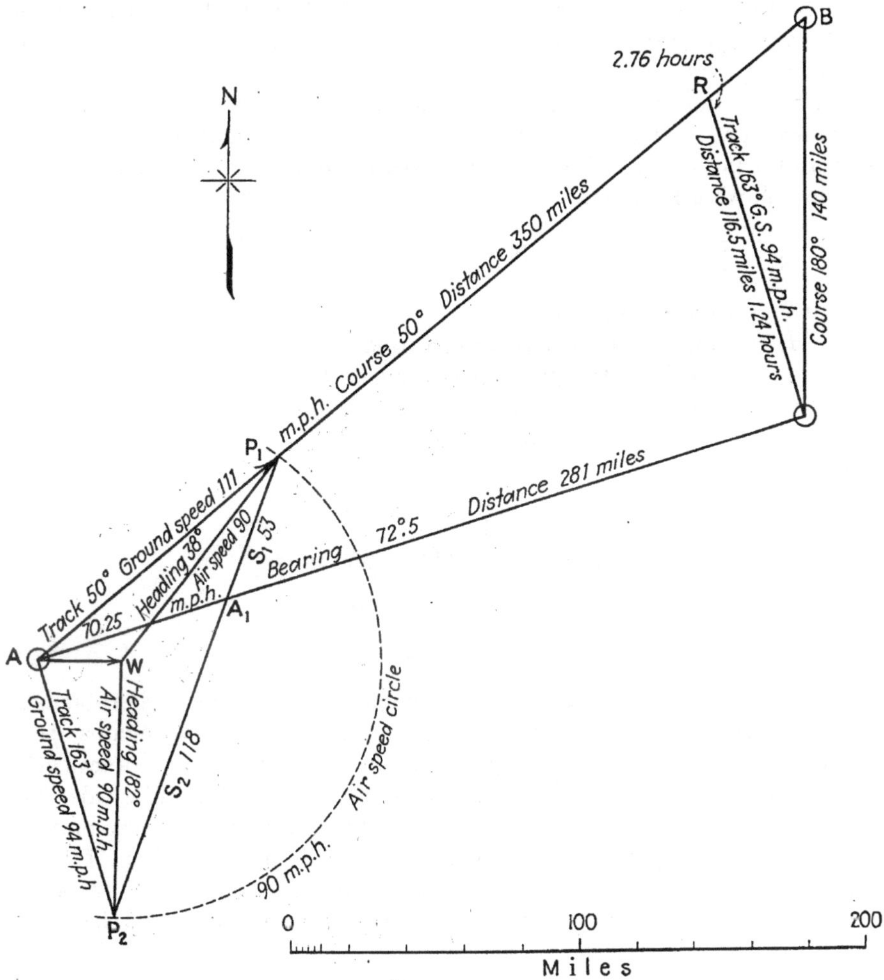

FIG. 120.—Alternate airport problem or radius of action from a moving base.

of turning, we may lay off this line from C. Where it intersects the course AB will give the point to turn, R.

By measurement to scale of the line AR we find the distance to the point of turning to be 306 miles. Dividing this by the ground speed, 111 miles per hour, gives us the time to point of turning, *i.e.*, 306/111 = 2.76 hr., or 2 hr. 46 min. By measurement to scale of the track from R to C we find the distance to C to be 116.5 miles from the point of turning. Dividing this by the ground speed, 94 miles per hour for this leg, gives the time to reach the alternate airport, *i.e.*, 116.5/94 = 1.24 hr., or 1 hr. 14 min.

Checking our results, we have the time to the point of turning plus the time from there to C = 2 hr. 46 min. + 1 hr. 14 min. = 4 hr., the total flying time.

After finding S_1 and S_2 in the 1-hr. triangle, the time to turn, the distance from A at turning, and the time to reach C and the distance to C from the point of turning may be solved in another way by substituting in the following formula:

$$t_1 = \frac{T \times S_2}{S_1 + S_2}; \qquad AR = t_1 \times g_1; \qquad t_2 = \frac{T \times S_1}{S_1 - S_2}; \qquad RC = t_2 \times g_2$$

where T = total flying time.
t_1 = time out from A before turning.
t_2 = time from the point of turning to C.
g_1 = ground speed for the first leg.
g_2 = ground speed for the second leg.

Thus

$$t_1 = \frac{4 \times 118}{171} = 2.76 \text{ hr.}$$

$$AR = 2.76 \times 111 = 306 \text{ miles}$$

$$t_2 = \frac{4 \times 53}{171} \doteq 1.24 \text{ hr.}$$

$$RC = 1.24 \times 94 = 116.5 \text{ miles}$$

6. In order to emphasize the *time* element in wind and speed problems, a simple problem where a plane is flying directly with and against the wind will be analyzed. Such problems as these are of a "practical" nature in that a pilot, unthinking, might essay something like this: "I have a 7-hr. fuel supply and the wind is off shore, blowing at the rate of 50 m.p.h.; therefore, to get the experience of overwater flying, I'll fly 3 hr. to sea and back. This will allow me an hour's supply of gas as a safety factor." The sad result in such an attempt would be that the plane would run out of gas 50 miles at sea.

In order to give the student a mental picture of what takes place, a diagram of several cases will be given. The essence of the problem might be summed up as follows: Since less time is required in going a given distance with the wind than when going the same distance against the wind, the gain in time going with the wind is less than the loss in time while going against the wind. The usual mistake will be in considering *distance* and not *time*. The wind effect of a given wind speed is proportional to the time it acts on the plane.

We will take the case of a plane with air speed of 100 m.p.h. flying 300 miles to sea with a wind along the track of 0, 50, and 100 m.p.h. With no wind, the round trip would obviously take 6 hr. It will be seen, by studying Fig. 121, however, that with a 50-mile wind the round trip will require 8 hr. Also, it will be clear that a plane cannot gain in a 100-mile wind if its air speed is only 100 miles.

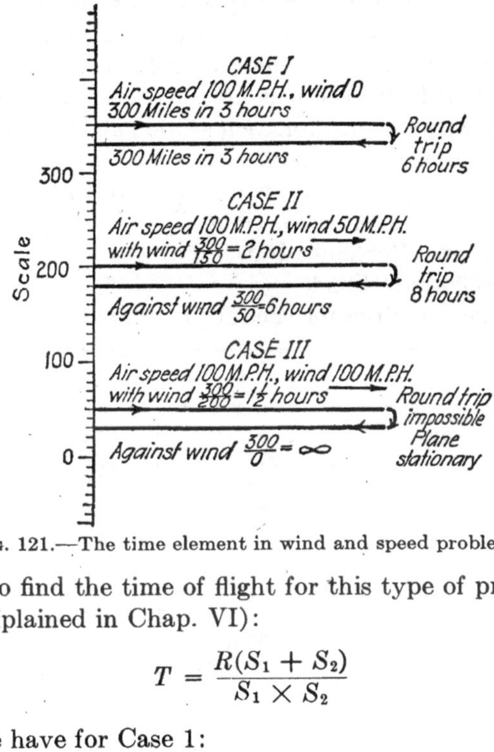

Fig. 121.—The time element in wind and speed problems.

We may also find the time of flight for this type of problem by using the formula (explained in Chap. VI):

$$T = \frac{R(S_1 + S_2)}{S_1 \times S_2}$$

From which we have for Case 1:

$$T = \frac{300(100 + 100)}{100 \times 100} = \frac{300 \times 200}{10{,}000} = 6. \quad \textit{Ans.}$$

For Case 2:

$$T = \frac{300(150 + 50)}{150 \times 50} = \frac{300 \times 200}{7{,}500} = 8. \quad \textit{Ans.}$$

For Case 3:

$$T = \frac{300(200 + 0)}{200 \times 0} = \frac{300 \times 200}{0} = \infty. \quad \textit{Ans.}$$

7. A flight is to be made from Pittsburgh-Allegheny County Airport to Chanute Field, Rantoul, Ill.

For this flight either the Cleveland and Chicago sectional charts, or regional chart 9M may be used. In this case the ship is fairly fast, dead reckoning (rather than piloting) will be employed, and the drainage pattern and larger cities will furnish sufficient check of position; therefore chart 9M is chosen.

Known data: Cruising speed of plane 165 m.p.h.; wind 20 m.p.h., from 165°.

Required: The distance, compass headings, and the total flying time.

A straight line between the two airports is drawn on the chart and, by means of the border scale of miles, the distance is found to be 434 miles.

When the route crosses more than 3° or 4° of longitude, the straight line should be divided into sections crossing approximately 2° of longitude each, and the true course for each section should be measured with the middle meridian of that section.

After a careful study the route is divided into three sections:

a. Pittsburgh—Mount Vernon, Ohio.

b. Mount Vernon—Portland, Ind.

c. Portland—Chanute Field.

The information required for each leg is tabulated below.

	a	*b*	*c*
Meridian nearest halfway	81°15′	83°45′	86°30′
True course	271°	270°	268°
Variation	+4°	+2°	−1°
Magnetic course	275°	272°	267°
Deviation	+1°	+1°	+1°
Compass course	276°	273°	268°
Wind	−7°	−7°	−7°
Compass heading	269°	266°	261°
Length	136 miles	132 miles	166 miles
Distance from Pittsburgh	136 miles	268 miles	434 miles
Time from Pittsburgh	48 min.	1 hr., 35 min.	2 hr., 34 min.

8. A flight is proposed from Pittsburgh-Allegheny Airport to North Platte Airport, Neb.

Required: The distance and compass headings.

The cruising speed of the plane in this case is relatively low, and the flight will be chiefly for pleasure. Navigation will consist in large measure of piloting, and the sectional charts will therefore be used.

Since the Lambert projection affords a perfect junction between any number of charts, if space is available, the charts required may be carefully fitted together and a straight line drawn across all of them, from starting point to destination.

However, when more than two or three charts are involved, it is often easier to plot the route first as a straight line, or series of straight lines, on a small-scale control chart, such as Coast and Geodetic Survey chart

3060a or 3074. The points at which the straight line crosses meridians and parallels on the small-scale chart are then measured and transferred to the large-scale charts, and connected on each of them with straight lines. The portion of the route appearing on each of the large-scale charts is treated as in the two preceding examples, in order to obtain the required distance and compass headings.

Following this procedure, we find that the straight line between the two airports on chart 3074 crosses longitude 84° (the western limit of the Cleveland sectional chart) at latitude 40°48′. This point is plotted on the Cleveland chart, using the marginal scale of minutes of latitude, and connected with Pittsburgh-Allegheny Airport by a straight line.

The portion of the route on the Cleveland chart crosses 4° of longitude, and is therefore divided into two sections crossing 2° of longitude each. The true course for each section is then obtained, and magnetic variation, compass deviation, and correction for the effect of wind are applied in order to find the required compass heading for each section. The total distance on the Cleveland chart is 215 miles.

In the same way the portions of the route crossing the Chicago, Des Moines, and Lincoln sectional charts are subdivided into sections of practical length, and the compass heading for each section determined. The distances from the various charts are totaled, of course, to obtain the distance for the entire route.

9. *a.* Assuming that the Hawaiian Islands extend 300 miles and that the distance from San Francisco to Hawaii is 2,100 miles, what maximum error in compass course may be tolerated and the islands still be reached?

Solution: If we aim for the middle we have 150 miles on each side.

Solving by simple algebra: Let x = maximum error in degrees from compass course that can be tolerated. Remembering that 1° error in course gives an error of 1 mile off course in 60 miles traveled we have:

$$x \times \frac{2{,}100}{60} = 150$$

$$x = \frac{60 \times 150}{2{,}100} = 4^\circ.3$$

When it is realized that some aircraft compasses are not marked closer than 5°, we see at once the necessity for steering a careful course.

b. If Colonel Lindbergh had steered a course 5° to right of the proper course on his 3,600-mile flight to Paris, how far from Paris would he have landed?

Solution: 5 × 3,600/60 = 300 miles.

c. If Admiral Coutinho had depended solely on his compass on his famous flight of 1,100 miles to St. Paul's Rock, and assuming he could see

25 miles, what maximum error could he have had in the course made good and still have picked up the rock?

Solution: Let x = maximum permissible error in degrees.

$$x \times \frac{1{,}100}{60} = 25$$

$$x = 25 \times \frac{60}{1{,}100}$$

$$x = 1^\circ.36$$

CHAPTER IX

INSTRUMENT FLYING

Instrument or *blind flying* is the term applied to the art of flying and navigating aircraft under those atmospheric conditions where neither the earth nor the sky is visible to the pilot. Such conditions are exemplified by thick fog, dense cloud, snowstorms, and the complete darkness of night, any of which would prevent a pilot from being able to orientate himself in space by reference to any known datum. The arts of piloting and navigating aircraft under circumstances so described are fundamentally connected, and must for this reason be studied as one subject.

Piloting and celestial navigation become impossible when neither the earth nor the sky is visible. The only methods of navigation applicable to blind flying are dead reckoning and radio position finding. It is, therefore, highly desirable for the navigator to become proficient in these methods before attempting to fly blind.

The Impossibility of Blind Flying by Instinct.—It is now a well-known fact that man cannot fly blind by instinct. The following extract from a British pilot's record of his experience in 1915 describes vividly the sensations and experiences of a pilot flying blind without special instruments and without special training for blind flying.

> A huge bank of black clouds loomed ahead. Our orders were to land if clouds were too bad, but as two machines pushed on ahead of me, I pushed on too. It started with a thin mist and then gradually got thicker. It continued so for about 10 min. and then I found that according to my compass I had turned completely round and was heading out to sea. The clouds got thicker and the compass became useless, swinging round and round. I was about 7,000 ft. up and absolutely lost. The next thing I realized was that my speed indicator had rushed up and the wind was fairly whistling through the wires. I pulled her up, but had quite lost control. I nose-dived, side-slipped, stalled, etc., time after time, my speed varying wildly. I did not get out of the clouds until I was only 1,500 ft. up. I came out diving headlong for the earth. As soon as I saw the ground, I of course adjusted my sense of balance and flattened out. I was, however, hopelessly lost—the sea was nowhere in sight. I steered by my compass (which had recovered, being out of the clouds) and after a short time picked up the coast.

An interesting description of the mechanics of blind flying is given by Donald Keyhoe in his book, "Flying with Lindbergh." Describing Lindbergh's blind flying between Boston and Portland, Keyhoe writes;

I could almost see the cockpit of the *Spirit of St. Louis*, and watch Lindbergh's eyes as they passed quickly but methodically from one instrument to another. Only by his perfect understanding of that set of instruments before him, and his calm vigilance in reading them correctly, could he win that battle with the elements.

It would almost be the same as the fight he had waged with the fog on the transatlantic flight. From the compass which kept him on his course, his eyes would have to go on rapidly to the bank-and-turn indicator. This would tell him whether he was flying straight or turning, and how steeply the wings of his ship were inclined, if he was not in level flight. Next, to the altimeter, so that he would not get dangerously close to the ground. With this, he must coordinate his knowledge of the particular terrain below, remembering whether it was rising or not, so that the sea-level altimeter would not betray him through a false sense of security.

From the altimeter his glance would have to go to the engine tachometer and the air-speed meter, so that he would be warned if the plane was climbing or diving, the first of which might lead him to a stall, the other perhaps to destruction if it were not quickly corrected.

At intervals his eyes would have to pass on to the clock, so that he could estimate the distance to be checked off on his map. Without this method of locating himself approximately he would indeed be lost. When he could find a spare second he would shoot a swift glance at the oil pressure and temperature gauges. Thus the cycle would end—to begin again, at once. And this must go on, over and over, until the grudging fog gave up and showed him the land below.

All of this while he hurtled along at almost a hundred miles an hour!

Blind flying such as this is the supreme test of any pilot. Some cannot stand this rapid movement above a hidden world, nor the haunting fear that they may have calculated erroneously and may be about to crash into some unseen obstacle. Sometimes their senses tell them that the instruments are wrong. They break under the strain imposed by their lack of confidence in their ability and realization of their own weakness. In desperation they climb up higher in the effort to pull out of the enshrouding fog, sometimes reaching clear air only at high altitude. At this height they cruise along miserably, afraid to come back through the mists, wondering where they are, and tortured by the knowledge that their gas is being used up and they soon must plunge back into that terrifying realm of blindness.

Or else they dive down with the hope of finding a clear spot close to the ground, where they can make a forced landing. Sometimes they succeed, but sometimes disaster comes without warning as the earth appears through the fog too late to avoid a crash.

Panic is fatal in this kind of flying. Only the man with utmost calmness and perfect understanding of his instruments can keep it up hour after hour. . . .

Physical Reactions to Blind Flying.—Under normal conditions of visibility a pilot controls his aircraft mainly by responding to his own sensory reactions. The senses most important to him are sight; the sensations of the vestibular labyrinths, from which is derived the sense of balance; deep muscle sensations consequent upon the shifting of the bodily weight; and, to a smaller extent, the sense of hearing. Sight is, of course, the chief guide, and while he is able to retain a view of the horizon or of the earth itself he can maneuver his aircraft accurately without reference to any of the instruments on his dashboard. With any diminution of visibility it may be necessary to make occasional reference to certain of his instruments, but it is only when he is completely deprived of external vision that he has to apply an entirely different technique in order to maintain control. It is now incontrovertibly

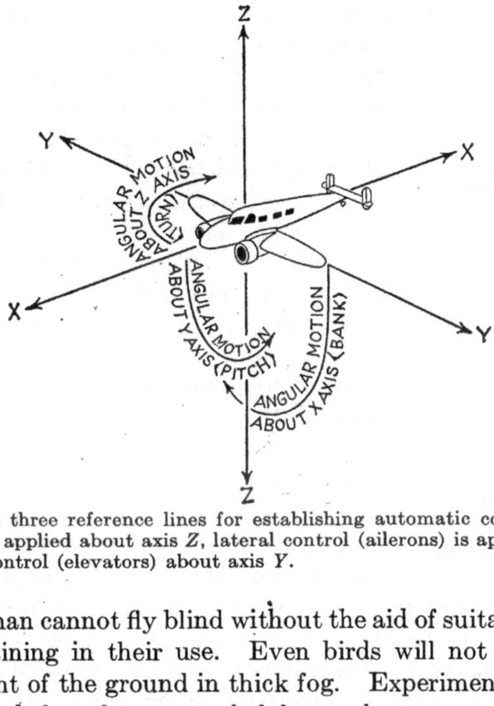

FIG. 122.—The three reference lines for establishing automatic control. Directional control (rudder) is applied about axis Z, lateral control (ailerons) is applied about axis X, and longitudinal control (elevators) about axis Y.

accepted that man cannot fly blind without the aid of suitable instruments and special training in their use. Even birds will not venture on the wing out of sight of the ground in thick fog. Experiments with pigeons, etc., have proved that they are as helpless as humans under blind conditions. We, however, have the advantage of mechanical aids. Since the mechanical aids used in blind flying usually include gyroscopes in one form or another, it is now necessary to give a brief description of them.

Gyroscopes.—A gyroscope is a spinning mass universally mounted; *i.e.*, mounted so that its axis may be pointed in any direction. This is accomplished by mounting a mass that is free to spin in a support carried on pivots whose axis is perpendicular to the spin axis, and finally supporting the whole on pivots whose axis is perpendicular to the other two, as shown in Fig. 122. Instruments that employ the gyroscope can be divided into two general classes: (*a*) those employing a free gyroscope and (*b*) those employing a controlled gyroscope. In the case of the free gyroscope the gyrowheel or spinning mass is supported in a mounting system that provides the least possible friction or resistance to movement about its supporting pivots. In the case of the controlled gyroscope certain forces are employed which, in a way, restrict the freedom of the gyroscope but make it do useful work.

In Fig. 123 is shown a simple gyroscope. It will be seen to contain a wheel or mass free to spin. This mass is mounted in a ring that is pivoted on axis AA'. The support for the pivots that provide axis AA' is in turn mounted to turn on a vertical axis.

Without the weight W this is a free gyroscope, since the gyrowheel may be set spinning and the spinning axis will hold its position in respect to space as the base is moved to various angles or rotated. Now add the weight W and it might be imagined that this would cause the gyroscope and inner ring to tilt about axis AA', thus allowing the weight to fall. This, however, would occur only if the gyrowheel were not spinning. The addition of the weight W will, in fact, cause the spinning gyrowheel and its supporting ring to rotate, or "precess" about the vertical axis B as shown by the arrow. It is this characteristic of the gyroscope that is employed in various ways in the controlled gyroscope. The torque need not necessarily be applied by a weight, but may be applied by a spring, friction, air reaction, or by any of several other means. It will be seen therefore that the least friction about any of the pivots must tend to reduce the accuracy of the gyroscope. If the gyroscope just described were frictionless, the weight W would cause the gyroscope and the inner ring to precess about the vertical axis B, and the weight W would not fall. Suppose, however, that considerable friction exists about the vertical axis B; this friction will clearly cause a frictional torque in the direction opposite to that shown by the arrow. Any such torque, frictional or otherwise, will cause the gyroscope to precess about the horizontal axis AA' and will allow the weight to fall after turning the gyroscope only a slight amount around axis B.

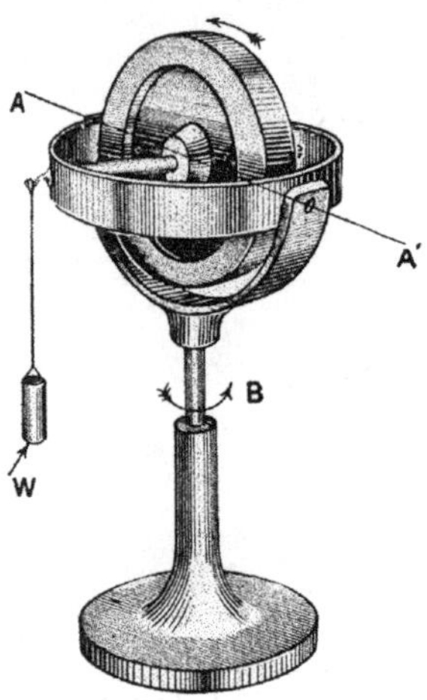

Fig. 123.—Simple gyroscope.

The following three simple rules will help in understanding the descriptions that are to be given:

1. Any torque about the horizontal pivots will cause a precession in azimuth.

2. Any torque about the vertical, or azimuth, axis will cause a precession about the horizontal or pitch axis.

3. All friction or restraint tending to oppose free movement about the vertical or horizontal axes will cause inaccuracy and must, therefore, be avoided as far as practically possible.

Applications of the Gyroscope in Aircraft Instruments.—The application of the gyroscope in aircraft instruments can generally be classified under the two types of gyroscopes: the free and the controlled.

1. The free types are those in which the gyroscope is used to indicate a set direction by means of the very low frictional value of its supports and its property of maintaining the direction of its axis in space. The directional gyro employs this type of mounting, which is provided with means of setting the spin axis in any desired direction then releasing it.

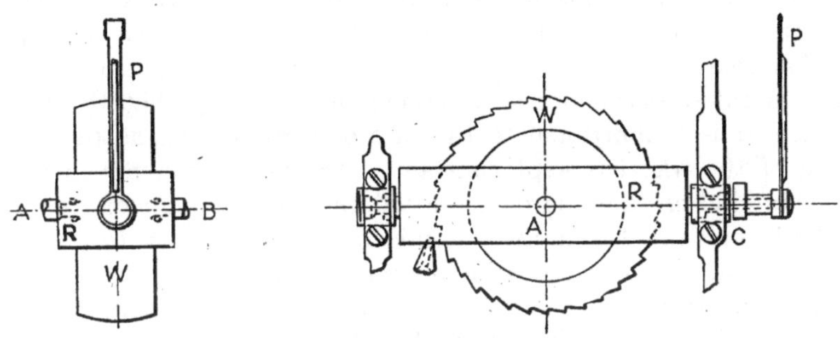

Fig. 124.—Schematic arrangement of turn indicator.

2. The controlled type is that in which a force is made to act on the gyroscopic system in order to produce the required results by precession. Under this heading is included the gyro turn indicator, and the gyro horizon. In the gyro turn indicator, the gyroscope is restricted by the omission of the axis represented by *B* in Fig. 123. From our knowledge of the gyroscope it will be seen that, if weight *W* were removed, the freedom by pivot *B* removed, and the base then rotated, the gyrowheel will be made to tilt, causing the top of the gyrowheel to move right or left depending upon which way the base is rotated. Mount this gyroscope in an airplane with axis *AA′* fore and aft, attach a pointer to the ring supporting the gyrowheel, and provide a dial for it, and we have a crude form of turn indicator. The gyro horizon has the full freedom of a universal mounting but is restrained in a position with its spinning axis vertical by one end of the gyrowheel being made heavier than the other. It is further controlled by air reaction that will be described later.

The *gyro turn indicator* consists of a spinning gyrowheel having a horizontal athwartship axle carried in a gimbal ring that is pivoted on a horizontal fore-and-aft line. Figure 124 shows two schematic views of the turn indicator, where *W* is the wheel having its axis *AB* horizontal and athwartships; this axis is mounted in the gimbal ring *R*, which is pivoted in a horizontal fore-and-aft line, the front pivot being shown at *C*. A pointer *P* is secured to the gimbal ring and moves over a dial.

When the airplane turns, a torque is applied through the gimbal-ring bearings to the gimbal ring and so to the wheel axis in a direction at right angles to the axis. As already explained, the wheel will then precess about an axis at right angles to the axis of spin (wheel axis) and to the axis of the applied torque (the vertical about which we are turning); this axis can be only the horizontal fore-and-aft axis. As the gimbal ring is pivoted on this axis, it will rotate on its bearings and so move the pointer over the scale.

In order to prevent this motion from becoming excessive and to ensure the pointer's returning to its central position when the turn causing the precession ceases, a centralizing spring is fitted. The instrument then becomes a rate-of-turn indicator having the property that the pointer displacement is proportional to the rate of turn so long as the rate of rotation of the wheel is kept constant. Actual instruments must also have some kind of damping device fitted and the indicator or cross level must be incorporated, but these details are not necessary to an understanding of the fundamental principles of the instrument.

Turn indicators are generally driven either by suction, when the incoming air is made to drive the wheel by striking buckets cut in the rim, or by electricity, in which case the wheel becomes part of a small electric motor.

Gyro Horizon.—The gyro horizon (Fig. 125) contains a controlled gyroscope, and its spin axis seeks the vertical when for any reason it has become displaced; it shows deviations from the vertical both in the fore-and-aft direction (pitch) and in the lateral direction (roll). It consists of a gyroscope with a universal mounting having the wheel axis vertical; the vertical-seeking quality may be imparted by any suitable means of detecting the deviation from the vertical and of applying the correct torque to precess the instrument back to the vertical.

The common peg top is a vertical-seeking gyroscope, as is known to any boy, but it does not seek the vertical directly although it finally gets there. The torque, due to the center of gravity of the top not being vertically above the supporting peg when the top is first spun, causes the axis of the top to precess sideways round the surface of a cone, the angle of which is steadily shrinking owing to other causes. The motion of the top axis is at right angles to the plane containing the vertical and the top axis.

Gyro horizons have been constructed that behave like this, but they are not so convenient or serviceable as those that move directly toward the vertical instead of following a shrinking cone. The preferred type of instrument will precess directly toward the vertical instead of reaching it ultimately through conical precession.

In order to replace the conical precession of the peg top by a direct precession it is necessary to apply the torque in a direction at right angles to both the wheel axis and the desired direction of return, which is the

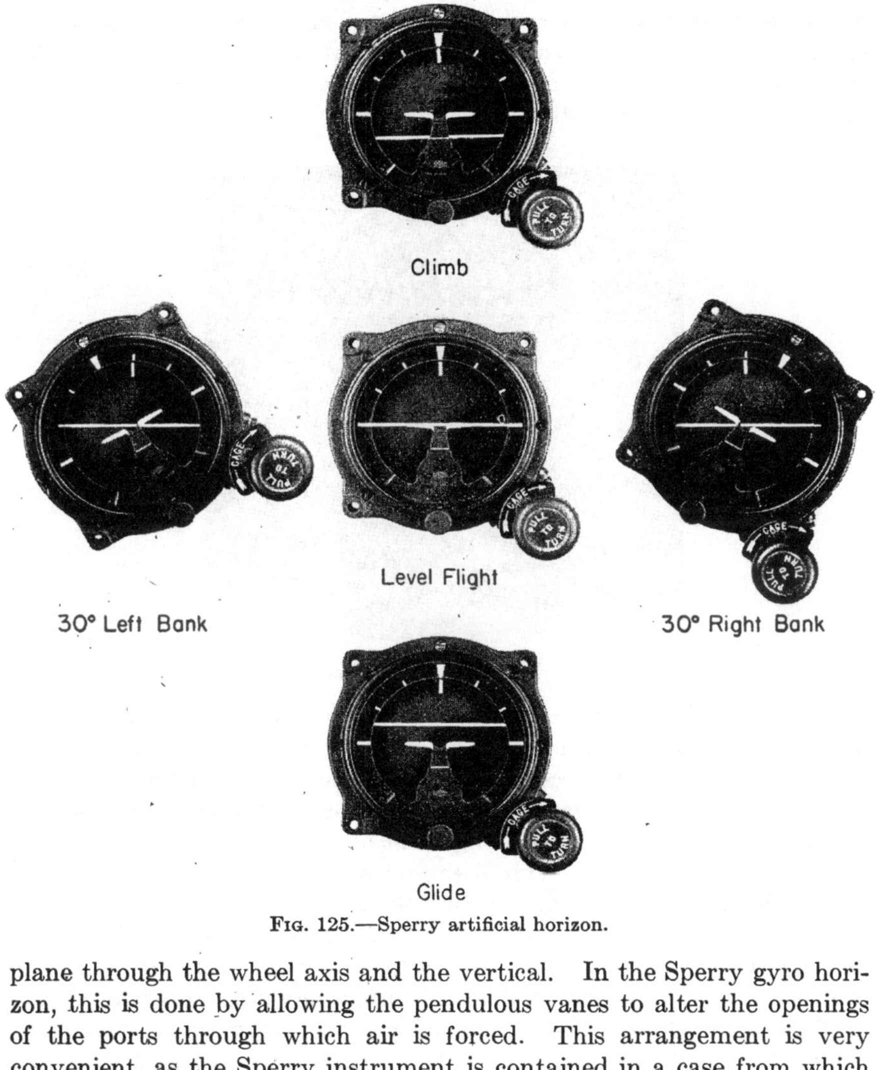

Fig. 125.—Sperry artificial horizon.

plane through the wheel axis and the vertical. In the Sperry gyro horizon, this is done by allowing the pendulous vanes to alter the openings of the ports through which air is forced. This arrangement is very convenient, as the Sperry instrument is contained in a case from which the air is evacuated.

In the Alkan electrically driven gyro horizon there is a channel in a plane at right angles to the wheel axis and, therefore, normally horizontal. A little, constant-torque friction drive is used to drive a steel ball around this horizontal channel at constant load. When the gyro axis is off the vertical, the steel ball runs uphill slowly, thus providing the correct imbalance to cause the gyro to precess back directly in the direction of the vertical; when the ball has reached the summit, it is allowed to run down the other side very quickly, so that the average imbalance is just what is required. The slight, permanently rotating imbalance, due to the ball's running round at constant speed when the gyro axis is vertical, is useful in causing a very slight nutation, which keeps the bearings from sticking.

Fig. 126.—Directional gyro.

The axis of the gyroscope of the gyro horizon seeks the vertical, but the vertical it seeks is not the true vertical; it is the apparent vertical, which is the resultant direction of the acceleration due to gravity and of any other accelerations that happen to be present. If a 180° turn is made, the erecting device will cause a slight tilt of the horizon a degree or two in the direction of turn. This is caused by centrifugal force acting on the pendulous vanes in the erecting device on the gyro, causing the gyro to precess in the direction of turn. If the turn is continued through to 360°, the error will cancel itself so that no error will exist. In any case, a few minutes of level flying will correct any errors introduced

Directional Gyro.—The directional gyro is a gyroscope universally mounted in neutral equilibrium. It will remain in whatever direction it is set until disturbed and made to precess. Consequently, if a perfect directional gyro could be made, it would maintain its direction in space undisturbed.

In the actual instrument (Fig. 126) the vertical gimbal ring is used to carry a strip compass card, and provision is made to set the instrument to agree with the magnetic compass. This must always be done, as the directional gyro has no azimuth-seeking properties, and its main purpose is to prolong a knowledge of direction during the intermittent periods that the magnetic compass is known to be unreliable because of turns and accelerations. It must therefore be checked at intervals and reset whenever necessary.

Gyropilot.—The gyropilot may be likened to the human body, but it detects smaller departures and acts on the controls with less delay

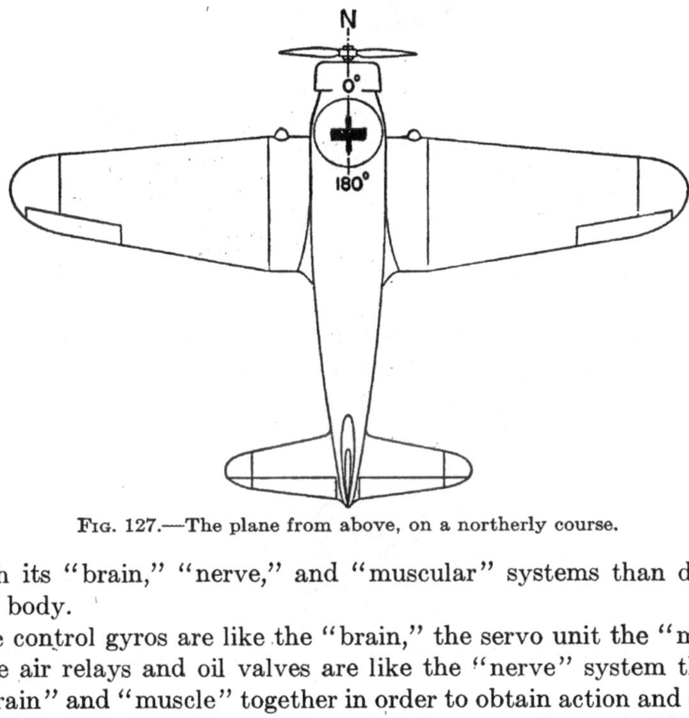

Fig. 127.—The plane from above, on a northerly course.

through its "brain," "nerve," and "muscular" systems than does the human body.

The control gyros are like the "brain," the servo unit the "muscle," and the air relays and oil valves are like the "nerve" system that ties the "brain" and "muscle" together in order to obtain action and control. The follow-up system is also part of the "nerve" system, carrying information back to the "brain" from the "muscle." The action of the follow-up is such that control is applied in proportion to disturbance, and overcontrolling of the aircraft is prevented.

The gyroscope that controls direction (the steering) is similar to the directional gyro but is arranged so that, in addition to indicating a departure from the course, it operates the necessary controls to return the airplane to the prescribed course. The gyroscope that provides lateral and longitudinal control is similar to the gyro horizon. It controls the positioning of the ailerons and the elevators (horizontal rudders).

The operating principles of the Sperry gyropilot that is in most general use can best be described with the aid of illustrations:

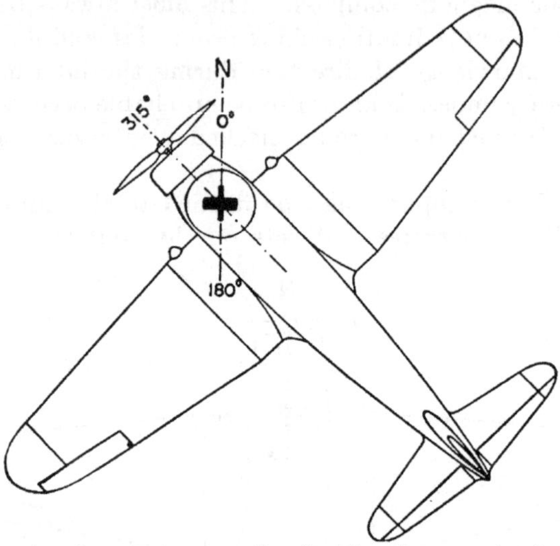

FIG. 128.—As the aircraft makes a turn, the directional gyro maintains its position and the card indicates the change of heading.

In Fig. 127, looking at the plane from above, the directional gyro is set to north, and the plane is flying north. If the plane turns left to a position as shown in Fig. 128, the gyro maintains its position, and the card reads 315°, a change in course of 45° west. This change in relation-

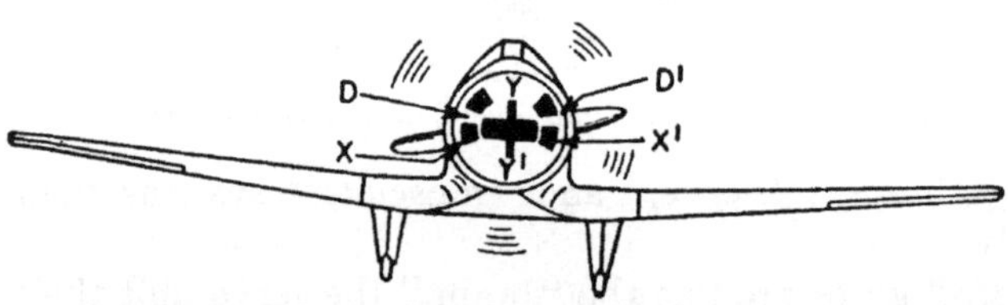

FIG. 129.—The air pick offs, added to the bank and climb gyro, are neutral when the plane is in level flight.

ship between the gyroscope and its supports is utilized to operate through an air relay a servo motor that moves the rudder to correct the departure from the course. The directional gyro, therefore, forms one unit in the "brain" of the gyropilot.

For the lateral and longitudinal controls, the departure of the airplane from the plane of the gryoscope operates the air valves for either or both axes and likewise moves either the lateral or longitudinal control, or both if the departure is about both axes.

In order to create the least possible amount of friction on the gyroscope the longitudinal and the lateral are arranged to provide control of their respective functions by means of "air pick offs"; therefore, no effort is required of the gyroscope.

Around the gyros are the air pick offs (XX′), as shown in Fig. 129. There are three sets of air pick offs—two connected with the bank-and-

Fig. 130.—In a bank (exaggerated for explanation purposes) the air pick offs are actuated, putting the system in operation to return the plane to level flight.

climb gyro for lateral and longitudinal control and one with the directional gyro for directional control. To illustrate the action of the gyropilot, only the aileron control will be used, as the rudder and elevator controls are operated in a similar manner.

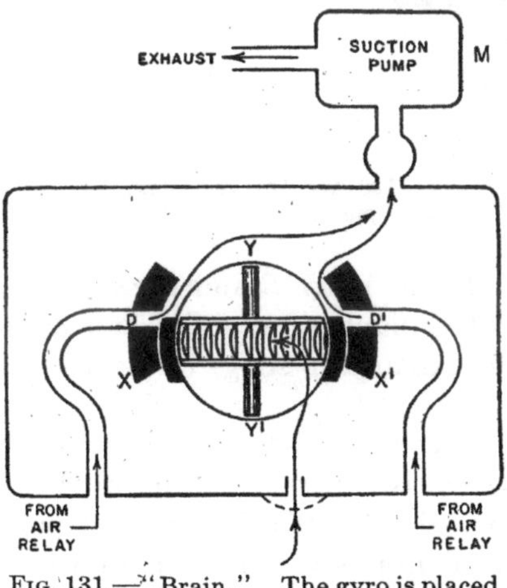

Fig. 131.—"Brain." The gyro is placed in a box with its air pick offs.

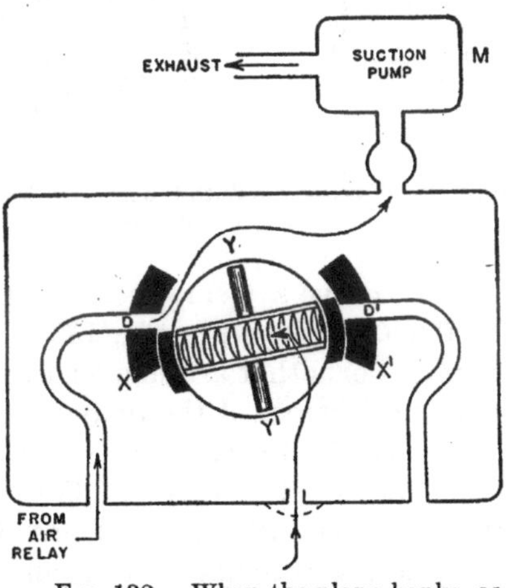

Fig. 132.—When the plane banks, as shown in Fig. 130, the gyro maintains its position and the box tilts, air being drawn from the air relay through the open port, *D*.

In Fig. 129 the attitude of the plane laterally is normal. In the air pick off (XX′) are two ports (*DD′*). In this position, air from the air relays flows equally through the ports as indicated by the arrows (*DD′*).

When the aircraft deviates from the level flight laterally as exaggerated in Fig. 130, the air pick offs, which from an integral part of the aircraft, take up the position as shown. Port *D′* is shut off, and port *D* is opened. Air, therefore, flows through port *D* and actuates the air relay-operated oil valve, which transmits oil to the servo piston to move the ailerons in the correct direction to bring the aircraft horizontal. The operation is reversed if the aircraft takes the opposite position.

Figure 131 shows the gyro and the aileron air pick offs placed in a box. Air is drawn into the bottom of the box by the suction pump *M* and directed to the gyro to spin it. Air is also drawn in from the air relay (through ports *D* and *D′* when the aircraft is level) by the suction pump, and exhausted at the top. With the aircraft in the position as shown in Fig. 130, the box is tilted and air is drawn through port *D* only, as shown in Fig. 132, port *D′* being closed.

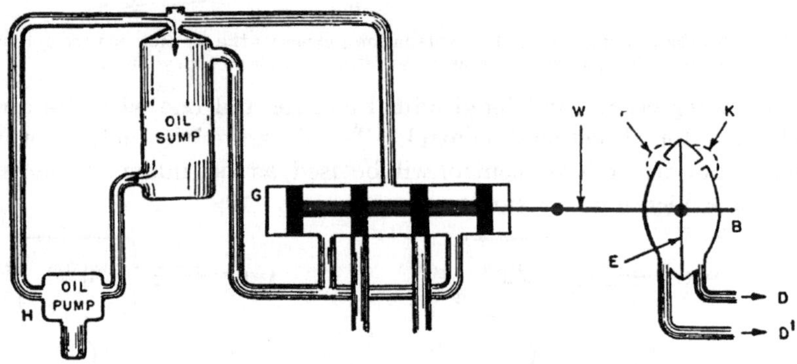

FIG. 133.—"Nerve." With the airplane level as in Fig. 129, the "nerve system" is neutral and no corrective action takes place.

The "nerve system," consisting of the air relay *B* and the balanced oil valve *G*, is shown in Fig. 133. *E* is the diaphragm, and *F* and *K* are two inlet ports that are smaller than the exhaust openings to the air pick offs at the bottom of the relay.

G is the balanced oil valve, which is connected to the air relay by the piston rod *W*. In Fig. 133 the system is neutral, the aircraft being level as in Fig. 129. Air is being drawn equally from the exhaust ports and, entering through ports *F* and *K*, maintains equal suction on both sides of the diaphragm. Therefore, there is no deflection of the diaphragm *E*, the oil valve piston is in the position shown, and no oil is permitted to flow to the servo cylinder.

If the aircraft changes attitude laterally, one of the ports (*DD′*) of the air pick off is opened fully and the other closed. This causes one side of the diaphragm in the air relay to receive increased suction while the suction on the other side falls off. Figure 134 shows the operation of the nerve system when there is suction at *D′*. The action of the diaphragm *E* moves the balanced oil valve to the left, permitting oil to flow to the servo unit through pipe 2. Oil from the other side of the piston returns through pipe 3 and flows back to the sump through pipe 4,

The "muscular system" consists of three hydraulic servo cylinders, one of which is shown in Fig. 135. Oil enters one end of the cylinder and moves the piston, an equal amount of oil being exhausted from the

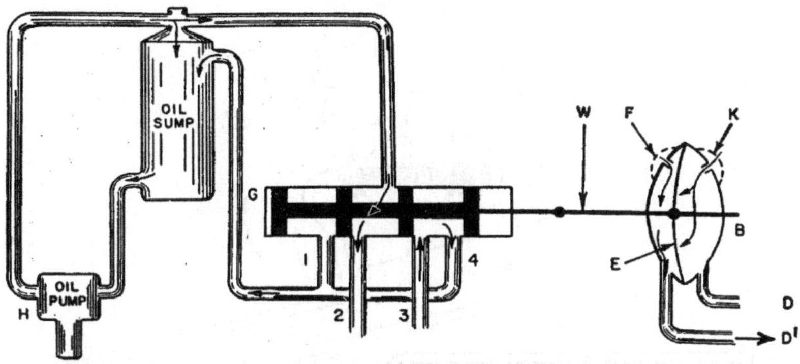

FIG. 134.—When the plane changes attitude laterally, the "nerve system" is energized, permitting oil to flow to the servo unit.

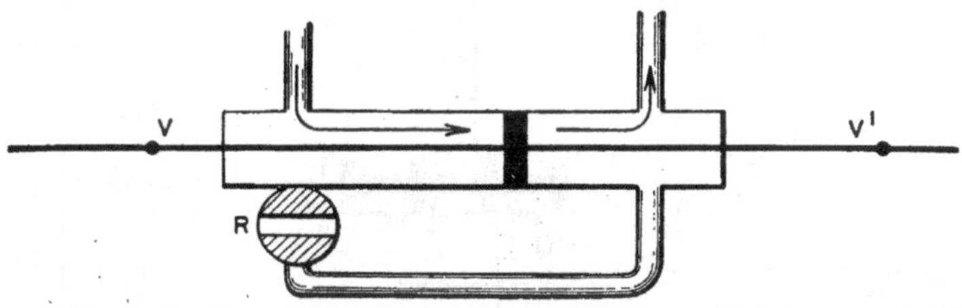

FIG. 135.—"Muscle." The oil moves the servo-unit piston, which is connected to one of the three control cables.

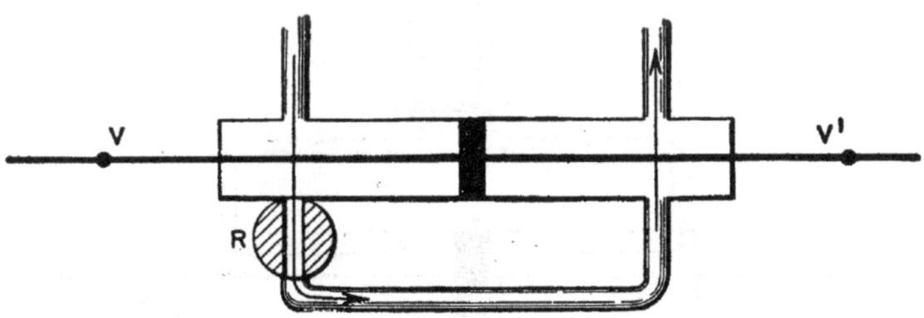

FIG. 136.—For take-offs and landings, the by-pass valve *R* is opened, permitting the oil to circulate around the piston in either direction.

other side of the piston and returned to the sump. The piston rod (*VV'*) is connected to one set of the control cables of the aircraft.

When the human pilot is flying the plane manually, the valve *R* (Fig. 136) is opened, oil flows through the by-pass tube, and the controls can be moved freely.

The three systems having been explained separately, Fig. 137 shows them combined. *O* is a suction regulator that keeps the vacuum at 4 in. of mercury regardless of the speed of the suction pump, which varies with the speed of the motor. The oil sump N carries the reserve oil. Q is a valve that regulates the oil pressure from the pump and permits it to circulate through the sump whenever the balanced oil valve

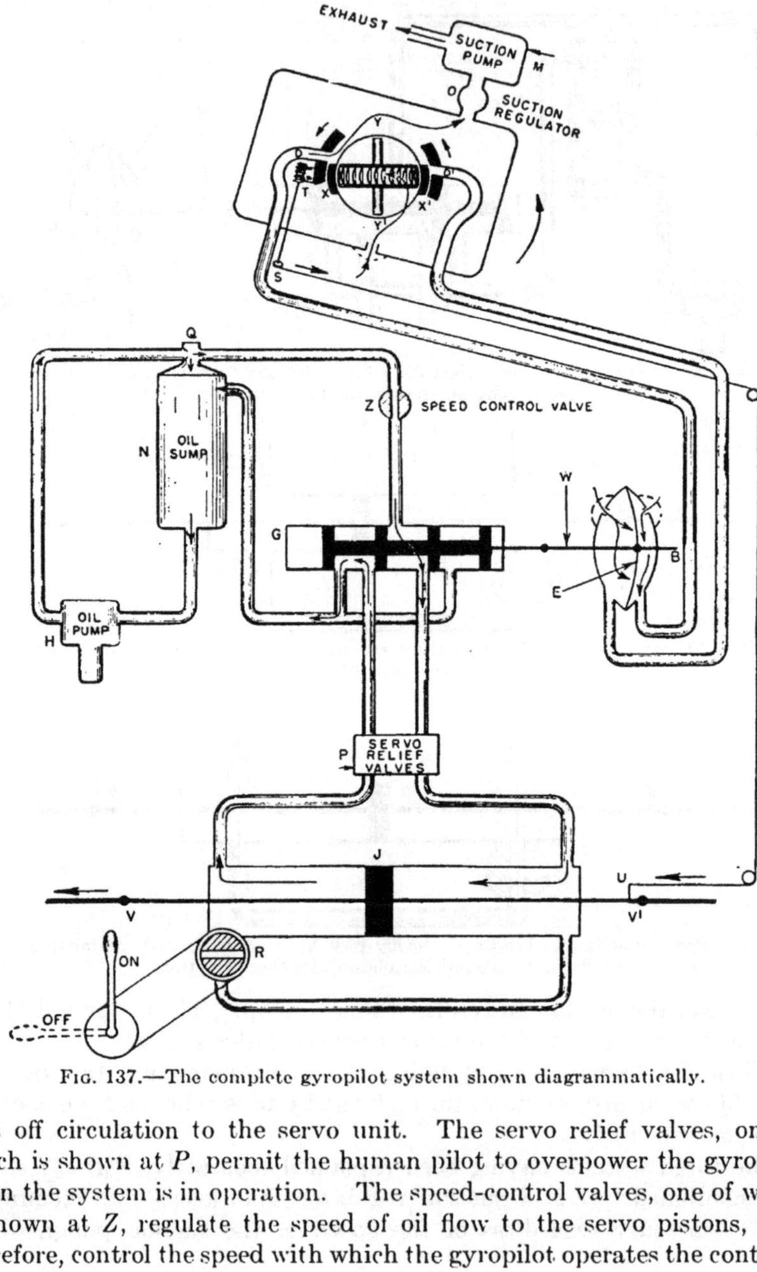

FIG. 137.—The complete gyropilot system shown diagrammatically.

cuts off circulation to the servo unit. The servo relief valves, one of which is shown at P, permit the human pilot to overpower the gyropilot when the system is in operation. The speed-control valves, one of which is shown at Z, regulate the speed of oil flow to the servo pistons, and, therefore, control the speed with which the gyropilot operates the controls.

The final part of the "nerve system" has been added in Fig. 137—the follow-up system. It must be remembered in controlling a plane that it is necessary not only to apply control to bring the plane back to level when it has been disturbed but also to begin to remove the applied control as the plane is returning to level so that the control surface will be back in neutral when the disturbance has been fully corrected. A further requirement is that the amount of control applied be in proportion to the displacement of the plane. All this is necessary to both manual and automatic control and in the latter is handled by the follow-up. The air pick offs XX′ are not fixed rigidly to the gyro box and the plane, but instead, can be moved in relation to them by the follow-up mechanism. A cable is connected to the servo piston rod at *U* and runs to the follow-up pulley *S* on the gyro box. The pulley controls a gear *T*, which is connected to a gear on the air pick offs XX′, which are commonly connected. When the piston *VV′* moves to the left, the follow-up cable at *U* moves likewise, and gear *T*, through the action of pulley *S*, moves X down and X′ up. When they reach a neutral position (both half open), the air relay and oil valve are centered and servo piston movement away from neutral is stopped. Now consider that the control-surface movement, which the servo has been producing, has been bringing the aircraft back to level flight. As the plane continues in toward level, the air pick offs that have been driven ahead of the gyro box pass beyond the neutral point and begin to cause servo movement in the opposite direction. This is not opposite control but is the removal of the control originally applied. The mechanism and its ratios are so arranged that the correct amount of control will be applied and also removed at the proper rate as the aircraft returns to level.

The Directional-gyro Control Unit.—This unit contains the directional gyro, which is the directional reference for both manual and automatic steering control. It also contains a ball bank indicator, as shown in Fig. 138, the air pick offs and follow-up mechanism for directional control, and a means for setting the gyropilot to steer any selected heading. The directional gyro control unit, together with the bank-and-climb-gyro control unit, is carried in the mounting unit and the whole is installed as a part of the instrument panel.

For military aircraft and others that require additional facilities for maneuvering, a turn control is supplied, which can be placed adjacent to the gyropilot, or wherever most convenient. The turn-control valve is a manually operated valve that controls the flow of air to a small air motor in the directional-gyro control unit. The motor moves the direction setting continuously, thus causing a turn. The control may be set for no turn or for a turn in either direction.

The Bank-and-climb-gyro Control Unit.—This unit contains the bank-and-climb gyro, which is used for lateral and longitudinal indication and control. It also contains the air pick offs for these two controls, together with the means for making manual adjustments. The aileron knob permits one set of air pick offs to be adjusted for the desired lateral attitude, and the elevator knob controls the setting of the other pick offs

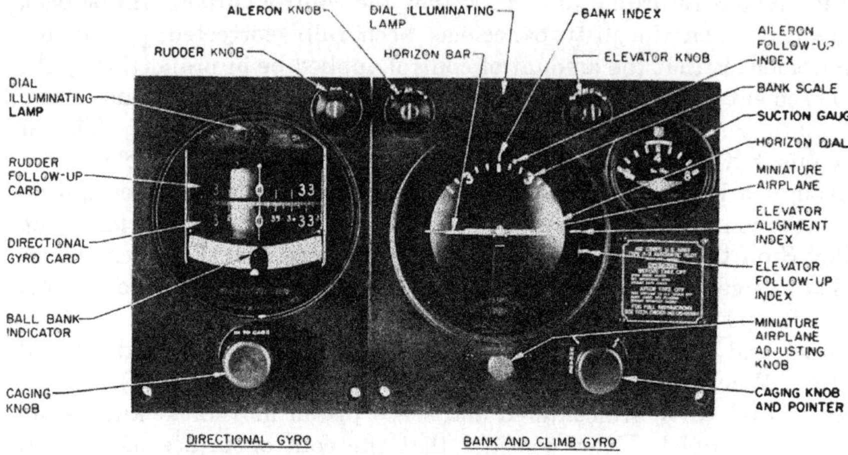

FIG. 138.—Gyro control units, front.

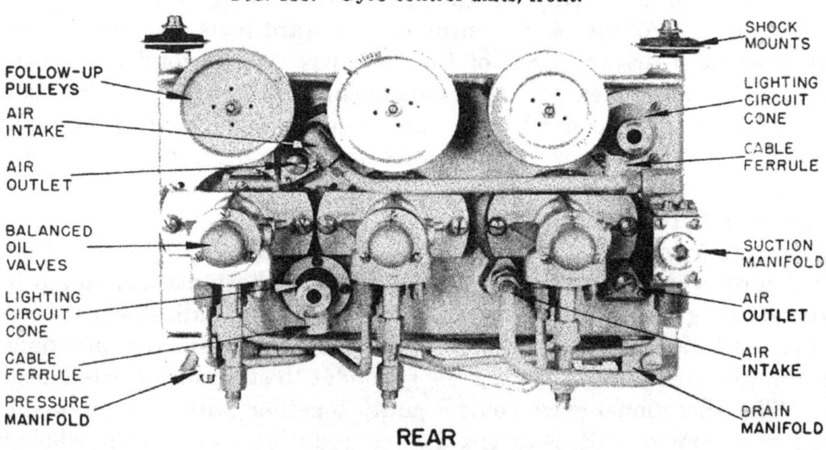

FIG. 139.—Gyro control units, rear, showing connections which engage with those on the mounting unit.

for longitudinal control. Follow-up indexes show the lateral and longitudinal settings that are made. The "level" knob permits the gyropilot to be set to fly the aircraft at any desired altitude. When adjusted to hold level flight, this control makes small shifts in longitudinal control to compensate for altitude changes caused by change of trim or by air currents. The bank-and-climb-gyro control unit is carried next to the directional-gyro control unit in the mounting unit.

The Mounting Unit.—The ease of maintenance of the gyropilot has been carefully planned. Tracks on the mounting unit permit the control units to be slid into place, where they are secured by four attaching bolts. Air, follow-up, and lighting connections are established automatically when the bolts are tightened, and disengaged when the control units are withdrawn. Either of the two control units can thus be replaced in a few minutes. Figure 139 shows the rear of the control units with the connections that engage with those on the mounting unit (Fig. 140).

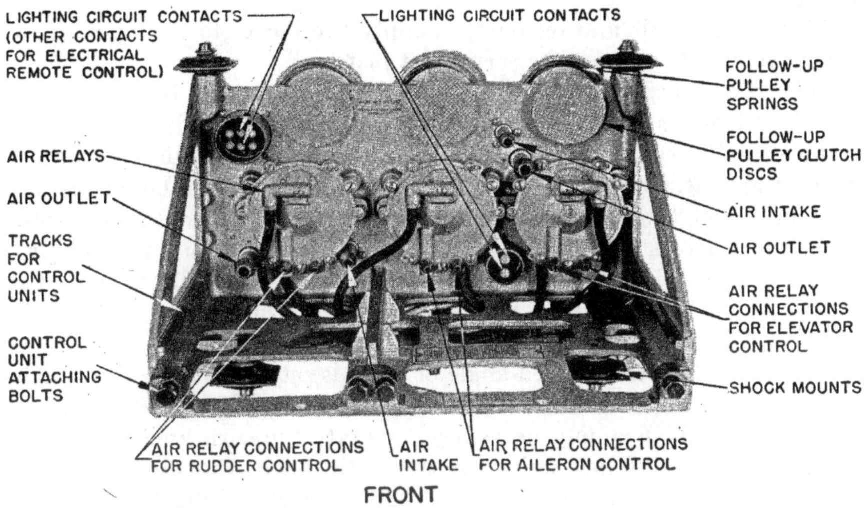

FIG. 140.—The mounting unit, supported on shock mountings at the back of the instrument panel and provided with tracks upon which the control units slide into place.

Other component parts that are carried on the mounting unit, such as the balanced oil valves, the air relays, and the follow-up pulleys, are easily adjusted in place or can be removed for inspection and servicing without having to disturb the rest of the equipment in any way. Standardized parts and the accessibility of the various units help to keep maintenance costs to a minimum.

Speed Valves.—The speed valves serve to distribute the main oil supply under pressure, through the balanced oil valves, to the three controls and permit setting the rate of flow of oil individually so that each control may be adjusted to operate to the best advantage. The speed valves may be mounted with the control units or placed wherever most convenient.

Servo Unit.—The servo unit consists of three cylinders cast en bloc, with pistons that are approximately to the center when the aircraft control surfaces are centered. The servo pistons are connected directly to the main control cables of the aircraft. They are also connected, through the follow-up cables, to the control units, so that corrective control will be removed at the proper rate, as previously explained. A manually operated "on-off" valve, connected to a lever accessible to the human pilot, by-passes the servo oil from one side of the pistons to the other when manual control is desired. Spring-loaded servo relief valves are also built into the servo unit so that the human pilot, by applying additional power manually to the controls, can blow the valves and thus fly the aircraft manually even while the gyropilot is engaged.

Motor-driven air and oil pumps supply the power for the gyros and their air pick offs and for the servo-unit pistons.

How the Gyropilot Is Used.—As soon as the aircraft is clear of the airport and on its course, the human pilot rotates the adjusting knobs on the gyropilot control unit so that the three follow-up indicators match the gyro indications for direction, bank, and climb. Then he moves the engaging lever "on" and takes his hands and feet from the controls. The climb knob is adjusted to obtain the desired rate of climb. Once this is set, the aircraft continues climbing steadily until the cruising altitude is reached, at which time another slight turn of the climb knob puts the plane in level flight.

For large turns the desired angle of bank is set in, and the turn knob rotated so as to turn the aircraft at the desired rate. If a small change of course is desired, it is only necessary to rotate the turn knob slowly to the right or left. When flying on the radio beam the precision with which these small changes in heading can be made while the gyropilot is in operation is an important factor in keeping the plane on course.

On long flights a glide is often started as much as 100 miles from the airport. A slight turn of the knob for glide is all that is necessary in order to obtain the desired gliding angle, and the gyropilot thereafter maintains a steady rate of descent until the human pilot is ready to take over the controls and make his landing.

To disengage the gyropilot, the human pilot takes over the controls and moves the engaging lever "off." If necessary, he can overpower the gyropilot while it is in operation.

Training.—As weather conditions are not always appropriate for this training, instructional aircraft are usually fitted with a translucent hood, which deprives the pilot of external view. The instructor, of course, occupies an open cockpit. In addition to the normal flying and navigational instruments the dashboards carry a turn-and-bank indicator and a pitch indicator. These are the only instruments necessary for basic training.

As successful blind flying depends to an appreciable extent on the self-confidence of the pilot, great care must be taken not to undermine this by forcing the issue in early stages. For this reason it is customary to make training flights of short duration and to master each control separately before any attempt to combine them is made. The pupil may begin with the rudder and should concentrate on keeping the aircraft straight and making gentle turns on to compass courses.

At this stage the vagaries of the compass should be fully explained. He will then pass on to the lateral control of the aircraft, and after flying the machine level, should manipulate this control while the instructor performs turns by means of the rudder. A similar process should apply to the elevator controls; then a gradual approach should be

Fig. 141.—Link-trainer ensemble showing the radio control table.

made to combining any two controls, and finally all three. It is here that particular care is essential on the part of the instructor, because the degree of unaccustomed concentration is a severe strain on the pupil, and an unsympathetic instructor can easily shatter the self-confidence of a promising pupil.

For example, it is not improbable that the pupil will maneuver the aircraft into an attitude from which he cannot extricate it; in this event the instructor should take control before the situation is apparent to the pupil. In the circumstances, his tendency is to revert to normal methods of flight by obeying his own reactions in preference to the information conveyed by the instruments. The conflict between what he sees and what he feels is perhaps the greatest psychological problem in the whole of his training, but he cannot hope for success until he is able to convince himself that, although his senses may play him false, the instruments will not mislead him.

Experience has shown that the average pupil makes very little progress in handling aircraft in all three dimensions during the first 3 or 4 hr. of his instruction in the air and, unless he receives proper encouragement, he is inclined to admit defeat. This frame of mind must be avoided if possible, because, although he will not believe it, he has reached the darkest hour before the dawn, and henceforth his progress should be considerably easier.

Throughout his training the pupil will be taught to make all control movements as gently as possible and to avoid either strenuous or jerky

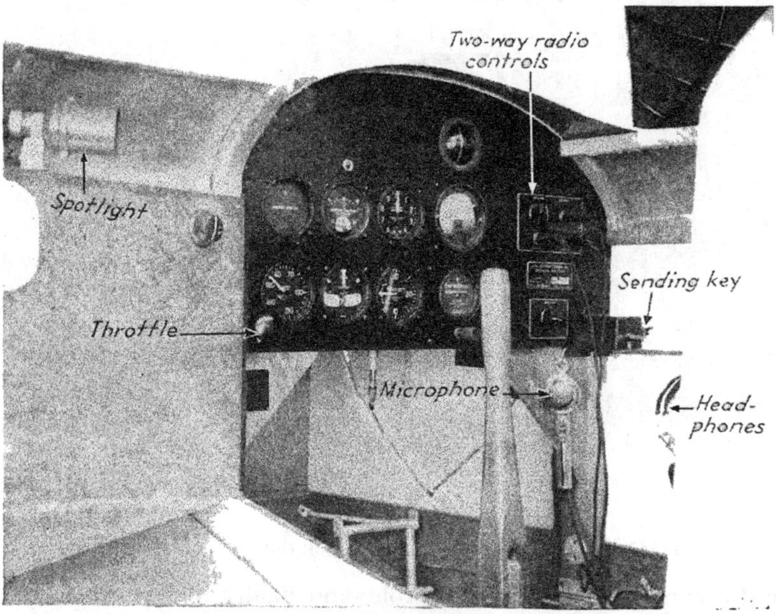

FIG. 142.—Cockpit of Link trainer.

movements of controls. These matters are important in defeating the condition of vertigo that occurs when the human body is deprived of sight and subjected to changing positions or velocities. One example will serve to illustrate this point. During the course of training the pupil is taught the various methods of extricating the machine from any attitude into which it may be put, but the most difficult maneuver to recover from is the spinning nose dive. While the spin is in process, the pupil is conscious of a velocity in a given direction, but as soon as the spin is stopped, and before the aircraft is out of the resultant dive, the change of direction has such a marked effect upon the pupil that he is convinced that he is spinning in the opposite direction. Systematic training will, however, soon defeat this bugbear, and there is nothing like spinning to convince the pupil that he *must* believe his instruments.

Training Devices.—With a view to reducing the cost of training, many attempts have been made from time to time to provide devices for use on the ground that would reproduce the same instrument indications as occurs in flight. An example of these is the *Link trainer*, which consists of a small model aircraft, the fuselage containing the pilot's seat, and standard stick and rudder, hood for blind flying, instruments, and wireless (see Figs. 141 and 142). An electric motor operates a vacuum turbine actuating a series of bellows. The air to these bellows is controlled by the stick and rudder, through a system of valves, in such a manner that the trainer banks, turns, climbs, and dives in response to the controls in the same manner as an airplane.

The instruments in the trainer consist of

1. Bank-and-turn indicator.
2. Air-speed meter.
3. Sensitive altimeter.
4. Rate-of-climb indicator.
5. Magnetic compass.
6. Directional gyro.
7. Artificial horizon.
8. Radio compass.
9. Marker beacon.
10. Two-way wireless, both voice and key.

The track the aircraft is flying is recorded and plotted by a relay system on a chart on the wireless desk so that a pilot may see the exact course he has been making good.

The sensitive altimeter shows changes of altitude when the trainer is in climbing, diving, or cruising attitude, exactly the same as in an aircraft. This permits problems to be flown at predetermined altitudes and makes it possible to practice landings and climbs by instruments. The altimeter, in conjunction with the radio compass, marker beacon, gyro compass, and horizon, is used for blind landings and instrument-approach practices. The trainer also has an automatic spinning device, which causes it to go into a spin when stalled. The apparatus is not intended to replace air training completely, but there is no doubt that it very greatly facilitates it.

Although the foregoing is not intended to be a complete survey of the problems appertaining to controlling an aircraft under blind conditions, it will serve to give a general idea of the difficulties involved. It was mentioned earlier in this chapter that the normal flying instruments, a turn-and-bank indicator, and some form of pitch indicator were the only instruments necessary for basic training. It is possible for a pilot who has received proper training with this combination to fly for several hours, but a great strain is imposed upon him, and for this reason commercial aircraft are fitted with additional aids that not only reduce the mental strain of blind flying but also make for greater accuracy in navigation. For this reason it is essential to provide basic training on a turn-and-bank indicator, which cannot be rendered useless by any maneuver of the aircraft, and to include this instrument on the dashboard of commercial aircraft. A point to note when using the artificial horizon is that the level-flight position of the miniature airplane in relation to the horizontal bar will depend to a small extent upon the loading of the aircraft. For example, a heavily loaded aircraft will fly tail down, and the instrument will indicate a slight climb; a lightly loaded aircraft will give the opposite indication. For this reason occasional reference should be made to the altimeter or rate of climb and descent instrument.

Navigation.—It is obvious that when the sky and earth are obscured from the pilot's view there is no opportunity to adopt normal methods of navigation. The pilot therefore has to rely upon dead reckoning or radio position finding in one form or another. It would be folly to rely exclusively on radio for the purpose of navigation, particularly when flying through fog, and it is therefore always wise to keep a continuous record of times and positions, so that up-to-the-minute data are available in case of necessity. A continuous plot of the dead-reckoning point of departure, the heading, and air speed, together with an allowance for the estimated wind, should also be kept.

The use of radio for navigational purposes is described in Chap. V.

Blind Landing.—The all-important operation of landing when visibility is nil has been taxing the minds of experts for many years. The Lorenz system has already been described in Chap. V. The conditions of visibility under which a pilot can land safely will depend to a large extent upon the experience of the pilot himself, but much training is necessary to make the fullest use of the system.

It will be clear from the foregoing that the ability to reach a desired destination depends first upon the knowledge necessary to control the aircraft, and secondly upon making intelligent use of the navigational aids available.

Once the plane has become enshrouded in fog, it is too late to record the last known point of departure accurately, and, without this knowledge, the record of dead reckoning is broken. The rules for safety in navigating during blind flying are:

1. Keep a continous record of times and positions of the plane so that the latest data will be available in case of necessity.

2. Keep a continuous plot of the dead-reckoning point of departure, heading, and air speed, together with allowance for the estimated wind.

3. Be prepared to check the dead reckoning by any of the three methods available, *viz.*, piloting, radio position finding, or celestial navigation.

4. Be prepared by adequate training to land by radio landing beam when necessary.

CHAPTER X

OUTLINE OF METEOROLOGY

By Dr. Sverre Petterssen

Introduction.—The navigator of a seagoing vessel is generally concerned with the tides, the currents, and the dangers to navigation of the ocean upon which he sails. The navigator of an aircraft is even more vitally concerned with the currents and the dangers to navigation in the broader ocean of the air in which he flies. The currents of the air are far greater in velocity, and both their direction and velocity are subject to far greater variation.

Moreover, the seagoing navigator who is confronted with a thick fog as he approaches his destination can slow down or come to a full stop and anchor. The pilot, however, who approaches a landing field with his fuel nearly gone, has no alternative but to land or to seek a near-by field where the fog may not be so thick, so that he can find his way down to a safe landing. A pilot who understands the behavior of the winds may be able to choose his altitude and route so as to have a 20-mile wind behind him, instead of a 20-mile wind against him.

The meteorological services of the various countries give the pilots information as to the direction and velocity of the wind, the weather, the visibility and ceiling likely to be encountered, and the dangers to air navigation such as fog, thunderstorms, snowstorms, and icing. Proper use of this information adds so much to the *safety of flight* that every pilot should have a knowledge of the fundamentals of meteorology.

The purpose of this chapter is to make the pilot acquainted with the terminology and the fundamental principles of meteorology to such an extent that he will be able to interpret and make adequate use of the weather charts, the airway forecasts, the airway reports, and other information furnished by the meteorologists, and also to understand the weather phenomena encountered en route. The terms and processes of direct importance for air navigation are explained fully in the text, and a selection of terms used frequently in meteorology is given in a glossary at the end of this chapter.

THE ATMOSPHERE

The atmosphere, which completely envelops the earth, may be regarded as a fluid sea at the bottom of which we live. The air, or the material of which the atmosphere is composed, is perfectly elastic and highly compressible. Although extremely light, it has a definite weight. At ordinary pressure and temperature the weight of a sample of air is 1/770 of the weight of an equal volume of water. Thus 1 cu. ft. of air weighs 1.22 oz. or, in metric units, 1 cu. meter of air weighs 1.3 kg.

In consequence of this weight, the air exerts a certain pressure upon the surface of the earth, amounting on the average to about 15 lb. per square inch. A column of air from the surface of the earth to the top of the atmosphere exerts a pressure on the surface that is approximately equal to the weight of a column of water 33 ft. (or 10 meters) high. This is equivalent to the weight of a column of mercury 30 in. (or 76 cm.) high. For this reason mercurial barometers are used to measure the pressure of the atmosphere.

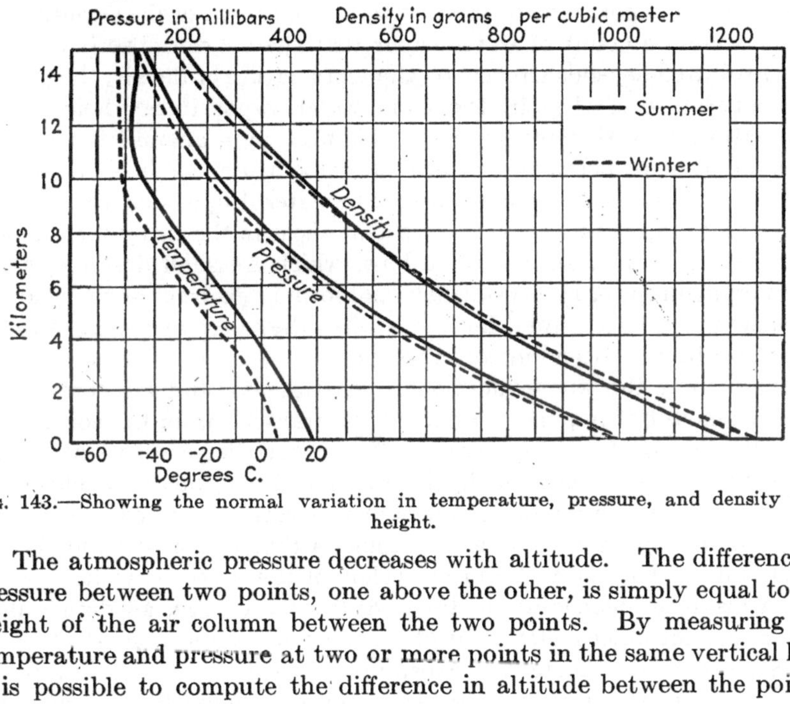

FIG. 143.—Showing the normal variation in temperature, pressure, and density with height.

The atmospheric pressure decreases with altitude. The difference in pressure between two points, one above the other, is simply equal to the weight of the air column between the two points. By measuring the temperature and pressure at two or more points in the same vertical line, it is possible to compute the difference in altitude between the points. Assuming normal distribution of temperature, pressure becomes a simple function of altitude, and special barometers (*altimeters*) have been constructed to record altitude instead of pressure.

Since the atmospheric pressure is equal to the weight of the air column, it follows that the pressure must decrease gradually and approach zero with increasing altitude. The same also applies to the density of the air. There is, therefore, no distinct upper limit to the atmosphere; it merges gradually into empty space. Figure 143 shows how pressure, density, and temperature normally vary with altitude in the lower 15 km.

Even though the atmosphere reaches to great altitudes, it is only the lower part of it that is of importance for the weather. The highest clouds (cirrus) are seldom more than 10 km. (33,000 ft.) above the earth's surface, while 50 per cent of the total weight of air and 90 per cent of the total moisture content are within about 5 km. (16,000 ft.) of the earth's surface.

Composition of the Atmosphere.—Air is a mechanical mixture of gases. A sample of dry and pure air contains about 78 per cent nitrogen, 21 per cent oxygen, and almost 1 per cent argon. There are also traces of other gases, but they occur in such minute quantities that they are of no practical importance in the study of the weather processes.

The air also contains a variable amount of water vapor. In many respects water vapor is the most important constituent of the atmosphere. The maximum amount of water vapor that the air can absorb without being saturated depends entirely on the temperature of the air; the higher the temperature of the air, the more water vapor it can hold. The air is saturated with moisture when this maximum amount is reached. When air is cooled below the saturation temperature, condensation takes place, the water vapor being condensed to water droplets or ice crystals. Small water drops and ice crystals are kept afloat in the air by the ascending currents. Under special conditions, which we shall describe later, these minute drops or ice crystals join and form large drops or snowflakes, which are precipitated from the clouds when they become too large to be kept afloat by the ascending air currents.

Apart from the above-mentioned constituents, the air contains a variable amount of impurities such as dust particles from the continents and salt particles from the oceans. These particles are so small that they cannot be seen individually by the naked eye, but their effect on visibility and on coloring of distant objects is easily observed. Through haze, distant objects are seen as if through a thin veil of pale blue if the object is dark, but the veil appears yellow if the object is white. At a certain distance, depending on the density of the haze, all details disappear, and the objects stand out like a silhouette against the sky. The denser the haze, the shorter is the distance at which the details disappear.

The presence of dust in the air is not only important because of its influence on visibility. If the air were perfectly pure, there would be no appreciable condensation of water vapor, because condensation takes place on certain nuclei. When air becomes saturated, condensation takes place on the active (hygroscopic) dust particles that form the nuclei of condensation. Not all dust particles are active as nuclei. Salt particles from the oceans are particularly active as condensation nuclei; observations show that such particles are present in the air in abundant quantities.

THE METEOROLOGICAL ELEMENTS

The meteorological elements observed at ordinary land stations are

1. Pressure.
2. Pressure changes (barometric tendency).
3. Temperature.
4. Wind direction and velocity.
5. Cloud forms.
6. Cloudiness.
7. Ceiling.
8. Weather.
9. Visibility.
10. Precipitations.
11. State of ground (field conditions).

Certain selected stations report wind velocity and direction in various levels above the ground; other selected stations report pressure, temperature, and humidity in the free air. We shall consider the most important of these elements briefly. The reader who desires fuller information is referred to instructions issued by the U. S. Weather Bureau, Washington, D. C.

Pressure.—The pressure is usually observed by means of a mercurial barometer. The direct reading of the barometer gives the *length* of the column of mercury whose weight balances the weight of the air column above the barometer. The length of the column of mercury depends on the temperature of the barometer. In order to render the observations of the various stations comparable, the readings are corrected for the thermal expansion of the metal tube and the mercury in the barometer, using 32°F. (or 0°C.) as the standard temperature. The weight of the air column depends also on the local gravity. A second correction is applied in order to obtain the pressure that would be observed at the station level if the local gravity were equal to the normal gravity at Lat. 45°N. The pressure thus obtained is the correct air pressure at the level of the barometer cistern. Since pressure varies with height, it is necessary to add a third correction in order to obtain the pressure that would occur if the barometer were placed at mean sea level. Hence the pressures given in meterorological reports are those which a correct barometer would show if it were mounted at sea level, if its temperature were 32°F. and if the local gravity were equal to the normal gravity at Lat. 45°N.

Up to about 1914, pressure was reported in units of length, either in inches of mercury or millimeters of mercury. In later years a new unit, called the *millibar* (mb.), has come into general use in all European countries, and is coming into use in this country. Normal pressure at sea level is roughly 30 in. or 762 mm., to which correspond 1,016 millibars.

Temperature.—The temperature of the air is measured by means of a mercurial thermometer. On land stations the instrument is hung in a louvered wooden screen in order to provide effective ventilation, and to protect the instrument from the influences of radiation and precipitation.

The thermometers are usually graduated either on the centigrade (C.) or the Fahrenheit (F.) scale. On the Fahrenheit scale the freezing point of pure water is 32° and the boiling point 212°. On the centigrade scale the freezing point is at 0° and the boiling point at 100°. To convert Fahrenheit degrees to centigrade degrees, subtract 32 and multiply by $5/9$. To convert centigrade degrees to Fahrenheit degrees multiply by $9/5$ and add 32.

Humidity.—The moisture content of the air can be expressed in terms of relative humidity, specific humidity, or dew-point temperature.

The *relative humidity* is the ratio (expressed as a percentage) of the moisture content of the air to the amount of moisture the air could contain if it were saturated at the same temperature.

The *specific humidity* is the number of grams of water vapor contained in one kilogram of air.

The *dew-point temperature* is the temperature to which the air must be cooled in order to become saturated. The closer the dew-point temperature is to the actual temperature of the air, the greater is the likelihood of formation of fog or clouds.

The moisture content of the air is obtained either from a hair hygrometer or from the readings of the "dry-bulb" and "wet-bulb" thermometers. The length of a hair varies with the humidity of the air. The hair hygrometer is an instrument for measuring variations in the length of a hair, or a bundle of hairs; it is graduated to indicate humidity instead of the length of the hair.

A wet-bulb thermometer is an ordinary thermometer whose bulb is kept wet by a piece of muslin and a wick that dips into a vessel containing pure water. If the air is not saturated with moisture, water will evaporate from the bulb of the wet thermometer. The evaporation cools the thermometer so that it shows a lower temperature than that of the dry thermometer. From the readings of the two thermometers the moisture content is readily obtained by means of humidity tables.

It is worthy of note that the specific humidity remains constant as long as no water vapor is added to or removed from the air, whereas the relative humidity varies with change of temperature.

Wind Direction and Velocity.—The wind direction is the direction from which the wind blows, a north wind being a wind blowing from true north. The direction of the wind is reported on a scale ranging from 00 to 32, where 08 = east, 16 = south, 24 = west, 32 = north (00 = calm).

The wind direction and speed are obtained from an instrument (anemometer or anemograph), and the speed is expressed in meters per second, kilometers per hour, or miles per hour. The relations between these units are

1 meter per second = 3.6 km. per hour = 2.24 m.p.h.

It is also possible to estimate the wind speed without the use of any instrument. The skilled observer will be able to estimate the force of

Beaufort number	Name	Miles per hour	Map symbol	Description
0	Calm	Less than 1		Calm; smoke rises vertically
1	Light	1– 3		Direction of wind shown by smoke but not by wind vanes
2	Light	4– 7	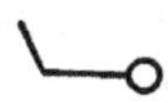	Wind felt on face; leaves rustle; ordinary vane moved by wind
3	Gentle	8–12		Leaves and small twigs in constant motion; wind extends light flag
4	Moderate	13–18		Raises dust and loose paper; small branches are moved
5	Fresh	19–24		Small trees in leaf begin to sway; crested wavelets form on inland waters
6	Strong	25–31	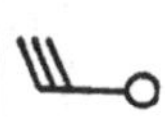	Large branches in motion; telegraph wires whistle; umbrellas used with difficulty
7	Strong	32–38		Whole trees in motion; inconvenience felt in walking against wind
8	Gale	39–46		Breaks twigs off trees; generally impedes progress
9	Gale	47–54		Slight structural damage occurs; chimney pots and slate removed
10	Whole gale	55–63	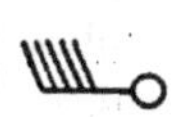	Trees uprooted; considerable structural damage occurs
11	Whole gale	64–75		Very rarely experienced; accompanied by widespread damages
12	Hurricane	Above 75		Devastation occurs

FIG. 144.—Beaufort scale.

the wind from the action that it has on certain objects. When this is done, the wind speed is referred to what is called the Beaufort scale. This scale, which was introduced by Admiral Beaufort in 1806, is still in international use, and wind velocities that are measured by instruments are converted to Beaufort numbers in order to have a uniform standard throughout the network of stations.

The wind varies with the distance above the ground, and the variation is particularly rapid close to the ground. Figure 144 gives the Beaufort scale of wind force, with the velocity equivalents at about 6 meters (20 ft.) above level ground.

The wind is not a steady current. The velocity varies in a succession of gusts and lulls of variable direction. These variations of short period are primarily caused by the friction along the surface of the earth, which creates eddies that travel with the general current, superimposed on it. Large irregularities in the wind are caused by convectional currents in unstable air. We shall discuss them in a later paragraph.

The turbulence of the wind is an important agent in carrying water vapor, heat, dust, etc., up to high levels.

Classification of Clouds.—Even though the number of forms that clouds may take is almost unlimited, it is a fact of experience that the number of *types* of clouds is rather limited. The international classification of clouds consists of 10 main types, which for convenience are arranged according to height above the ground in the following manner:

Cirrus (Ci)	High clouds
Cirro-stratus (Cs)	High clouds
Cirro-cumulus (Cc)	High clouds
Alto-stratus (As)	Medium clouds
Alto-cumulus (Ac)	Medium clouds
Strato-cumulus (Sc)	Low clouds
Nimbo-stratus (Ns)	Low clouds
Cumulus (Cu)	Low clouds
Cumulo-nimbus (Cn)	Low clouds
Stratus (St)	Low clouds

Cirrus (Ci) is the highest of all clouds. It has a typical fibrous (threadlike) structure and a delicate silky appearance (Fig. 145). Cirrus clouds are sometimes arranged irregularly in the sky as detached clouds without connection with cirro-stratus or alto-stratus. They are then called fair-weather cirrus. If the cirrus clouds are arranged in bands, or connected with cirro-stratus or alto-stratus, or otherwise systematically arranged, they are usually the forerunners of bad weather. In thundery or squally weather a special kind of cirrus (*cirrus densus*) is frequently observed, which originates from the anvils of cumulo-nimbus. These clouds are often called *false cirrus*, because they are denser and usually lower than the real cirrus.

Cirro-stratus (Cs) is a thin whitish sheet of cloud, sometimes like a veil covering the whole sky and merely giving it a milky appearance, at other times showing signs of a fibrous structure like a tangled web. Cirro-stratus often produces halo around the sun or moon. It is often

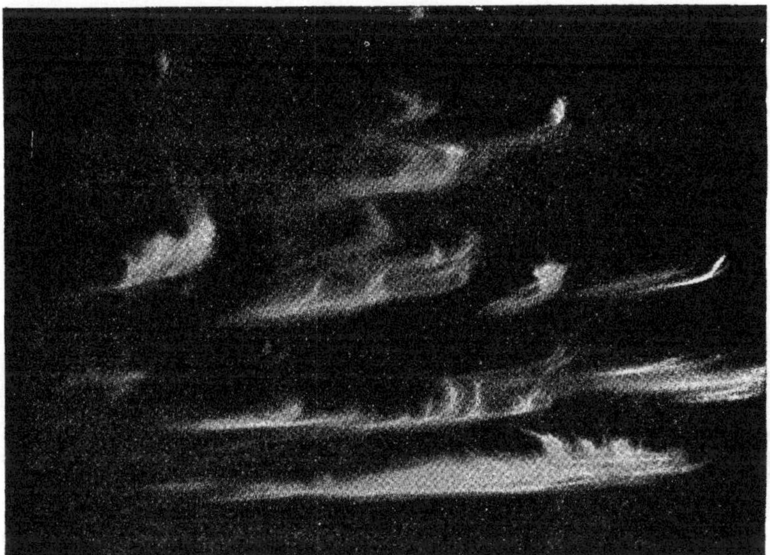

FIG. 145.—Cirrus, tufted form. (*International Atlas of Clouds. Photograph by Fundacio Concepcio Rabell.*)

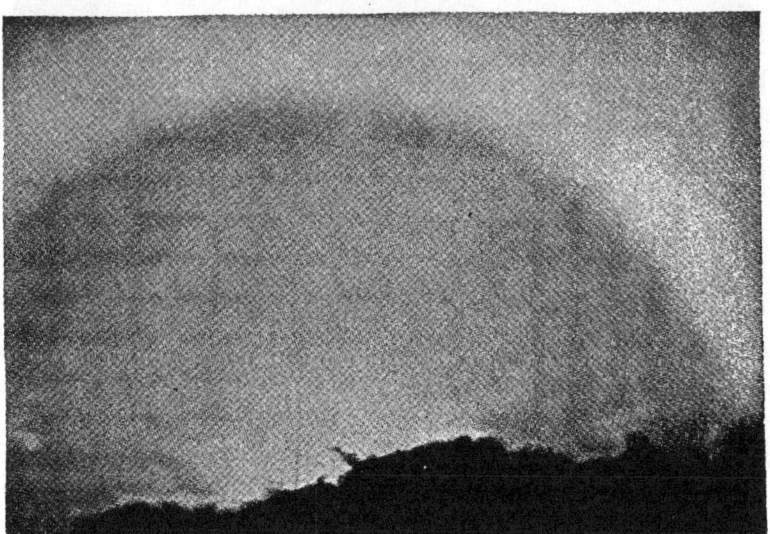

FIG. 146.—Cirro-stratus with halo. (Below, tops of cumulus.) (*International Atlas of Clouds. Photograph by G. A. Clarke.*)

a sign of approaching bad weather. A typical example is shown in Fig. 146.

Cirro-cumulus (Cc) is usually small white flakes of clouds without shadow, arranged in a regular pattern. Cirro-cumulus develops from

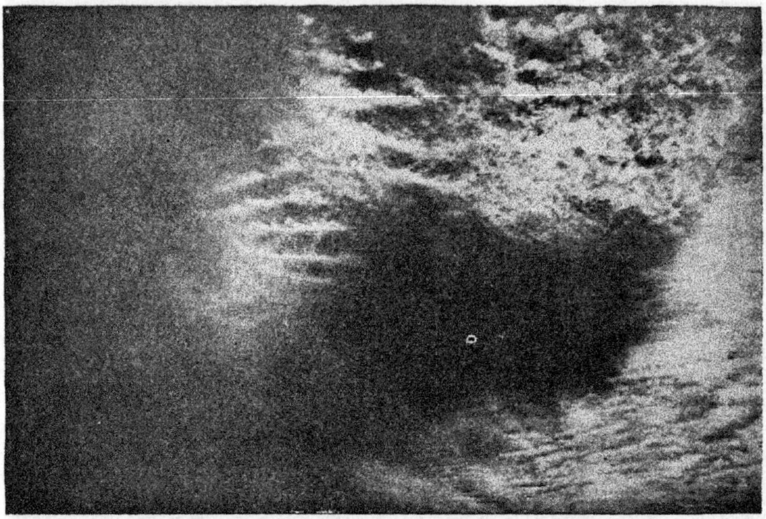

FIG. 147.—Cirro-cumulus. (*International Atlas of Clouds. Photograph by M. Loisel.*)

FIG. 148.—Alto-stratus (thin) with fracto-stratus underneath. (*International Atlas of Clouds. Photograph by G. A. Clarke.*)

cirro-stratus. The pattern is due to a single or a double undulation of the cloud sheet (Fig. 147).

Alto-stratus (As) is a dense sheet of gray or bluish color, often showing a fibrous structure (Fig. 148). It often merges gradually into cirro-stratus. Increasing alto-stratus is usually followed by precipitation of a continuous and lasting type.

FIG. 149.—Alto-cumulus. (*International Atlas of Clouds. Photograph by G. A. Clarke.*)

FIG. 150.—Strato-cumulus (thin, with breaks). (*International Atlas of Clouds. Photograph by G. A. Clarke.*)

Alto-cumulus (Ac) differs from cirro-cumulus in consisting of larger globules, often with shadows, whereas cirro-cumulus clouds show only indications of shadows, or none at all (Fig. 149). Alto-cumulus often develops from dissolving alto-stratus. An important variety of alto-cumulus is called *alto-cumulus castellatus.* In appearance it resembles ordinary Ac, but in places turreted tops develop which look like miniature cumulus. Alto-cumulus castellatus usually indicates a change to a chaotic thundery sky.

Strato-cumulus (Sc) is a cloud layer consisting of large lumpy masses or rolls of dull gray color with brighter interstices (Fig. 150). The masses are often arranged in a regular way resembling alto-cumulus.

Nimbo-stratus (Ns) is a dense shapeless and rather ragged layer of low clouds from which steady precipitation usually falls. It is usually con-

FIG. 151.—Cumulus humilis, or fair-weather cumulus (very flat). (*International Atlas of Clouds. Photograph by Meteor. Magn. Observatorium, Potsdam.*)

nected with alto-stratus, which is present above the nimbus. Fragments of nimbus that drift under the rain clouds are called *fracto-nimbus* or *scud.*

Cumulus (Cu) is a thick cloud whose upper surface is dome-shaped, often of a cauliflower structure, while the base is usually horizontal. Cumulus clouds may be divided into two classes. Flat cumulus clouds without towers or protuberances are called *cumulus humilis* or fair-weather cumulus (Fig. 151). Towering cumulus clouds with typical cauliflower structure showing internal motion and turbulence are called *cumulus congestus* (Fig. 152). These latter cumuli may develop into cumulo-nimbus.

Cumulo-nimbus (Cn) thunder clouds or shower clouds are great masses of cloud rising like mountains, towers, or anvils, and having a base that looks like a ragged mass of nimbo-stratus. The tops are often anvil-shaped or surrounded by false cirrus. Figure 153 shows a cumulo-nimbus without anvil, and Fig. 154 shows one with anvil. The cumulonimbus clouds are accompanied by showers, squalls, or thunderstorms

FIG. 152.—Cumulus congestus. (*International Atlas of Clouds. Photograph by Meteor. Magn. Observatorium, Potsdam.*)

FIG. 153.—Cumulo-nimbus calvus. (*International Atlas of Clouds. Photograph by C. J. P. Cave.*)

and sometimes hail. The *line-squall cloud* is a variety of cumulonimbus that extends like a long line or arch across the sky (Fig. 155).

Stratus (St) is a uniform layer of low cloud like fog, but not lying on the ground (Fig. 156).

The heights of the various types of clouds vary within wide limits. The high clouds are usually above 20,000 ft. and below 30,000 ft. The medium clouds are most frequently between 8,000 and 20,000 ft., and the low clouds below 8,000 ft. The tops of cumulus clouds, notably those of cumulo-nimbus, may reach to great heights, while their base on the average is about 3,000 ft. above the ground.

The preceding classification of clouds is based on their appearance rather than on the nature of the processes that form them. In order

FIG. 154.—Cumulo-nimbus with anvil. (*International Atlas of Clouds. Photograph by Fundacio Concepcio Rabell.*)

to make full use of cloud observations it is necessary to discuss the processes leading to their formation. We shall, therefore, return to this discussion in later sections.

Fog.—Fog is formed when the air near the earth's surface is cooled below its dew point. Fog is, therefore, nothing but a cloud that touches the ground. By international agreement such a cloud is called *fog* when the visibility is less than 1 km.; if the visibility is greater than this, it is called *mist.* However, in the United States the word *mist* is commonly used to designate the falling from clouds of fine water droplets. Thus, the American word *mist* designates about the same phenomenon as the English word *drizzle.* In order to avoid confusion, we shall here use the phrase *thin fog* instead of the English word *mist.*

The following scale may be used to indicate the intensity of the fog:

Description	Visibility, Meters or Yards
Dense fog	Less than 50
Thick fog	50–200
Fog	200–500
Moderate fog	500–1,000
Thin fog	Greater than 1,000

Fog may be formed when the air travels over a surface that is colder than the air itself. This type of fog, which is called *advection fog,* is prominent at sea in spring and summer, when warm air from the continents streams over cold water. In winter, advection fog is most frequent inland when warm and moist air from the oceans invades cold continents. Advection fog is a frequent phenomenon throughout the

FIG. 155.—Cumulo-nimbus arcus (line-squall cloud). (*International Atlas of Clouds Photograph by Fundacio Concepcio Rabell.*)

year over cold ocean currents (for example, the Labrador Current) when the air comes from warmer regions.

Fog may also form in stagnant air after clear weather when the ground is cooled by outgoing radiation to such an extent that it cools the air below its dew point. This type of fog, which is called *radiation fog*, is most frequent in winter over level country. It has a marked diurnal variation with a maximum about sunrise.

The height to which the fog reaches varies considerably. Radiation fog, which forms overnight in calm air, is usually very shallow, but if it persists for days, it may be very deep. Advection fog, which occurs in moderate or strong winds, is usually very deep.

Both kinds of fog have a tendency to lift or "burn off" during the day because of the diurnal heating and the mixing of air. Increasing wind velocity and turbulence often dissipate the fog, or cause it to rise and become a layer of stratus. This process will be described in a later section when comments will be made on the stratus and fog that occur along the coast of California.

FIG. 156.—Stratus. (*International Atlas of Clouds. Photograph by M. Gain.*)

Ground fog is a fog that is so shallow that an observer on the ground can discern the sky.

Ice fog (colloquially termed *frozen fog*, *frost in the air*, *ice needles*, etc.) consists of spicules of ice crystals which are so light that they seem to be suspended in the air. It occurs usually under conditions of clear, windless weather and low temperatures. Its occurrence is confined to polar regions and mountains. The sun is usually visible in an ice fog, but the horizontal visibility may be considerably restricted.

Arctic smoke is a kind of fog that forms when very cold winter air streams over water, the temperature of which is considerably higher than that of the air. The cause of the arctic smoke is the intense evaporation of water vapor. Arctic smoke occurs most frequently in the polar regions in winter.

Haze.—Haze consists of exceedingly fine particles of foreign matter from the continents, or of salt particles from the spray of the oceans. These particles, which are very evenly distributed in the air, are too small to be seen individually by the naked eye, but their effect on visibility and on coloring of distant objects is easily observed. Through haze, distant objects are seen as if through a thin veil of pale blue if the object is dark, but the veil appears yellowish if the object is white (snow-covered mountains). At a certain distance, depending on the density of the haze, all details of the landscape and all details of color disappear, and the objects stand out like a silhouette against the sky. The characteristic distance at which the details disappear is, in fair weather with sunshine on the object, one-third of the distance at which the contours of a mountain can be distinguished; in dull overcast weather the distance is only one-sixth of the total range of visibility.

Dust.—Dust consists of finely divided earth (clay, loam, humus) that is whirled up by the wind. It is coarser and not so evenly distributed in the air as are the particles of haze. It imparts a tannish or grayish hue to distant objects. The sun's disk is pale and colorless or perhaps of a yellowish tinge. The dust is usually of local origin; in strong wind it is blown about in clouds or sheets (blowing dust).

Sandstorm.—A strong wind carrying heavy particles such as coarse sand, usually mixed with dust and extending over a considerable area, is called a sandstorm. Owing to the coarseness of most of the material, it is seldom carried more than 100 to 200 ft. into the air, although the fine particles may be carried to great heights. The blowing sand is not carried to any appreciable distance from its source. A severe sandstorm may cause considerable damage to aircraft and motors, and radio operation is exceedingly difficult on account of static. Conditions favorable for the formation of sandstorms are extreme dryness of the ground, strong winds, and steep lapse rate of temperature. Severe sandstorms may occur in connection with cold fronts, like a line squall, associated with thundery conditions.

Smoke.—Smoke consists of a suspension in the air of particles of foreign matter resulting from combustion. It is a frequent obstruction to vision in industrial regions and where there is or has been brush or forest fire. In light amounts it may be confused with haze or dust; and at a distance it may be confused with a fog, but usually it can be differentiated from these by its odor. At sunrise and sunset, the sun's disk is very red; during the daytime, it has a reddish tinge.

Hydrometeors.—Hydrometeors are bodies of solid or liquid water falling through the air. The following distinctions should be noted:

1. *Rain* is liquid precipitation of large drops or small drops that are not numerous.

2. *Freezing rain* is rain that instantly freezes to objects it strikes in the open (glazed frost).

3. *Mist*[1] (in the United States meaning) denotes the falling from clouds of fine water droplets that individually do not make wet spots over 1/16 in. in diameter on pavements, boards, etc. It often resembles fog but is identified by the occurrence of drops of a size appreciable to the face or hands.

4. *Freezing mist* is mist (United States meaning) that instantly freezes to objects it strikes in the open.

5. *Drizzle* consists of small liquid droplets that are so numerous they seem to fill the air. Drizzle originates in fog or stratus. Thus *mist* (in the United States weather code) and *drizzle* (in the international weather code) are essentially the same.

6. *Snow* is precipitation in the form of ice crystals, usually of hexagonal or star structure. The crystals may fall singly, or a large number may be matted together in large flakes.

7. *Granular snow* is precipitation of opaque small grains (diameter 1 mm. or less) falling from stratus (frozen drizzle).

8. *Sleet* (United States meaning) is precipitation consisting of clear ice pellets, formed by the freezing of rain drops that fall from warm air aloft into a layer of cold air near the ground (dangerous icing conditions).

To the English the word *sleet* denotes precipitation consisting of *melting snow* or a *mixture of snow and rain.* In this meaning the word *sleet* is used in the international weather code.

It is worthy of note that granular snow and ordinary dry snow do not form appreciable deposits on aircraft, while sleet, wet snow, and subcooled water drops may form heavy deposits.

9. *Soft hail* is precipitation of opaque grains (diameter 1 to 5 mm.). They are crisp and bounce off the ground and disintegrate easily. They sometimes consist of a core of snow surrounded by a (wet) crust of opaque ice. They often fall with rain.

10. *Hail* is precipitation of round or irregular lumps of solid and fairly transparent ice. They occur almost exclusively in thundery weather and are rare in high latitudes.

The above classification is mainly based on the appearance of the hydrometeors. The processes leading to their formation are mainly of three kinds, *viz.,*

a. More or less continuous precipitation from continuous cloud cover (alto-stratus and nimbo-stratus). This kind of precipitation is caused by *slow* upward movement of a *large* mass of air, due to convergence in the horizontal motion of the air (*frontal precipitation*).

[1] To the English, "mist" stands for a thin fog whose visibility is greater than 1 km.

b. Showers, or precipitation of short duration that begins and ends suddenly, usually with fair periods between. This kind of precipitation, which originates in clouds of the cumulus family, is caused by the fairly *rapid* rising of *small* bodies of air through the atmosphere (instability precipitation).

c. Drizzle,[1] or numerous small droplets falling from fog or stratus. This kind of precipitation is not connected with any appreciable ascensional velocity; on the contrary, the small drops fall out of the cloud because of the absence of any appreciable upward movement.

We shall return to the discussion of precipitation in later sections on air masses, fronts, and cyclones.

Cloudiness.—At the hour of observation a note is made of how many tenths of the sky are covered with clouds.

Owing to the importance of the low clouds for aviation, a special note is made of how many tenths of the sky are covered by low clouds.

Ceiling.—The height of the base of the low cloud above the ground is an important element for aviation. It is usually estimated by the observers. Some stations measure the heights by means of clinometers or by small free balloons of known ascensional velocity.

Visibility.—The term *visibility* is used to describe the horizontal transparency of the air. In some countries the visibility has been defined as the distance at which the outline of objects seen against the sky disappears. In other countries it has been the distance at which objects seen against a background of the surface of the earth become indistinguishable. In other countries again it has been the distance at which an object such as, for example, a tree or a house can be recognized as such. The International Meteorological Organization has recently agreed that the last definition of visibility should be adopted for future use. The discrepancies between the various definitions are of slight consequence when the visibility is bad, but they appreciably affect the measurements when the distance to the object is large.

The preceding definition refers to the horizontal daylight visibility. In darkness it is necessary to use lights of known candlepower to determine the horizontal visibility.

In the United States the visibility is reported in miles and/or fractions thereof when the visibility is less than 4 miles, to the nearest whole mile from 4 to 15 miles, inclusive, and to the nearest 5 miles when greater than 15 miles. Fractions of miles are reported as follows: 1/8, 1/5, 1/4, 1/2, 3/4, 1, 1 1/4, 1 1/2, 1 3/4, 2, 2 1/2, 3, 3 1/2 and 4 miles.

In the countries that use the international weather code the visibility is reported on a scale ranging from 0 to 9. The following table gives the scale numbers and the corresponding daylight visibility,

[1] *I.e.*, mist in the United States meaning.

and also the corresponding distance at which a light of 100 c.p. becomes indistinguishable.

TABLE OF VISIBILITY

Scale number	Corresponding daylight visibility		Night observations			
			Distance of object		Distance at which a light of 100 c.p. becomes indistinguishable	
0	Less than 50 m.	Less than 150 ft.	50 m.	150 ft.	100 m.	330 ft.
1	50– 200 m.	150– 700 ft.	200 m.	700 ft.	330 m.	1,000 ft.
2	200– 500 m.	700–1,600 ft.	500 m.	1,600 ft.	740 m.	2,500 ft.
3	500–1,000 m.	1,600–3,200 ft.	1 km.	0.6 mile	1,340 m.	4,400 ft.
4	1– 2 km.	0.6– 1.2 miles	2 km.	1.2 miles	2.3 km.	1.4 miles
5	2– 4 km.	1.2– 2.4 miles	4 km.	2.4 miles	4.0 km.	2.4 miles
6	4– 10 km.	2.4– 6 miles	10 km.	6 miles	7.5 km.	4.5 miles
7	10– 20 km.	6 –12 miles	20 km.	12 miles	12.0 km.	7.2 miles
8	20– 50 km.	12 –30 miles	At greater distances a 100-c.p. light is not suitable.			
9	Over 50 km.	Over 30 miles				

From this table we see that a light of 100 c.p. is visible at greater distances than such objects as are used to determine the daylight visibility when the visibility is less than 4 km. For greater visibilities the reverse is true.

The visibility depends greatly on the weather. The table that appears below shows the normal relation between visibility and weather as defined in the explanations to the international weather code.

Scale number	Fog, mist or haze	Snow or sleet	Drizzle	Rain
0	Dense	Very heavy		
1	Thick	Very heavy or heavy	....	Tropically heavy
2	Medium	Heavy		
3	Moderate	Moderate	Thick	Very heavy
4	Mist	Light	Moderate	Heavy
5	Slight mist or haze	Very light	Slight	Heavy
6	Slight mist or haze	Very light	Slight	Moderate
7	....	....	....	Light
8	....	....	....	Very light
9				

The following table gives a summary of corresponding visibilities and weather phenomena in terms used in the United States weather code:

Visibility	Weather
1/5 mile or less	Heavy snow, dense fog
3/4 mile or less	Heavy mist
1/5–3/4 mile	Moderate snow, moderate fog
Over 3/4 mile	Light mist, light snow, light fog
Less than 1 mile	Thick blowing snow (blizzard), thick blowing dust (duststorm), thick blowing sand (sandstorm), thick smoke, thick haze
1–6 miles	Blowing dust, blowing sand, blowing snow

The haze is often arranged in layers in the atmosphere. In such cases the vertical visibility may vary greatly in different directions. A pilot flying in sunshine above a haze layer may not be able to see the landing field, while the aircraft is perfectly visible from the ground. This condition is due to the reflection and scattering of light from the top of the haze layer. It is also worthy of note that the "pilot's visibility" may be considerably less than the real visibility owing to dew, rain, or ice that gathers on the wind shield.

Use of Meteorological Observations and Frequency Tables.—It is essential that the meteorological observations should be made simultaneously at all stations within large areas, so that they give an adequate picture of the state of the atmosphere at a given moment. Such observations plotted on a map furnish the basis for the safeguarding of the airways as well as for general weather forecasting, storm warnings, and other forecasts. The weather maps also furnish an invaluable means for scientific research in the causes of weather phenomena.

A series of observations covering several years may be used to evaluate the *average* or *normal* values of certain factors. A set of such mean values gives us what we call the *climate* of the station in question.

For aviation purposes the mean values are of but little interest. For the planning of new airways, landing fields, and airports, it is of but little use to know, for example, the mean visibility. It is far more important to have tables showing the frequencies of the various values of visibility, cloud heights, cloudiness, wind forces and directions, etc. The International Commission for Air Navigation has prescribed what information should be accumulated in the form of frequency tables. These tables of frequencies are published by the meteorological services of the various countries.

TEMPERATURE VARIATIONS AND SOME OF THEIR PHYSICAL EFFECTS

Stability and Instability.—We say that the atmosphere is in a stable state of equilibrium if an air particle that is moved a small distance upward or downward has a tendency to return to its original level. If such a particle, when moved a small distance and then left to itself, has a tendency to move farther away from its original level, the atmosphere is in an unstable state of equilibrium. If the particle can be made to rest at any level, the atmosphere is in an indifferent or neutral state of equilibrium.

A rocking chair at rest is in a stable state of equilibrium, because if it is rocked slightly it will return to its original position. A pencil balanced on its point is unstable, because a slight movement will cause the pencil to move away from its original position. A perfectly round ball on a highly polished and perfectly horizontal slate is in an indifferent state of equilibrium.

Since any minute disturbance will upset the unstable systems and bring them to a stable state, it follows that only stable or indifferent conditions can exist for any appreciable length of time. The transition from unstable to stable states of equilibrium involves a reduction of the potential energy, and all systems left to themselves will try to avoid instability and obtain a minimum of potential energy.

The principles of stability and instability also govern the stratification of the atmosphere, but the conditions are here more complicated owing to the fact that the density of the air depends on both the temperature and the pressure. A second complication arises when the air becomes saturated with moisture. The latent heat of vaporization is then made available to increase the temperature of the air.

Let us imagine a volume of non-saturated air, say 1 cu. meter, rising through the air without heat being supplied to or withdrawn from it from the surroundings. As it rises it comes under lower pressure, and it expands. The work done against the external pressure during the expansion must be compensated by a loss of internal energy, or, in other words, the temperature of the rising air will decrease. Taking the physical constants involved into consideration, we find that the rising air must cool 1°F. per 180 ft. (1°C. per 100 meters) of elevation. The process described is called an *adiabatic* process, the qualifying adjective "adiabatic" indicating that no heat is supplied from without. When a body of air moves down through the atmosphere it comes under higher pressure and is compressed. The temperature of the descending air then increases at the rate of 1°F. per 180 ft. (1°C. per 100 meters) of descent. A lapse rate of 1°C. in 100 meters is called the *dry-adiabatic lapse rate.*

The above deductions hold for non-saturated air only. If the air cools to its saturation temperature, condensation takes place, and the heat energy, which once was used to evaporate the water, is made free. This heat is called the *latent heat of vaporization.* For each gram of water that is condensed, a certain amount of heat is made free, and this heat is used to increase the temperature of the air.

If saturated air rises, it will cool because of the expansion and will simultaneously be heated because of the condensation. The balance between these two processes comes out in the negative, with the result that the rising air cools about half as quickly as it would if it were non-saturated. The amount that saturated air cools per 100 meters of

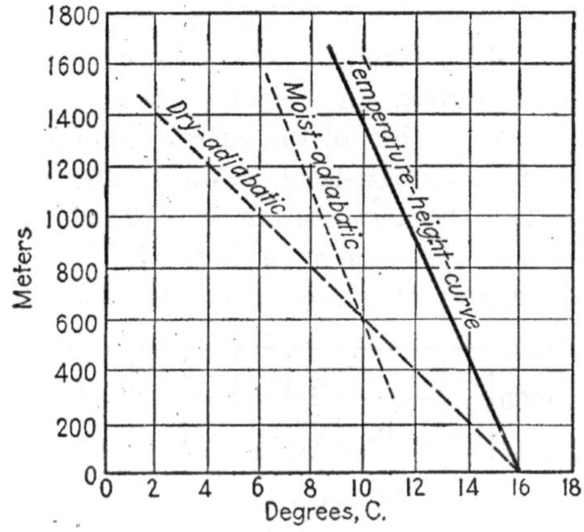

FIG. 157.—Illustrating stability.

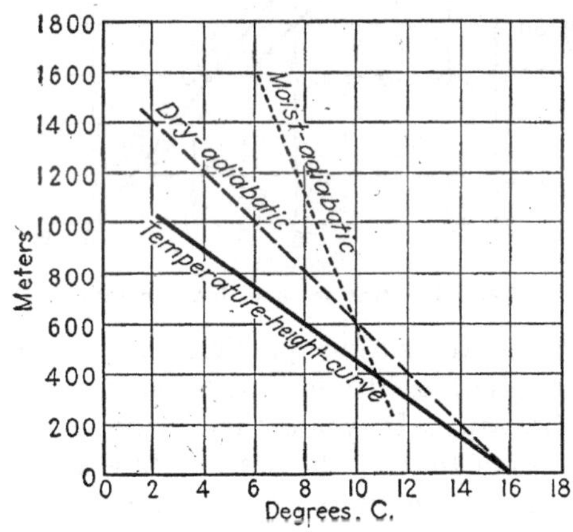

FIG. 158.—Illustrating instability.

ascent is called the *moist-adiabatic lapse rate*, and the process of cooling is called *moist adiabatic*.

With the above principles in mind, we shall turn to the above graphs in order to demonstrate in greater detail the conditions of stability and instability.

Stability.—Figure 157 shows a hypothetical distribution of temperature as a function of height. Let us consider the air particle at the surface of the earth whose temperature is, say, 16°C. Let us, to have a concrete example, say that its moisture content is such that the air becomes saturated at 10°C. If this particle is lifted through the atmosphere, it will cool dry adiabatically till its temperature falls to 10°C., and, if it is lifted farther, it will cool moist adiabatically. In the diagram the broken line represents the dry-adiabatic cooling, and the dotted line the moist-adiabatic cooling.

From this diagram we see that, if the said particle were lifted through the atmosphere, it would arrive at every level with a temperature lower than that of the surrounding air. The particle would, therefore, also have a density greater than that of the surrounding air. Hence it would resist the lifting, and, if left to itself, would sink to its original level. In this case we say that the atmosphere is *totally stable.*

Instability.—Let us next consider Fig. 158. The full line is the temperature-height curve; it shows that the temperature decreases

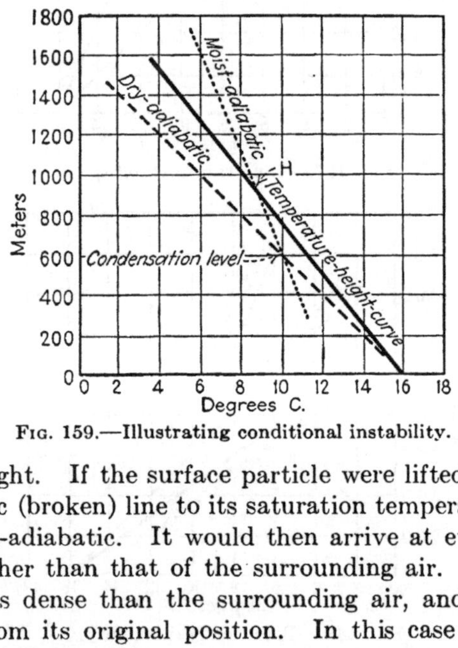

Fig. 159.—Illustrating conditional instability.

rapidly with height. If the surface particle were lifted, it would follow the dry-adiabatic (broken) line to its saturation temperature, and thence follow the moist-adiabatic. It would then arrive at every level with a temperature higher than that of the surrounding air. It would, therefore, also be less dense than the surrounding air, and so would move farther away from its original position. In this case we say that the atmosphere is *totally unstable.*

Conditional Instability.—There is a characteristic state of equilibrium between the two cases just discussed, in which the temperature lapse rate is less than the dry-adiabatic and greater than the moist-adiabatic. We consider the temperature-height curve in Fig. 159. If we now lift the surface particle, it will cool dry adiabatically to its saturation temperature (10°C.) and then cool moist adiabatically while it rises farther. At a certain height (*H* in Fig. 159) the temperature of the ascending particle is just equal to that of the surrounding air. If the particle rises beyond this point, it becomes warmer and lighter than the air around it, and it will be accelerated upward if it is left to itself.

We thus see that, if the particle is lifted, it will first resist lifting, but, if it is lifted to a certain height, it will move farther by itself. In such cases we say that the atmosphere is *conditionally unstable.*

In the above examples we have discussed only the behavior of the air at the surface of the earth. The same principles hold for any particle, and all characteristic points of the temperature-height curve should be considered to make sure how deep the stable or unstable layers are.

Heating and Cooling of the Atmosphere.—The direct radiation of the sun is absorbed to only a slight extent in the atmosphere, because the air is almost transparent to high-temperature radiation (short waves). The radiation emitted from the sun is partly scattered in the atmosphere and partly reflected from clouds and the surface of the earth, but a considerable portion is absorbed by the earth's surface. The latter emits low-temperature radiation (long waves) toward the atmosphere, and a portion of this radiation is absorbed by the water vapor in the air or by the water of the clouds.

The atmosphere thus acts as the glass in a greenhouse. It lets through practically all incoming short-wave radiation from the sun, and it prevents the outgoing long-wave radiation from the earth from getting back to the universe. The difference between incoming and outgoing radiation goes to heat the earth's surface and the air that comes into contact with it.

In the lower layers of the atmosphere the greater part of the heating comes from direct contact with the earth. In the higher atmosphere the influence of the earth's surface decreases with elevation; above a certain height, the heating and cooling of the air are completely controlled by radiation.

The heating of the lower atmosphere is different over land and over sea. We shall, therefore, in due course discuss the following cases separately: the heating of the air over land, the heating of the air over sea, and the heating of the upper atmosphere.

Heating and Cooling of Air over Land.—The absorption of heat by the earth, after sunrise, increases the temperature of the earth's surface. This heat is accumulated in the upper few inches of the earth, owing to the very slow conduction of heat in the earth. The air that comes into contact with the earth becomes heated, and the lapse rate of temperature in the very lowest layer of air increases rapidly. When the lapse rate has reached or surpassed the dry-adiabatic rate, the layer becomes unstable, and vertical currents set in, carrying heat and moisture, picked up from the surface, to higher levels. As the sun gets higher in the sky, the heating increases, and the unstable layer of air increases in thickness, the heat obtained from the surface being transported to higher and higher levels. The diurnal heating of the earth's surface results in a steep lapse rate and stirring of the lower atmosphere.

As the temperature of the earth's surface increases, the outgoing radiation increases too, and, some 2 hr. after the culmination of the sun. there is balance between the loss and the gain of heat. The temperature of the earth reaches its maximum and begins to decrease. The cooling of the earth affects the air temperature, and, as before, the influence is greatest near the surface, with the result that the air cools more quickly along the earth than above. The temperature lapse rate decreases, and the air becomes stable again. Figure 160 shows three typical temperature-height curves, illustrating the diurnal variation of temperature in the lower atmosphere.

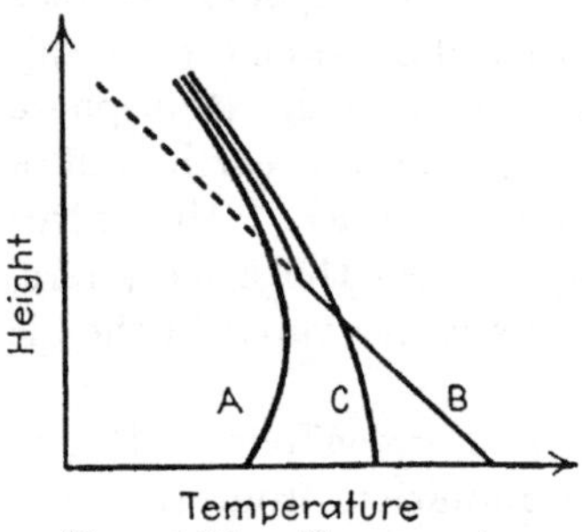

FIG. 160.—Temperature-height curves illustrating the diurnal variation in temperature over land. *A* = early morning; *B* = midday; *C* = evening; broken line = dry-adiabatic.

The annual temperature variations are similar to the diurnal ones, only on a larger scale. The result is stirring of the lower atmosphere and maintenance of a steep lapse rate of temperature which favors convectional currents.

Heating and Cooling of Air over Oceans.—It is important to note that the absorption of solar radiation by the sea surface does not appreciably affect its temperature. This condition is due to a variety of causes. If the sun is not overhead, or nearly so, the greater part of the incoming radiation is reflected from the surface of the sea, and is lost as far as heating of the sea or the lower atmosphere is concerned. If the sun is near the zenith, the amount reflected by the sea is small, but in this case the radiation penetrates to a considerable depth before it is completely absorbed. Since the specific heat of sea water is very large, the effect of radiation on the temperature will be small, the heating being spread over a deep layer of water. Other effects help to reduce the diurnal temperature change in the sea. Part of the heat gained by the sea is used to evaporate water. Through evaporation the salinity of the surface layer increases; it becomes denser than the water underneath and sinks to be replaced by colder water from beneath.

The outgoing radiation at nighttime cools the sea surface. But when the top layer cools, it becomes denser and sinks, to be replaced by warmer and lighter water from underneath. Thus both at day and night the sea surface keeps a fairly constant temperature. The air that is in contact with the sea will also have a fairly constant temperature.

Observations show that the diurnal range of temperature at the sea surface is less than 1°F.

The regulating influence of the sea surface on the air temperature decreases with elevation. A few hundred meters above the sea surface the temperature variation is controlled by radiation, and the air has a diurnal variation in temperature greater than the variation at the surface of the sea. At still greater altitudes the effect of radiation decreases, and the temperature remains fairly constant day and night.

Figure 161 shows two characteristic temperature-height curves over sea. It is readily seen that the lower part of the atmosphere has a tendency to be stable in the daytime and unstable at nighttime. At greater heights the reverse is true.

The diurnal variation of stability in the lower atmosphere over oceans is in sharp contrast to the conditions prevailing over land, and this has a marked influence on the weather.

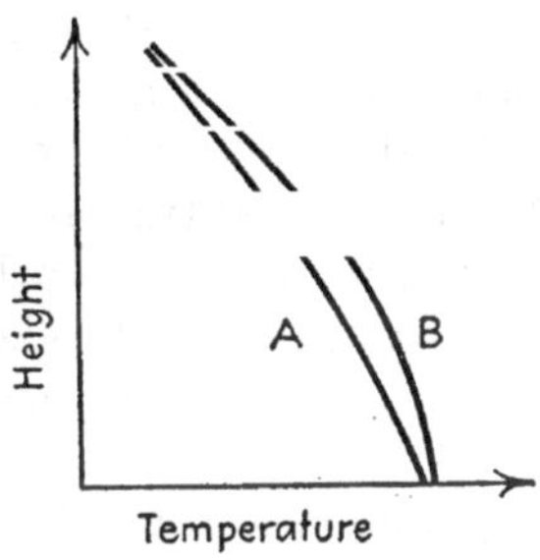

FIG. 161.—Temperature-height curves illustrating the diurnal variation in temperature over oceans. A = night; B = midday.

Heating and Cooling of Traveling Air Masses.—An air mass that is colder than the surface over which it travels is called a *cold air mass.* It will be heated from below and, by continued traveling over a warmer surface, instability will develop in the lower layers and gradually spread upward. The vertical currents resulting from instability will carry heat and moisture to higher and higher levels. The changes in such an air mass are completely analogous to those in air over land that is heated by sunshine on the earth's surface, which we have described in a previous section.

If such an air mass travels over land, it will have the effect of the diurnal temperature changes superimposed on the effect originating from its traveling toward warmer regions. The instability of the air will then vary during the day, having a maximum in the early afternoon and a minimum in the early morning. If the difference in temperature between the surface and the air is large, the air will remain unstable day and night. If the difference is small, the air may become stable in the night.

An air mass that is warmer than the surface over which it travels is called a *warm air mass.* Through continued traveling toward colder regions it will be cooled from below and will acquire pronounced stability in the lower layers. The stability hinders vertical currents, and the cooling will be limited to the lower layer. If such an air mass travels over land, it will have the diurnal variation in stability superimposed on the effect originating from its traveling toward colder regions. If the temperature difference between the surface and the air is large, it will remain stable day and night. If the difference is small, it will be stable at night and unstable by day.

Over oceans the conditions are different, inasmuch as the diurnal variation of stability is small. Over oceans, therefore, stability and instability are determined mainly by the travel of the air masses, and there is but slight diurnal variation. This is particularly true in high and intermediate latitudes. In the equatorial oceanic regions the sea-surface temperature is fairly constant over large areas. The travel of the air masses in tropical oceanic regions has, therefore, less influence on stability and instability than have the diurnal heating and cooling.

Convection.—We have seen in the foregoing paragraphs that instability is created in the lower layer of the atmosphere either through the diurnal heating of the earth's surface by the sun, or through heating of the air when it travels toward warmer regions. A number of phenomena, such as gustiness, turbulence, cumulus clouds, showers, and thunderstorms, are directly caused by instability. We shall, therefore, discuss the consequences of instability in greater detail.

As soon as the temperature lapse rate near the earth exceeds the dry-adiabatic rate, the slightest disturbance will upset the stratification. Air from the earth's surface rises, and air from higher levels sinks to replace the ascending air. This process of turning over unstable air is called *convection.*

The ascending currents reach only to the top of the unstable layer, or slightly beyond it. The height to which the air must ascend in order to be cooled to its saturation temperature is called the *condensation level.* The weather phenomena that convection will produce depend on the depth of the unstable layer, the height of the condensation level, and the distribution of temperature above. We shall discuss the following typical cases separately:

1. The condensation level of the unstable air is essentially above the top of the unstable layer.
2. The condensation level of the unstable air is lower than the top of the unstable layer, and the air above the unstable layer is completely stable.
3. The air is unstable up to the condensation level and conditionally unstable above.
4. The conditions are the same as mentioned under 3, and the convective currents reach to levels where the temperature is below the freezing point.

Case 1.—Since the condensation level is well above the unstable layer, it is obvious that convection will not produce clouds. This case is typical of the conditions in the early morning after clear nights. By continued heating from below, the depth of the unstable layer will grow, and when it passes the condensation level condensation will take place and cumulus clouds will begin to form. The type of cumulus clouds and the weather that will develop depend entirely on the temperature conditions above the condensation level.

Case 2.—We consider Fig. 162. The rising air from the earth's surface will cool dry adiabatically to its condensation level C, and thence moist adiabatically. It is seen from the figure that the ascending air, when it passes the level H, will be colder than the surrounding air. The ascending currents, when they pass this level, will be retarded, and they cannot penetrate to greater heights. The base of the clouds will be at the level C, and the top will be slightly above H.

Owing to the pronounced stability above the level H, the cumulus clouds will be flat, and they will show no tendency to produce towers or protuberances that grow

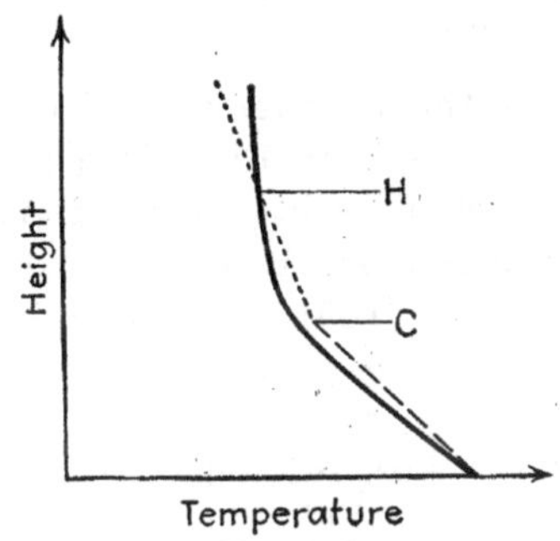

FIG. 162.

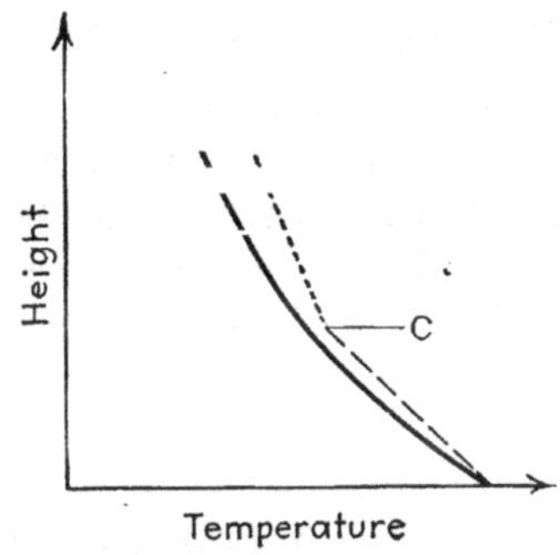

FIG. 163.

FIG. 162.—Temperature conditions in shallow convective layer with cumulus humilis. Full line = temperature-height curve; broken line = dry-adiabatic rate; dotted line = moist-adiabatic rate; C = condensation level; H = top of convective layer.

FIG. 163.—Temperature conditions in deep convective layer with cumulus congestus, or cumulo-nimbus. (Symbols as in FIG. 162.)

upward. The type of cumulus that forms under these conditions is the fair-weather cumulus (cumulus humilis, Fig. 151).

Case 3.—We consider Fig. 163. The ascending air cools dry adiabatically to its condensation level C and then moist adiabatically. Owing to the conditional instability above the condensation level, the cumulus clouds that form will grow upward. Such clouds will have towering heads and protuberances showing internal motion (see Fig. 152). These clouds are called *cumulus congestus*. Neither the cumulus humilis nor the cumulus congestus produces precipitation, but the latter may develop into cumulo-nimbus.

Case 4.—This case differs from case 3 only in that the convective clouds have reached to levels where the temperature is below 0°C. The water droplets in the top of the cloud begin to freeze. This is an important change in the cloud, because the water in the cloud will gather on the ice crystals, which then become so large and heavy that they usually cannot be kept afloat by the ascending currents in the cloud. The presence of ice crystals in the top of the clouds releases the precipitation, and the cumulus congestus changes rapidly into a cumulo-nimbus.

Figures 164 to 169 illustrate the characteristic features of the convective clouds. The cumulus humilis (Fig. 164) are rather flat and well separated from one another. The base is well defined, and there are no towering heads or protuberances from the upper part of the clouds.

Figure 165 shows the characteristic features of a cumulus congestus. Towers of typical cauliflower structure grow up from the main cloud, whose base is usually well defined. Internal motion and turbulence are noticeable in cumulus congestus; the cumulus humilis has a rather "dead" appearance. Both these types of cumulus consist of water droplets without admixture of ice crystals in their tops.

Figure 166 shows a towering cumulus with a fine silky veil around its top. The veil sometimes consists of ice crystals. If such a cloud develops further, it changes into a cumulo-nimbus. Figure 167 shows a cumulo-nimbus that has developed from such clouds as are shown in Figs. 165 and 166. The top of this cloud consists of a mixture of ice crystals and water droplets. The valleys and crevices in the towering parts are disappearing and the cloud is about to lose its cauliflower structure. This type of cloud is called *cumulo-nimbus calvus.* (See also Fig. 153.)

Figure 168 shows a cumulo-nimbus with anvil. (See also Fig. 154.) Ice crystals are present in abundant quantities in the top part of the cloud and therefore give heavy precipitation and frequently also thunderstorms.

It is possible with practice to forecast the weather for a few hours ahead by looking at the cumulus clouds and observing their development. It is then most important to observe the changes in the upper part of the clouds. If there are no towers (as in Fig. 164), there is no chance of precipitation. If there are towers (as in Fig. 165), it is possible that precipitation may develop. If some of the clouds show signs of presence of ice crystals, the precipitation is sure to be released soon. Figure 166 often occurs as a transition from Fig. 165 to Fig. 167, but Fig. 167 may also develop directly from Fig. 165.

The cumulus clouds sometimes dissolve by general shrinking and disappear gradually. This usually occurs when a sheet of high clouds develops above them. In dissolving in this way, they pass through the state of fair-weather cumulus (Fig. 164). Most frequently the cumulus clouds flatten out into rolls or bulging layers; this development is shown in Fig. 169. This often occurs in the evening when the atmosphere is settling down after the diurnal heating. In any case, the dissolution of cumuli shows that the atmosphere is developing toward a stable stratification.

If the convection is caused by diurnal heating over land, it has a pronounced diurnal period with a maximum of cumulus clouds in the afternoon and clearing in the evening. Over oceans the diurnal convection is only slight (except in tropical regions), and the maximum of cloudiness has a tendency to occur in the night.

If the convection is caused by the travel of air towards warmer regions, there is but slight diurnal variation, and cumulus and cumulonimbus may develop both day and night.

The precipitation caused by convection is always of a showery character; it begins and ends suddenly owing to the rapid transition from ascending to descending currents. The sky is variable with frequent changes from a dark and threatening appearance to clearing.

FIG. 164.

FIG. 165.

FIG. 166.

FIG. 167.

FIG. 168.

FIG. 169.

FIGS. 164–169.—Various types of cumulus clouds.

Thunderstorms.—Thunderstorms develop from clouds of the cumulo-nimbus type in excessively unstable air. The mechanism of their formation may be described briefly as follows:

The falling velocity of raindrops depends on their size. If the drops grow larger than 1/6 in. (4 mm.) in diameter, they will fall with a velocity exceeding 24 ft. (8 meters) per second. By such high falling velocity, the drops breaks up into smaller drops, which then fall less rapidly.

If the ascending currents in the cumulo-nimbus exceed 8 meters per second, the largest raindrops will be split up into smaller drops, and will be carried upward. The ascending current in a cumulo-nimbus is not steady; it consists of a succession of gusts and lulls, so that the drops may rise and fall, grow and break up repeatedly.

Each time a drop breaks up into smaller drops, the negative and the positive electricity will be separated, the air taking up a negative charge and the drops a positive charge. By repeated splitting up of drops, enormous electric charges are made available for the thunderstorms. Since the air ascends much more rapidly than the drops that break up, it follows that the positive charge is accumulated in the part of the cloud where the ascending current is strongest, while the rest of the cloud becomes negative or neutral. The lightning discharge starts from the positive part of the cloud toward the negative part, or toward the negative part of some neighboring cloud; or it starts from the ground toward the negatively charged portion of the cloud.

Severe thunderstorms are often accompanied by hail. The structure of the hailstones shows conclusively the existence of large ascensional velocities in the thunderstorms. The hailstones often consist of concentric shells of clear ice and snow, which shows that the hail must have been moved repeatedly from the liquid to the snow part of the cloud.

A fully developed thunderstorm is accompanied by heavy rain or hail, lightning, and thunder. The wind freshens during the approach of the storm, blowing at first toward the advancing storm. As the thundercloud arrives overhead, the wind changes in direction, blowing out from the storm in a forward direction. The barometer falls while the storm approaches, but when the wind changes a brisk rise amounting to a few millibars occurs. The precipitation, which began as a sudden heavy downpour, changes into a more continuous rain, which gradually decreases in intensity.

The passage of a thunderstorm is frequently accompanied by strong gusts that may cause complete loss of control of an aircraft.

Mixing of Air.—The wind is never a steady current; it consists of a succession of gusts and lulls of short period. This condition is mainly due to friction along the surface of the earth, which creates eddies that are forced up to higher levels by the irregular motion. If the air is in a stable state, the turbulence will be damped; in unstable air the eddies may travel to considerable heights. What we are here concerned to emphasize is the effect of turbulence on the distribution of temperature and humidity along the vertical.

We consider first a vessel partly filled with cold salt water and with warm fresh water on top. The cold and salt water is heavier than the warm and fresh water, and the stratification is a stable one. If turbulent motion is created and maintained in the vessel, the two layers will mix into a homogeneous liquid showing constant temperature and constant salinity.

In the air the conditions are somewhat different owing to the fact that air is compressible. The ascending air will cool adiabatically,

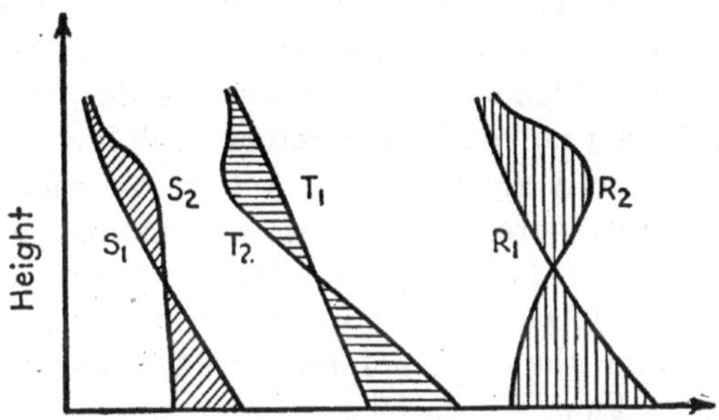

FIG. 170.—Showing the effect of turbulent mixing. S = specific humidity; T = temperature; R = relative humidity. Subscripts 1 denote the conditions before the mixing commenced, and subscripts 2 the conditions after complete stirring.

and the descending air will be heated adiabatically. The result of turbulent mixing in the air is to create an adiabatic lapse rate of temperature.

The specific humidity of the air, being independent of the adiabatic processes, will mix, and the final result of the stirring of the air will be a distribution of specific humidity that is constant with height.

We consider next a stable layer of air along the surface of the earth in which the specific humidity decreases with elevation. (This case occurs frequently in nature.) We suppose that the wind velocity, for some reason or other, increases so that the layer becomes mixed by turbulence. Figure 170 shows the changes that occur in the vertical distribution of temperature, specific humidity, and relative humidity.

The relative humidity will increase rapidly with height because of the cooling and increasing of specific humidity in the upper part of the stirred layer. It needs no further explanation that this process of stirring often leads to condensation of the water vapor in the top part of the stirred layer. The clouds that form under such conditions are of the stratus type.

Inversions.—The air temperature normally decreases with height. Under special conditions a reversal of the lapse rate may develop, showing a layer of air in which the temperature increases with height. This is called an *inversion.*

Inversions develop easily near the earth's surface during calm and clear nights, or when warm air travels over a cold surface. They may also develop above the ground at the top of the layer which is stirred by turbulence. Inversions may also develop in the free atmosphere as a result of descending air spreading out laterally.

The essential feature of an inversion is the pronounced stability of the air in the layer that has increasing temperature. A well-developed inversion acts as a lid through which no convection or mixing can take place, the ascending currents being repulsed by the excessively stable layer.

The higher the inversion is above the ground, the deeper is the layer that can be stirred by convection and turbulence. The height of the inversion above the ground has a marked influence on the diurnal range of temperature. The higher the inversion, the deeper the layer to be heated, and the diurnal range of temperature at all levels under the inversion will be correspondingly smaller. A low inversion favors a large diurnal range of temperature. Observations from San Diego, Calif., show this very distinctly. The following table gives in the left-hand column the observed heights of the base of the inversion above the ground, and in the right-hand column the corresponding average diurnal ranges of temperature, in degrees Fahrenheit.

Height of Inversion, Ft.	Diurnal Range, °F.
0– 150	23
150– 300	17
300– 600	14
600– 900	10
900–1,200	9
1,200–1,500	7

Low inversions are easily destroyed by diurnal heating and mixing of the air, while a high inversion usually persists day and night. Dust and smoke from the earth gather under the inversion, forming a distinct layer of haze whose top coincides with the base of the inversion.

Formation of Stratus.—We have seen in previous paragraphs that air cooled to its dew point along the surface of the earth will produce fog. This cooling along the surface of the earth is counteracted by the effect of the turbulent mixing, which causes heating of the surface layer of air and cooling of the air at some distance above the ground. The result is often that fog forms in stagnant air, while stratus forms in air whose velocity is large enough to create sufficient turbulence.

If air from warm and moist regions travels a considerable distance over a colder surface, it may cool so much that the fog fills the whole layer that is under the control of turbulent mixing. For this reason fogs that occur in strong winds are usually deep fogs. Such fogs often occur in the warm sectors of depressions that travel from lower to higher latitudes.

Shallow fogs have a marked tendency to "burn off" after sunrise. Through the combined effect of diurnal heating and turbulent mixing, such a fog may dissolve and reappear as a stratus, or fragments of stratus. Deeper fogs often do not dissipate after sunrise but, through the combined effect of diurnal heating and turbulent mixing, may rise and become a layer of stratus.

FIG. 171.—Surges on a layer of stratus (or fog) indicating the topography beneath.

Low layers of stratus in the morning often dissolve or break during the day owing to the diurnal heating, but high layers may persist day and night. This condition is due to the fact that a low layer of stratus is more easily affected by the diurnal heating and the turbulent mixing than is a high layer.

The stratus layers often show surges like a large swell on an ocean of cloud. These surges often indicate the topography of the terrain (see Fig. 171).

Fog and stratus are limited to the lower layer of the atmosphere. They are both water clouds without admixture of ice crystals. The precipitation that may fall from these clouds is of the drizzle type.

Stratus often develops into strato-cumulus. This happens when a single or a double undulation breaks up the stratus layer into lumps or rolls.

The California Fog.—In previous sections we have seen that fog forms either through advection of warm air to colder regions, or through outgoing radiation that cools the air below its dew point. On the other hand, stratus is mostly formed through mixing of air whose temperature is sufficiently close to the dew point. All these processes combine to make the so-called California fog.

Air from the Pacific Ocean is, during the summer and autumn months, cooled over the cold upwelling water along the coast, whereby the air temperature is brought close to its dew point. The diurnal variation in radiation causes the air temperature to fluctuate around the dew point with the result that fog forms in the evening and dissolves in the morning. Usually there is sufficient turbulent mixing to change the fog into a low stratus whose upper limit coincides with the base of the strong quasi-permanent temperature inversion between 500 to 2,000 ft. above the sea.

If the base of the inversion is close to the ground, the cloud layer is shallow and the fog (or stratus) burns off early in the morning; but, if the inversion is high above the ground, the fog or stratus layer is deep and burns off later. In pronounced cases the fog (or stratus) persists throughout the day.

Troposphere and Stratosphere.—We have seen in the previous paragraphs that the temperature changes in the lower atmosphere are due mainly to influences from the surface of the earth, the conduction of heat to higher levels being effected by turbulence and convection.

The influence from the earth's surface decreases with elevation and vanishes at a certain height above the ground. Above this level the temperature distribution is completely controlled by the balance between incoming and outgoing radiation. The effect of radiation is to smooth out temperature differences. The result of this process is that a constant temperature is maintained.

The lower part of the atmosphere, in which convection is a prominent feature, is called the *troposphere.* The upper atmosphere, in which the temperature remains constant with height, is excessively stable and has no convective currents; this part of the atmosphere is called the *stratosphere.* Owing to its stable and cloudless state, the stratosphere offers the nearest approach to ideal flying conditions.

The base of the stratosphere is called the *tropopause.* Its height above the ground varies with latitude, season, and the weather situation. It is higher over anticyclones than over depressions. Figure 172 shows the average height of the tropopause and the normal distribution of temperature. It is of interest to note that the temperature of the stratosphere decreases from the poles toward the equator.

Icing on Aircraft.—Icing is one of the greatest dangers to air navigation. Ice formation cannot as yet be forecast with satisfactory accuracy, owing partly to lack of adequate observations from the free atmosphere and partly to insufficient knowledge of the physical processes involved.

Ice usually forms on the forward edges of wings and struts, or on the propeller, but sometimes it forms also on the horizontal faces. The ice that forms on the forward edges of the wings changes the wing profile. Often the cross section of the ice deposit is of a mushroom shape. This causes a general change in the streamlines around the wings, whereby the airplane may lose so much in dynamic lift that flying becomes disastrous.

Ice that forms on the propeller is dangerous because of the irregular rotation that results from the asymmetrical distribution of the weight of the ice. There is no force perpendicular to the axis of the propeller in the case of perfect symmetry. But when ice gathers unevenly, or when a lump of ice breaks off from one blade of the propeller and not from the other, a force proportional to the asymmetry of the weight and

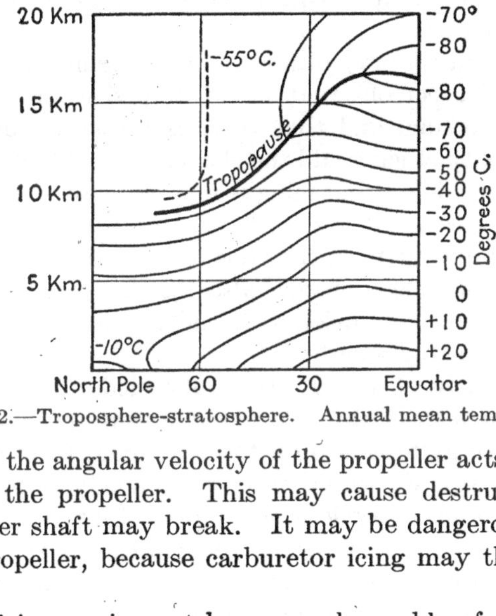

FIG. 172.—Troposphere-stratosphere. Annual mean temperature.

to the *square* of the angular velocity of the propeller acts perpendicularly to the axis of the propeller. This may cause destructive vibrations, and the propeller shaft may break. It may be dangerous to reduce the speed of the propeller, because carburetor icing may then cause engine difficulties.

Modern de-icing equipment has proved capable of overcoming icing of slight or moderate intensity on the propellers and on the forward edges of the wings, but no adequate means has been invented for eliminating intense icing, or the icing that takes place on the horizontal faces. Icing, therefore, remains a potential danger to air navigation.

It is known that icing may occur (1) outside the clouds, (2) within the clouds, and (3) in subcooled rain that falls from warmer air aloft into a layer of cold air.

The icing that occurs in cloudless air is caused by direct condensation of the water vapor on the aircraft. This kind of icing is of only slight intensity. The ice is usually crisp and breaks easily.

The icing that occurs within the clouds may be of any intensity. On the average the accumulated ice on the forward edge of the wing amounts to $\frac{1}{30}$ in. per minute, and the maximum icing intensity observed within the clouds amounts to $\frac{1}{4}$ in. per minute. It has been observed frequently that no icing occurs in clouds with temperature below freezing. No icing occurs when the drops are small and the water content low. This is due to the fact that small drops are swept away by the air current around the wing, while larger drops strike the wing. The observed maximum of icing intensity is well below the theoretical maximum.

Both the size of the drops and the water content vary considerably even within one cloud. It happens, therefore, quite frequently that two airplanes flying through the same cloud experience widely different icing intensities. When intense icing starts, it is advisable to change altitude in order to get into regions where the cloud droplets may be smaller or the water content lower.

The difficulties in the way of forecasting icing are due mainly to the circumstance that direct observations of drop sizes and the amount of liquid water are lacking, and that there is no outstanding correlation between the intensity of icing and the elements usually observed. There is, however, some evidence that icing is more intense when the lapse rate of temperature is great. In the United States intense icing occurs most frequently in unstable air from the Gulf of Mexico.

The intensity of icing depends also on the speed of the aircraft. An airship, therefore, can drift with the air current and change its altitude in order to avoid icing.

Icing is frequently observed when subcooled rain falls from warm air aloft into a cold layer of air. This kind of icing is particularly dangerous because the ice gathers also on the horizontal faces of the aircraft. In such cases the aircraft should ascend as quickly as possible into the warm air aloft in order to avoid a heavy load of ice.

Icing may occur in clouds whose temperature is slightly above the freezing point, especially when an airplane descends, because it will then be colder than the air, and the water droplets may freeze when they strike the aircraft. But even in a level flight, ice may form on the aircraft. This happens when the relative humidity in the cloud is less than 100 per cent. Part of the water that is deposited on the aircraft will then evaporate, and this may cool the remaining water sufficiently to produce icing. It should also be mentioned that wet snow or sleet may gather and freeze on aircraft, while dry snow is usually swept away by the air current.

AIR CURRENTS

The Causes of Air Currents.—As far as motion is concerned, the atmosphere may be regarded as an engine that creates kinetic energy from heat energy, the difference in temperature between the poles and the equator and between the upper and lower atmosphere being the significant sources of energy.

The forces that move the air particles depend mainly on the distribution of pressure. Let us consider a cube of air with horizontal and vertical faces. Since the atmospheric pressure decreases with altitude, it follows that the pressure on the lower face is greater than that on the top face. This represents a force that is directed upward. This force is counteracted by the weight of the air within the cube. Usually there is balance between the two forces, but occasionally one exceeds the other and vertical accelerations are created. In this way the convective currents are created. The mean vertical velocity over large areas is, however, small, and rarely amounts to more than 1 or 2 in. per second. The large wind systems are, therefore, mainly horizontal currents.

The pressure also varies horizontally. The pressure on one of the vertical faces of the said cube will, therefore, be greater than that on the opposite face. The difference in pressure is equivalent to a force tending to drive the cube of air toward lower pressure.

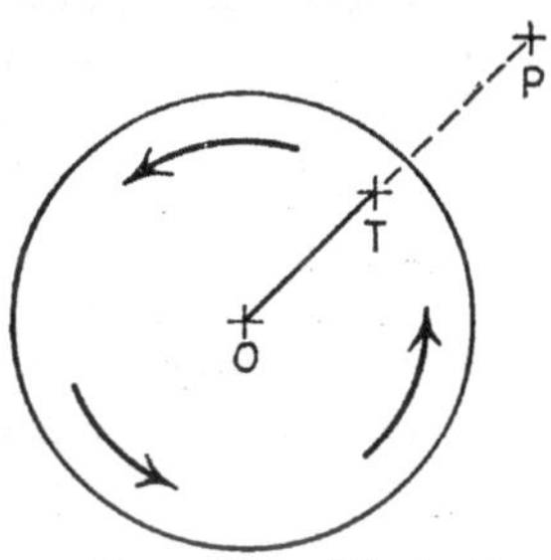

FIG. 173.—Illustrating the deflective force due to the rotation of the earth.

If the earth did not rotate, our cube of air would move directly toward the deficit of pressure, which would be immediately nullified. However, since the earth rotates on its axis, deflective forces originating from the rotation of the earth will act on the moving air and tend to balance the pressure gradient. In order to understand the wind systems of the earth, it is necessary to understand also the deflective force of the earth's rotation.

The Deflecting Force.—The nature of the deflecting force of the earth's rotation may be demonstrated in the following manner. We consider a circular disk which rotates as indicated by the arrows in Fig. 173. A bullet is shot from the center *O* of the disk at a target *T* on the edge. At the moment the bullet is shot, the line from the center toward the target runs through a point *P* which, being outside the disk, does not take part in the rotation. The bullet will move on the straight line *OP*, but in the meantime the target *T* will have moved toward the left. An observer on the rotating disk, not seeing the point *P*, would get the impression that the bullet curved off to the right of the target *T*; while another observer outside the disk, not seeing the target *T*, would think that the bullet moved on the straight line. Both observers would be right in their statements. The one observes the *relative* motion, and the other observes what we may call the *absolute* motion.

The movement of the bullet relative to the rotating disk would be deflected and curved off to the right of the target, and this deflective movement can be accounted for by a force that produces an acceleration equal to $2V\omega$, V being the velocity of the bullet and ω the angular velocity of the rotating disk. The acceleration would be at right angles to the path of the bullet.

The Geostrophic Wind.—The above example of a rotating disk illustrates the conditions at the north pole. A similar effect is present also in lower latitudes. We consider next the movement of an air particle relative to the earth in latitude ϕ. The acceleration due to the deflective force of the earth's rotation would then be $2V\omega \sin \phi$, where V is the wind velocity and ω the angular velocity of the earth's rotation.

If an air particle, originally at rest, starts to move toward lower pressure, the deflective force of the earth's rotation will drive the particle more and more to the right until it moves in a direction perpendicular to the pressure gradient. The air particle will then have attained a velocity that is determined by the balance of the pressure force and the deflective force. These two forces will then act in opposite directions, and the air particle will move in a direction perpendicular to both these forces.

A wind velocity that corresponds to complete balance between the pressure gradient and the deflective force is called the *geostrophic wind.* If V denotes the velocity of the geostrophic wind, the following formula holds:

$$V = \frac{G}{2\rho\omega \sin \phi}$$

where G is the pressure gradient and ρ the density of the air. In the above formula, ρ is almost constant in the horizontal, so that the geostrophic wind is directly proportional to the pressure gradient.

In the southern hemisphere the rotation of the earth is opposite to that in the northern hemisphere. The wind is therefore deflected to the *left* of the pressure gradient.

The rotation of the earth prevents the air particles from moving directly toward the deficit of pressure. On a rotating earth, therefore, pressure gradients can be maintained. The actual wind observed near the earth's surface is less than the geostrophic wind, owing to the retarding influence of friction along the earth. The influence of friction decreases with altitude, and the wind in the free atmosphere approaches the geostrophic value.

The most convenient way to represent the distribution of pressure is to draw lines through points of equal pressure. Such lines, which are called *isobars*, are most conveniently drawn for each $\frac{1}{5}$ in. or for each fifth millibar. A map prepared in this way will show centers of high pressure, or anticyclones; centers of low pressure, or depressions; troughs of low pressure; wedges of high pressure; and cols.

The pressure gradient is everywhere inversely proportional to the distance between the isobars. Since the geostrophic wind is directly proportional to the pressure gradient, it follows that the wind is strongest where the isobars are most crowded.

The geostrophic wind blows along the isobars with low pressure to the left in the northern hemisphere, and to the right in the southern hemisphere. We may say that the air streams between the isobars in the same way as water streams in a river bed; the narrower the river the greater the velocity.

The geostrophic wind is a good approximation to the actual wind if we are concerned only with the large-scale movement of the air. In problems dealing with the details of the wind, it is necessary to discuss the behavior of the actual wind. We shall discuss both things briefly in the following paragraphs.

The General Circulation.—Maps of the mean pressure for representative winter and summer months are shown in Figs. 174 and 175 for the northern hemisphere. Certain features of the maps are common to both winter and summer. Around the equator there is a region of almost uniform pressure, in which the winds are light and variable. This belt of light and variable winds with frequent showers and thunderstorms is called the *doldrums*. On the average it is slightly to the north of the equator, and it moves northward in the northern summer, and southward in the southern summer.

Farther away from the equator are belts of high pressure with easterly winds on their equatorial sides, and westerly winds on their poleward sides. These belts of high pressure are called the *subtropical anticyclones*. The easterly winds on their equatorial sides are called the *trade winds*. They have a slight component toward the equator, and converge from both hemispheres into the doldrums. The convergence of air into the doldrums results in ascending currents and precipitation. The belts of high pressure are regions of divergence and descending currents with dry weather. On the poleward side of the high-pressure belts, the winds blow from a westerly or southwesterly direction.

In the southern hemisphere there is a pronounced belt of low pressure on the poleward side of the subtropical high, and there is but little difference between winter and summer. In the northern hemisphere the conditions are more complicated in high latitudes, owing to the distribution of land and sea. A mere glance at the maps will show that there is a marked tendency for anticyclones to form over the continents in winter and over the oceans in summer. This condition is due to the accumulation of cold and heavy air over the coldest regions. The general circulation in these regions therefore shows a pronounced seasonal variation.

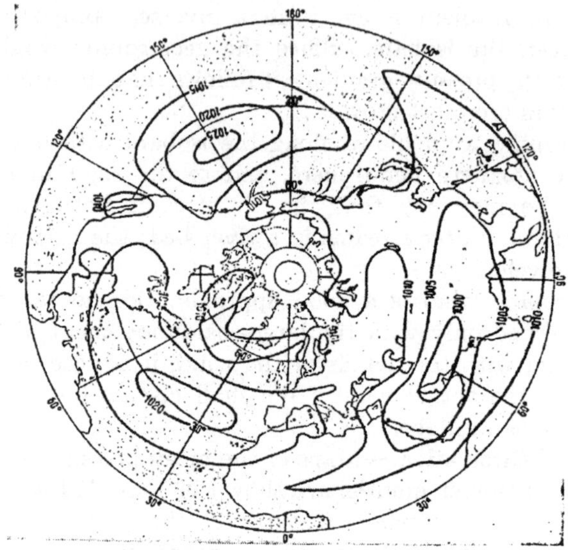

FIG. 174.—Mean pressure in February.

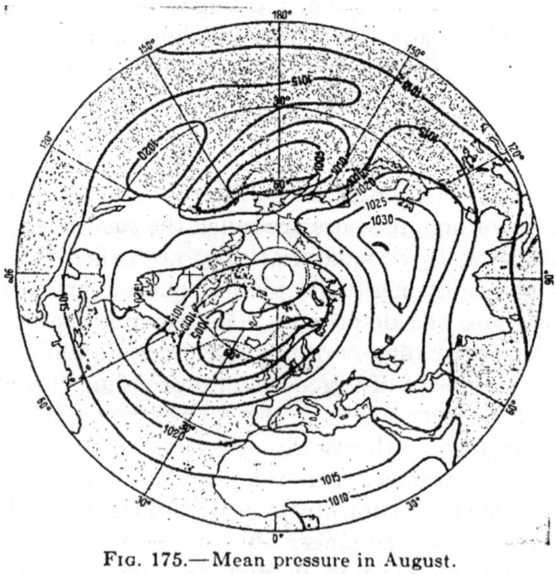

FIG. 175.—Mean pressure in August.

The Monsoons.—Owing to the difference in temperature between continents and oceans, wind systems develop that blow with great persistence from sea to land in summer and from land to sea in winter.

Such winds are called *monsoons,* even though the word usually refers to the monsoons in the Indian Ocean only.

The circulation in Asia affords the most striking example of monsoon winds. In winter the subtropical anticyclone is intensified by the cold of Central Asia. Cold offshore winds prevail over the east and south coast of Asia, and westerly winds prevail along the coast of Russia and Siberia. As the summer approaches, the anticyclone dies away and is replaced by an immense low-pressure area, which covers almost the whole of Asia and parts of Africa and Europe. The wind system is completely reversed, the winds blowing along the isobars with a component inward to the continental low-pressure area. The Asiatic monsoon system affects the wind circulation over a vast area, particularly in summer season. During this period the doldrums and the subtropical high of the Indian Ocean are both completely submerged.

Land and Sea Breezes.—In addition to the seasonal contrast of temperature and pressure over land and water there is a daily contrast that exercises a similar but more local effect. In summer the land is warmer than the sea by day, and colder than the sea by night. The slight variations in pressure thus established cause a system of breezes with a component landward during the daytime and seaward during the night. These breezes are shallow and do not penetrate far inland. The day breeze may attain the strength of a fresh or a strong breeze, while the night breeze is usually gentle.

In the tropics the land and sea breezes blow with great persistence. In higher latitudes they are often overshadowed by stronger winds of more general character.

The Circulation of the Free Atmosphere.—The wind velocity increases rapidly with height in the immediate vicinity of the earth and then increases slowly so as to approach the geostrophic wind about 1,500 to 3,000 ft. above the ground.

The pressure distribution itself varies with height, and above 3,000 meters the distribution of land and sea has practically no influence on the general circulation, except in India, where the monsoon disturbance is still noticeable.

The general circulation in the free atmosphere is controlled by an immense low-pressure area centered over the poles, a belt of high pressure in the subtropics, and a belt of uniform pressure around the equator. As a result of this pressure distribution, the mean wind in the free atmosphere is westerly everywhere on the poleward sides of the subtropical high-pressure belts. The westerly winds increase with altitude and attain their maximum velocity at the tropopause. On the equatorial sides of the subtropical high-pressure belts the mean wind blows from an easterly direction.

The various circulations described above represent the *mean* state of motion of the atmosphere. On this mean state of motion are superimposed many disturbances that account for the variability of the weather.

Turbulence.—The main source of turbulence in the atmosphere is the friction along the surface of the earth. The roughness of the surface creates eddies or whirls of air which are forced up to higher levels. Records of the details of the wind structure show that the turbulent flow consists of a succession of gusts and lulls, the period of which is irregular and of the length of a few seconds. The strength of the gusts is somewhat proportional to the roughness of the ground and the velocity of the wind.

Turbulence caused by friction along a rough surface is called *mechanical turbulence.* It is usually limited to a layer of air of a few hundred meters in thickness. This turbulent layer along the earth's surface may be visualized from Fig. 176*a*, which shows eddies along the surface of the earth and a fairly steady flow above.

Another source of turbulence in the atmosphere is the irregular distribution of temperature. The warmer lumps of air will rise and the colder will sink, thus leading to irregular flow. This kind of turbulence may be called *thermal turbulence.* The mechanical and the thermal turbulence combine to produce the resultant turbulence.

The thickness of the turbulent layer along the earth's surface depends mainly on the stability of the air. The mechanical turbulence is damped in stable air and intensified in unstable air. The thickness of the turbulent layer is, therefore, greater in unstable air than in stable air.

In unstable air the convectional currents are superimposed on the general flow. They cause fluctuations in the wind similar to those characteristic of pure turbulence, but with periods ranging from a few minutes to an hour or more. Such wind variations are called wind squalls if they are strong. The wind squalls are usually connected with convective clouds, and the approach of a towering cumulus or a cumulonimbus may be taken as a warning of the arrival of a wind squall.

Although the pure turbulence is most pronounced near the surface of the earth, the vertical wind squalls are strongest in the free atmosphere, notably in the convective clouds and in their immediate vicinity. A skilled pilot will be able to estimate the chances of severe gusts and squalls from the appearance of the convective clouds.

The intensity of the thermal turbulence and the convectional gusts and squalls depends largely on the nature of the surface. The amount of heat absorbed by the earth's surface depends to a large extent on its moisture content and on its color. The midday temperature on sunny summer days is higher over macadam and sandy fields than over grasslands, and is lower over wet ground and woods, and still lower over water. These differences in temperature give impulses of considerable intensity to local convectional currents. Local currents of this kind are the main cause of the bumpiness experienced above level country on warm days. The convective currents also form easily on the sunward slopes of hills and mountains. In hilly country, therefore, turbulence, local eddies, and convection combine to make the air very bumpy.

Turbulence and Obstacles.—Figure 176*a* illustrates the formation of eddies in the air flow above level ground. If such a current meets an obstacle, the distribution of the turbulence is greatly modified, the

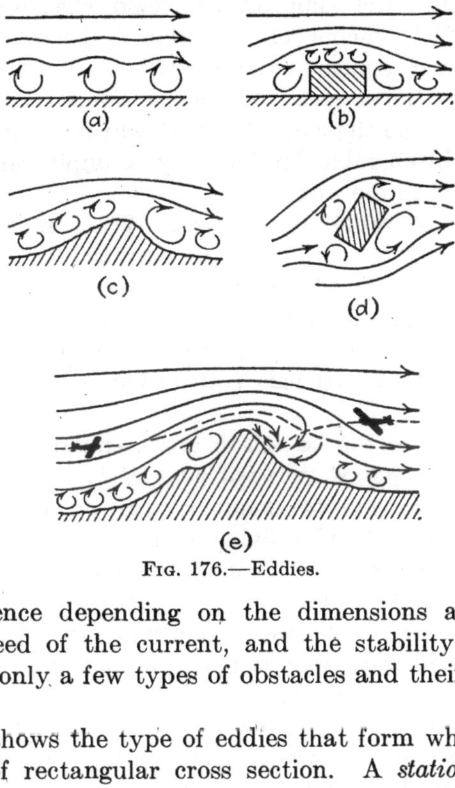

FIG. 176.—Eddies.

resulting turbulence depending on the dimensions and shape of the obstacle, the speed of the current, and the stability of the air. We can here discuss only a few types of obstacles and their influence on the air current.

Figure 176*b* shows the type of eddies that form when an air current passes a ridge of rectangular cross section. A *stationary* eddy forms on the windward side of the ridge. Smaller irregular traveling eddies develop above the roof of the obstacle, re-forming continuously along the windward edge. On the lee side of the ridge, eddies form and travel downwind. The lee eddy is usually stationary while it develops; when it has attained a certain intensity, it leaves the obstacle and travels with the wind while it dissipates gradually.

If the horizontal dimensions of the obstacle are small, the air current has a tendency to stream around it. In this case eddies with vertical axes form at the edges, as shown in Fig. 176*d*. Whether the air will stream around the obstacle or across it depends mainly on the length of the obstacle and the stability of the air. The resulting turbulence caused by large buildings, hangars, etc., is usually a combination of horizontal and vertical eddies. If the air is unstable and the wind speed high, landing should not be attempted too close to a hangar. The eddies on the lee side may cause a complete reversal of the wind direction and make landing unsafe. The eddy on the windward side is less dangerous because it is stationary and does not extend far into the wind.

Figure 176*c* shows the eddies that develop when the current crosses a small ridge. If the incline on the windward side is not too steep, the stationary eddy disappears and there are only the usual eddies caused by the roughness of the surface. On the lee side there are usually larger eddies that form on the slope and travel downwind. Again, if the wind is strong and the air unstable, landing may be unpleasant or even unsafe on the lee side close to the ridge.

The Influence of Mountain Ranges.—The influence of mountain ranges on air currents is, in general, the same as the influence exercised by the obstacles discussed in the previous paragraph. Figure 176*e* shows a cross section through a mountain range. The most striking feature is the well-developed eddy on the lee side of the mountain. On the windward side there may be a stationary eddy or not, according to whether the incline is steep or not.

The eddies that form on the lee side are often dangerous to air navigation. *A pilot flying against the wind may, if he does not keep sufficient altitude, be forced down into the mountain side, or he may lose control of the plane in the eddy.* A pilot flying with the wind will usually gain altitude while he approaches the mountain range. If the mountain is very steep, there may be a stationary eddy on the windward side, which may cause difficulties. Under these conditions, the pilot must carefully consider the effect of ascending and descending air currents in making turns or banks with a heavy load.

The eddies around mountain ranges reach up to some altitude above the range, causing intense mixing of the air. With a favorable distribution of humidity and temperature, this mixing will lead to formation of clouds (stratus) around and above the mountain range. The process of formation of stratus has been described in a previous paragraph. What we are here concerned to emphasize is that stratus has a tendency to form in mountainous country owing to the increased turbulence. A general layer of stratus will be lower on the mountain sides than in the free air. Oftentimes stratus forms only on the mountain sides and not in the free air.

Apart from forming eddies and local clouds, the mountain ranges affect the streamlines of the general flow on a large scale. At some considerable altitude the streamlines are unaffected by the mountain range, and the current is mainly horizontal. It follows then that the cross section of the current is diminished by the range. The speed of the current will, therefore, increase in proportion and attain a maximum velocity above the range. The general flow has an upward component on the windward side and a downward component on the lee side. This large-scale influence on the main current is noticeable at great distances. The downward component of the general flow dissolves clouds, and this effect often reaches 150 miles or more to the lee. The ascending current is a frequent cause of general cloudiness and precipitation on the windward slope. If the air is stable, it flows more smoothly and has a tendency to stream around mountain ranges. Unstable air, on the other hand, streams easily across, and the upward movement favors the formation of convective clouds and showers.

Bumpiness.—Most frequently bumpiness is caused by upward or downward currents in the air. These currents may be caused by turbulence, convective currents, or eddies caused by obstacles. The bump is felt when the aircraft flies into or out of a rising or descending current.

Another cause of bumpiness is sudden horizontal variations in the wind. If the aircraft flies with the wind, a sudden lull will cause a sudden increase in dynamic lift, which is felt as an upward bump. Likewise, a sudden horizontal gust will cause a sudden decrease in dynamic lift, which is felt as a downward bump. If the aircraft flies against the wind, a lull would give a downward jolt, and a gust an upward jolt.

Bumpiness is also experienced when the aircraft passes a temperature inversion. Temperature inversions are usually wind discontinuities, and both the velocity and the direction of the wind are different above and below the inversion. The following example suffices to demonstrate the cause of this kind of bumpiness: An aircraft flies under the inversion with the wind. Its air speed is v, and its ground speed is V, V being greater than v. If the air above the inversion is calm, the aircraft will arrive there with an air speed equal to V. This sudden increase in the air speed gives increasing dynamic lift, which is felt like an upward bump. If the aircraft flies under the inversion against the wind and passes through the inversion into calm air above, it will experience a sudden loss in dynamic lift and drop down again under the inversion. Under the control of the elevator the plane may rise, only to receive another downward bump, or a series of bumps. If the aircraft changes its direction sufficiently, it may pass through the inversion with hardly any bumping.

It happens sometimes that ripples or billows develop along the inversion. The aircraft, flying horizontally close to the inversion, may then be exposed to a series of bumps in a regular period. These can be avoided by a slight change in altitude.

Temperature inversions have a very pronounced effect on lighter-than-air craft flying through them. An airship flying in the air above the inversion will have acquired a temperature close to that of the surrounding air. When diving down through the inversion it moves into colder air. The superheat of the ship may then be sufficient to overcompensate the rudder control, and the ship rises above the inversion. Likewise, if an airship flies in the cold air under the inversion and attempts to rise through it, it will receive a downward jolt, because the gas of the ship is colder than the warm air above. In Southern California, where a strong inversion is present most of the year, it is known that airships have experienced serious difficulties in attempting to pass through the inversion. The difficulties are particularly great if the inversion is close to the ground because the downward jolt received by the ship when entering the inversion may be sufficient to bump the ship against the ground.

AIR MASSES

Life History of Air Masses.—The ideas underlying the principles of air-mass analysis are based on the fact that the general circulation of the atmosphere has a tendency to produce vast masses of air whose physical properties are more or less homogeneous within large areas, the transition from one such mass to another being rather abrupt. The region where an air mass is formed is called its *source.*

When a mass of air leaves its source, it begins to change its physical properties, and the weather phenomena that develop within the mass depend entirely on its *life history.* In studying the life history of an air mass, the following factors should be considered: (1) the source whence the air mass obtained its fundamental properties, (2) the path it has traveled, and (3) the time it has spent on its journey.

1. *Source.*—The air mass absorbs the properties of the source (as determined by the temperature of the surface and the radiation characteristic of the latitude) if it remains at rest or moves for a considerable time over regions of quasi-homogeneous source properties. Examples of such sources are the subtropical anticyclones, and the polar or continental anticyclones over snow-covered regions. In these anticyclones the air movement is slight or moderate, and the air masses, therefore, get time to acquire the typical properties. Therefore, stationary or quasi-stationary anticyclones are usually effective air-mass sources.

Traveling depressions rarely produce air masses, because the air is moved so quickly that it keeps changing its characteristics.

2. *Path.*—The second important feature to consider is the path that the air mass has traveled since it left its source. The air mass will change its properties and its structure en route from the source to the destination according to whether it travels toward colder or warmer, moister or drier regions.

3. *Age.*—The third important feature to consider is the time the air mass has spent on its journey from its source, or, in other words, the age of the air mass. The amount of change in the air-mass properties depends largely on the nature of the surface over which it travels and on how long it has been in contact with the surface. The absorption of properties proceeds from the surface upward. The maximum of modifying influence is, therefore, always near the earth's surface when the air mass is young. How far upward the influence penetrates depends entirely on the nature of the influence and the age of the mass.

The transition from one air mass to another is usually rather abrupt. It is along the border between adjacent air masses that the greatest contrasts in energy are found, and it is there that the traveling depressions develop. These phenomena will be discussed in a later section. What we are here concerned with are the weather phenomena that develop *within* the air masses. It is then useful to classify air masses according to types of life history.

Classification of Air Masses.—Air masses may be classified according to the source whence they acquired their fundamental properties. There are only four main sources, *viz.*, the doldrums, the subtropical anticyclones, anticyclones in high latitudes, and the regions occupied by arctic (or antarctic) snow and ice. Accordingly, we may distinguish between the following types of air masses:

1. *Equatorial air* (air from the doldrums).
2. *Tropical air* (air from the subtropical anticyclones).
3. *Polar air* (air from anticyclones in high latitudes).
4. *Arctic air* (air from the arctic fields of snow and ice).

The difference between equatorial air and tropical air is not considerable; the same is true of the difference between polar air and arctic air. Disregarding these minor differences, we may consider only two outstanding types of air mass, namely, *polar air masses* and *tropical air masses.*

The above classification is based on the nature of the source of the air mass without regard to the influences sustained on the journey. These influences depend on whether the air has traveled toward colder or warmer regions. Accordingly, we distinguish between a *cold air mass* and a *warm air mass*. A cold air mass is one that is colder than the surface over which it travels, and a warm air mass is warmer than the surface over which it travels. In this classification it is indifferent whether the air temperature is high or low, or whether the one mass is colder or warmer than the other. The classification is based on the difference between the temperature of the air near the surface of the earth and that of the surface itself.

The amount of heat that the air absorbs depends on the difference in temperature between the air and the surface over which it travels. The amount of moisture that it picks up depends on the aridity of the surface. It is, therefore, convenient to supplement the above classifications by the qualifications *maritime* and *continental.*

The classifications described above supplement one another. Thus, the term "a cold mass of maritime polar air" indicates a vast body of air of polar origin which is traveling over an ocean that is warmer than the air itself. Taking into consideration the season and the time that this air mass has been under maritime influence, the above term is descriptive of a definite type of weather phenomena characteristic of the mass as a whole.

The Properties of a Cold Mass.—The source of the cold masses is normally in the subpolar or arctic regions, but, in winter, masses of cold air may develop over continents down to Lat. 25° or 30°N. While in their sources, these masses are cooled from below and are characterized by:

1. Stable stratification, notably in the lower layers.
2. Low specific humidity.
3. Low temperature.

When such a mass, for some reason or other, moves toward warmer regions, it will arrive there with a temperature lower than that of the surface over which it travels. The mass will be heated from below, and thermal instability will soon develop in the lower layers and gradually spread upward. If the air originally contained inversions, these will be destroyed by continued heating from below with the result that a uniform steep lapse rate develops throughout the mass; this results in convective currents.

If the cold mass travels over water, it will pick up moisture, which is brought up to higher and higher levels by the convective currents. Convective clouds form and soon develop into cumulo-nimbus. If the cold mass travels over land, it will be heated from below, but it will not absorb much moisture. In this case, therefore, convective clouds do not easily form until the instability has reached up to very great altitudes. Continental cold masses are, therefore, often cloudless.

The following examples will show the outstanding difference between a cold continental and a cold maritime mass: (1) Summer air from, say, Saskatchewan streams southward toward Texas. In spite of being heated 30°F. or more, it remains clear or perhaps produces a few scattered cumuli, because it does not absorb much moisture while it is heated. (2) Cold and dry winter air from, say, Texas streams across the coast toward the Gulf whereby it may be heated 15°F. At a distance of about 100 miles off the coast, towering cumulus and cumulo-nimbus develop, because the air has been heated and has absorbed moisture from below.

A typical *maritime* cold mass is recognized by the presence of several or all of the following characteristics:

1. Increasing temperature and humidity.
2. Steep lapse rate, and instability.
3. Turbulence, gusts, and squalls.
4. Cumulus and cumulo-nimbus.
5. Variable sky, changing from dark and threatening to bright or clearing.
6. Showery precipitation that begins and ceases abruptly.
7. A slight diurnal variation in cloudiness and precipitation with maximum in the early morning.
8. The visibility between the showers is good because marine polar air is fairly pure in its source and has not picked up dust on its journey.
9. The height of the base of the clouds is moderate and rarely less than 1,000 ft. This condition is due to the heating of the air from below whereby the condensation level is kept at some distance above the ground.

A typical *continental* cold mass is recognized by the presence of several or all of the following characteristics:

1. Increasing temperature and fairly constant humidity.
2. Steep lapse rate and instability.
3. Scattered cumulus clouds and occasionally cumulo-nimbus.
4. Pronounced diurnal period in cloudiness with a maximum in the afternoon.
5. If precipitation occurs, it is of a showery character with considerable bright intervals, and it occurs generally in the afternoon.
6. The visibility is variable but on the whole good, except when the air has traveled a considerable time over dusty land or industrial regions.
7. The height of the base of the clouds is considerable and rarely less than 2,000 ft.

When a maritime cold mass invades a continent in summer, the instability is intensified; the showers increase in both frequency and intensity. When a maritime cold mass invades a continent in winter, the instability decreases and the showers decrease in intensity, and the cumulus clouds begin to flatten out into bulging layers resembling stratus, strato-cumulus, or nimbo-stratus. This influence is felt first in the lower layers and gradually spreads upward.

When a cold continental mass invades an ocean in summer, it develops into a stable state. When a cold continental mass invades an ocean in winter, its instability increases and the showers increase rapidly in intensity and in frequency.

The shower activity in cold air masses counteracts the instability that is caused by the heating from below. This condition is due to the liberation of the latent heat of vaporization above the condensation level. Therefore, old cold masses gradually develop toward a stable state of equilibrium. It happens, therefore, quite frequently that a cold mass that has traveled from the United States across the Atlantic Ocean arrives in western Europe in a fairly stable state. The maximum instability occurs in air masses that have moved rapidly directly from the polar regions.

The Properties of a Warm Mass.—By far the most important sources of warm air masses are the subtropical anticyclones in oceanic regions. In summer, warm masses may also develop over southern continents, notably in anticyclonic situations.

The air in the maritime subtropical anticyclones is warm and fairly stable, and the moisture content is high, particularly in the lower layers. When such a mass travels toward colder regions, it arrives there with a temperature higher than that of the surface over which it travels. The air becomes cooled from below, and the lower layer of air becomes increasingly stable. The pronounced stability that develops in this way hinders turbulence and completely prevents convectional currents. The result of this process is that the cooling is mainly limited to the lower layer of the air mass. The cooling from below results in a cooled surface layer of air, while the air above the inversion is mainly unaffected, except for the slow cooling caused by outgoing radiation.

By continued cooling from below, the air along the surface of the earth may be cooled below its dew point, in which case fog forms. If the wind velocity is high, there may be sufficient mechanical turbulence to stir the air under the inversion, in which case stratus, and not fog, would develop, as has been explained in a previous section.

A typical maritime warm mass is recognized by the presence of several or all of the following characteristics:

1. Stable lapse rate and inversions in the lower layer.
2. Slight turbulence. Steady wind.
3. Poor visibility.
4. High relative humidity.
5. Stratus, mist or fog, and haze.
6. Drizzle.

Air from the doldrums, or from the western part of subtropical anticyclones, and warm air that develops over continents in summer are usually conditionally unstable. When such air moves toward colder regions it will be stable in the lower layer but will remain conditionally unstable at greater altitudes. It will, therefore, have properties characteristic of a warm mass near the earth's surface, but cold-air-mass properties above. This is typical of the tropical air that invades the United States from the Gulf of Mexico.

When a warm air mass invades a warm continent in summer, instability develops rapidly, whereas the air changes from a warm mass to a cold mass. In winter, on the other hand, the stability is increased and deep layers of fog (advection fog) may cover large areas.

TRAVELING DISTURBANCES

We have already discussed the types of weather that develop within the air masses, and we have seen that the physical causes of these weather phenomena can be ascribed to the heating and cooling of the air that is in contact with the earth's surface. The aim of this section will be to discuss the weather phenomena that develop along the border between two adjacent air masses of different temperature, as well as other phenomena that can be classed as traveling disturbances.

The motion of the air is mainly controlled by the pressure distribution. It is, therefore, convenient to discuss separately the following types of pressure distribution: cols, depressions (or cyclones), and anticyclones.

The Col.—A *col* is the saddle-backed region between two anticyclones and two depressions arranged as shown in Fig. 177. Since the air streams mainly along the isobars, it follows from the diagrams that, in the region of a col, there are two main currents of air which blow against one another and deviate sideways. In the regions of a col, therefore, air masses from widely different regions may be brought toward one another.

If the isotherms are more or less parallel to the broken line through the col (*i.e.*, the axis of outflow), the isotherms will sooner or later be concentrated in the vicinity of this line, and great temperature contrasts and energy contrasts will be created. If the isotherms are more or less parallel to the dotted line (the axis of inflow), it is easily seen that the isotherms will move away from one another, the air in the region of the col becoming more and more homogeneous.

Let us now imagine that the axis of outflow runs east and west, and that the air to the south is warm and the air to the north cold. The air movement around the col would then bring warm and cold air into juxtaposition. In this case a marked discontinuity in temperature would develop, and it would move toward the axis of outflow. The process of creating such a discontinuity is called *frontogenesis*, and the discontinuity created is called a *front*.

Let us next imagine that the anticyclone to the northwest of the col is a cold polar anticyclone, and that the one to the southeast is a warm subtropical anticyclone. The circulation in the region of the col would then bring tropical air and polar air into juxtaposition and create a front between the two masses. The front that separates the polar air from the tropical air is called the *polar front*. As we shall see later, the polar front has a far-reaching influence on the weather in intermediate and high latitudes.

The preceding does not fully explain the intricacies of the process that leads to the formation of fronts. Frontogenesis may also take place in other wind systems, notably in troughs of low pressure. But

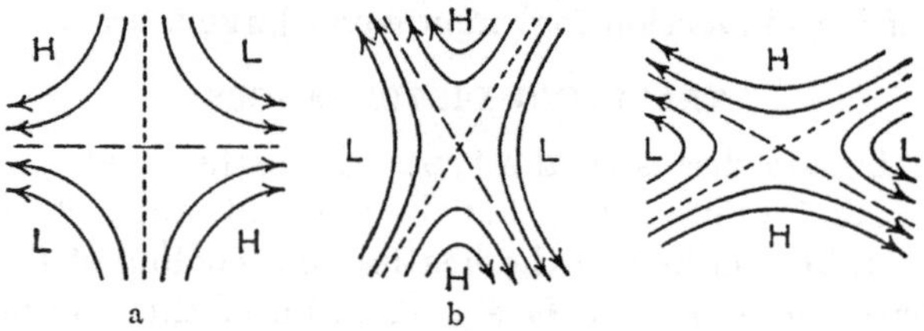

FIG. 177.—Types of cols.

the most active frontogenesis occurs in the vicinity of cols, and the fronts have a tendency to be parallel to, and not very far from, the axis of outflow.

Whether a front develops or not depends entirely on the distribution of temperature, and often, when the isotherms are unfavorably arranged, the cols destroy temperature contrasts instead of creating them.

The cols are very perfidious systems, and almost anything may happen in them. In summer, afternoon showers and thunderstorms develop frequently in the cols over warm continents. In winter, fog is a frequent occurrence. In each particular case we must turn to the weather map to see what surprises the col may bring. As a general rule we may say that the type of col shown in Fig. 177*b* gives fair and settled weather, while the type shown in Fig. 177*c* usually gives a rapid development toward worse weather.

Frontal Zones.—We have seen previously that the general circulation of the atmosphere has a tendency to produce huge air masses of more or less homogeneous properties. The general circulation also tends to produce frontal zones between the various air masses. Such frontal zones, or fronts, are maintained when the air currents converge toward a line. On the whole there are three main frontal zones.

The *equatorial frontal zone* is situated in the doldrums and separates the equatorial air of the winter hemisphere from that of the summer hemisphere. Since the temperature difference between the two equatorial air masses is but slight, this frontal zone is not a pronounced one.

The most important frontal zone is the *polar front*, which separates the tropical air from the polar air. In the Atlantic area this frontal zone extends generally from the United States toward Norway; it moves northward in summer and southward in winter. In the same way a prominent polar front extends generally from the east coast of China toward the west coast of the United States; this zone, too, moves northward in summer and southward in winter.

Often a third principal frontal zone develops between the maritime polar air and the arctic air farther north. This frontal zone is called the *arctic front*.

These are the main frontal zones of the world. Fronts of less extent, persistence, or intensity may form within the air masses, notably between old and young maritime polar masses. Such fronts are called *secondary fronts*.

The importance of the fronts is most easily understood when we consider the energy of storms. Along the fronts tremendous amounts of energy (temperature differences) are concentrated and used by the air to create kinetic energy. This explains, as we shall see presently, how the traveling depressions (or cyclones) develop along the fronts and feed on the heat energy that is there accumulated.

The Traveling Depression.—A *depression* is an area where the pressure is low relative to that of its surroundings. The pressure is lowest in the center of the depression, where the winds usually are light and variable. The winds blow around the depression in a counterclockwise direction in the northern hemisphere, and in a clockwise direction in the southern hemisphere. The wind also has a drift across the isobars toward the center of the depression.

Through studies of a number of weather charts, J. Bjerknes was led to the conclusion that the traveling depressions often consisted of a sector of warm air surrounded by colder air, the two air masses being separated from each other by a front. Figure 178 shows the structure of such a depression, or the so-called *cyclone model*. The sector of warm air extends to the center of the depression, which moves in the direction of the current in the warm sector. The part of the front where warm air replaces cold air is called the *warm front*. In the rear, cold air replaces warm air along the *cold front*.

The lower part of the diagram shows a cross section through the depression to the south of the center. The warm air overruns the wedge of cold air that forms the warm-front surface, and the cold air behind the cold front cuts under the warm air in advance of it.

The inclination of the warm-front surface is on the average $\frac{1}{100}$, and the inclination of the cold front is usually greater than $\frac{1}{100}$, and varies with the speed of the front and the difference in temperature.

It is important to note that the warm air is lighter than the cold air; it will, therefore, rise above the cold air along the frontal surfaces. The air in the warm sector blows up the slope of the warm-front surface, whereby it comes under lower pressure and expands and cools. At a certain distance above the ground, depending on the moisture content, the warm air has cooled so much that it becomes saturated with moisture. Above this level a huge cloud system develops in the warm air above the sloping warm-front surface.

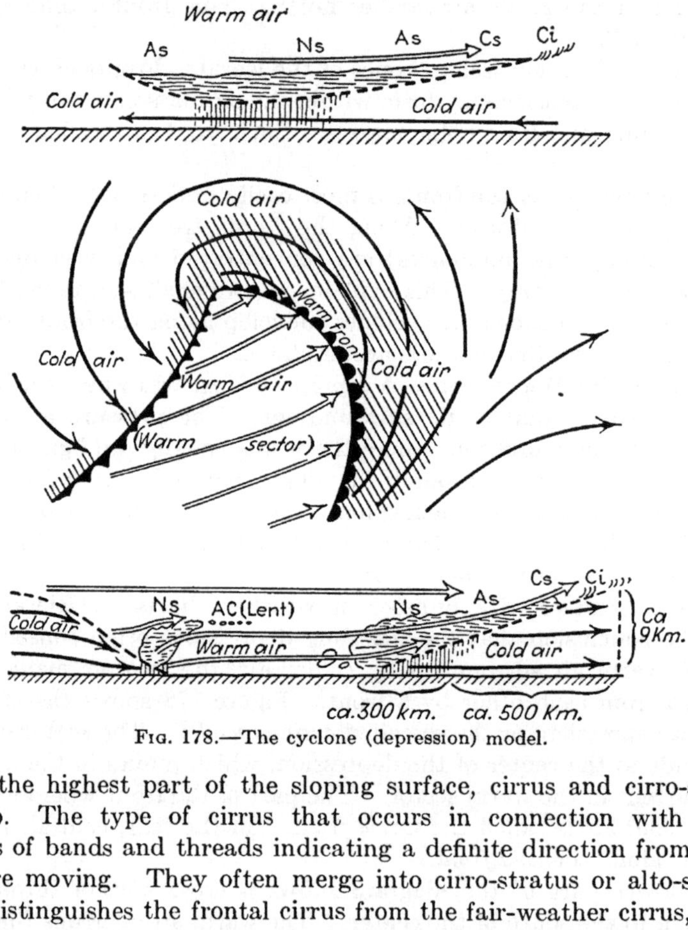

FIG. 178.—The cyclone (depression) model.

At the highest part of the sloping surface, cirrus and cirro-stratus develop. The type of cirrus that occurs in connection with fronts consists of bands and threads indicating a definite direction from which they are moving. They often merge into cirro-stratus or alto-stratus. (This distinguishes the frontal cirrus from the fair-weather cirrus, which show no systematic arrangement, and no connection with cirro-stratus.)

The cirro-stratus again merges gradually into alto-stratus, which gradually becomes lower and denser, and eventually changes into nimbo-stratus. The top part of the warm frontal cloud system consists of ice crystals, whereas the lower part of it usually consists of water droplets. In the zone of transition the water will gather on the ice crystals, which then grow and become so heavy that they fall out of the cloud as precipitation.

Seen from below, the cloud system has an even appearance that is distinctly different from that of convectional clouds. The precipitation that falls from the frontal cloud system is usually continuous without much variation in intensity. At weak fronts the precipitation may be intermittent, but the variations are not so rapid as in the case of showery precipitation.

The cold air in the rear of the cold front cuts under the warm air and lifts it. Condensation of water vapor and precipitation occur in the same way as along the warm front, but, owing to the current above the cold-front wedge, the cold-front cloud system gets a forward tilt, and the upper part of the cold-front surface remains cloudless. The rain area behind the cold front is fairly narrow, and it often extends in a forward direction into the warm sector.

The upper part of the diagram (Fig. 178) shows a section through the depression north of the center. The cold air forms a valley that is partly filled with clouds.

The above diagram represents the "ideal" or "model" depression. In nature many complications that are not contained in the diagram may occur. Moreover, the diagram represents only a single stage in the development of a depression.

The Development of Depressions.—Further investigations have shown that a depression originates as a *slight* wave on the front, and that the wave moves along the front, steadily growing in amplitude. After some time the wave has reached the stage represented by Fig. 178. During the further development, the cold front overtakes the warm front, and the warm air is squeezed between the two cold wedges and disappears aloft.

During this development the depression grows in intensity, and the wind around it increases. Some time after the cold front has reached the warm front, the depression begins to decrease in intensity and gradually dies away as a frontless vortex of air. This development is shown in Fig. 179. In Fig. 179*a* the front is a straight line, and the frontal surface slopes up from the warm side toward the cold side, having an inclination of about 1 in 100. The warm air blows from a westerly direction and the cold air from an easterly direction. An aircraft that ascends through the frontal surface will experience a sudden wind shift from an easterly to a westerly direction. The sheering motion along the frontal surface is comparable with a strong wind that blows along the surface of the sea and creates waves. The waves that form on the surface between cold (heavy) air and warm (light) air are huge waves the length of which may be 1,000 miles or more.

The waves that form on the surface of the sea are usually *stable* waves, *i.e.*, waves that travel with a fairly constant amplitude. A

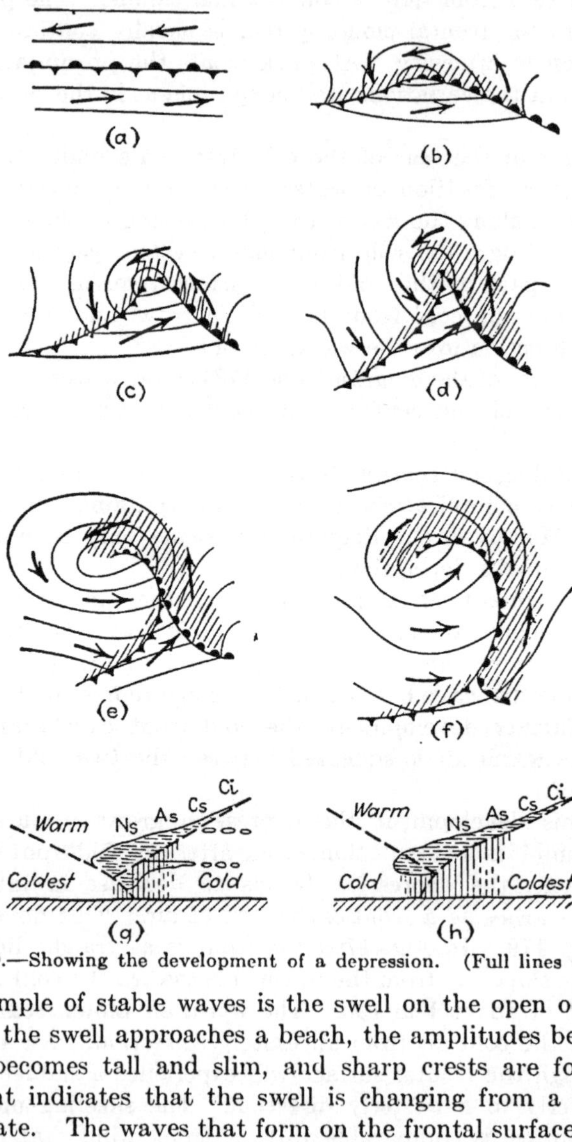

FIG. 179.—Showing the development of a depression. (Full lines = isobars.)

typical example of stable waves is the swell on the open ocean. However, when the swell approaches a beach, the amplitudes begin to grow; the wave becomes tall and slim, and sharp crests are formed. This development indicates that the swell is changing from a stable to an unstable state. The waves that form on the frontal surfaces are usually *unstable* waves. Their amplitudes grow, and simultaneously the cold front overtakes the warm front. This process is called the *occlusion* of the depression, and the front that results when the cold front has reached the warm front is called an *occluded front.*

The development described above is illustrated in the diagrams of Figs. 179*a* to 179*f*. Figure 179*b* shows the beginning of a wave, Fig. 179*c* a young wave, Fig. 179*d* a wave that is nearly occluded, Fig. 179*e* a newly occluded depression, and Fig. 179*f* an old occlusion.

If the cold air in the rear of the cold front is colder than the air in advance of the warm front, the rear air will cut under the warm front as shown by the cross section in Fig. 179*g*. If the air in advance of the warm front is colder than the air in the rear of the cold front, the latter will climb up the warm-front surface as shown by the cross section in Fig. 179*h*. Thus an occluded front may act either as a cold front or as a warm front. An occlusion that approaches a warm continent will change into a cold-front type, an occlusion that approaches a cold continent will change into a warm-front type.

Some Further Remarks on Fronts and Depressions.—The frontal cloud system shown in Fig. 178 illustrates the conditions when the air of the warm sector is stable. Warm air that originates from the doldrums or from the western extremities of the subtropical anticyclones, for example, the Gulf of Mexico, or from warm continents may be in a conditionally unstable state. When such air is lifted along the frontal surfaces it becomes unstable. In such cases turreted tops like cumulus congestus or cumulo-nimbus grow up from the alto-stratus layer, and showers or even thunderstorms may be superimposed on the warm-front area of precipitation. This is quite a frequent occurrence in the eastern part of the United States, because the warm air from the Gulf of Mexico is usually conditionally unstable. The warm air from the Atlantic subtropical anticyclone arrives in Europe in a stable state, but, in summer, tropical air that develops over northern Africa or southern Europe is sometimes conditionally unstable.

If the warm-sector air is stable, fog, stratus, or strato-cumulus are the characteristic clouds in the warm sector. If the air is unstable or conditionally unstable, convective clouds may develop in the warm sector. This happens quite frequently in the United States in summer.

If the warm-sector air is unstable or conditionally unstable, the cold-front cloud system breaks up into convective clouds with showers and thunderstorms. This, too, happens frequently in the United States in summer.

Usually there is a slight descending motion (subsidence) in the cold air some distance ahead of the warm front. This descending motion is apt to dissolve the clouds in the cold air under the cirro-stratus and alto-stratus layer. Cumulus clouds in the cold air will therefore shrink and disappear when the warm-front cloud system approaches. Likewise, there is often a descending current immediately after the cold front. Therefore the passage of a cold front is often followed by a clearing of brief duration, after which follow the weather phenomena characteristic of the cold-air mass.

A *trough of low pressure* is an elongated area of relatively low pressure that extends from the center of a depression. The trough may have either U-shaped or V-shaped isobars. The U-shaped troughs have no fronts in them, and the meteorological elements vary continuously while the trough passes. Even though there is no front in such a trough, the weather is usually unsettled or bad. Moreover, secondary fronts may form in the U-shaped troughs.

The V-shaped troughs are always connected with fronts. The V points away from the center of the depression, and the front runs through the angles of the V-shaped isobars. When such a trough passes a station, the meteorological elements vary abruptly.

Wind Variations and Fronts.—It is of interest to note that a front is a wind-shift line. When a front passes a station, the wind direction changes more or less abruptly in a clockwise sense (for example, from south to southwest, or from southwest to northwest). This applies to all fronts, whether warm, cold, or occluded.

If an aircraft rises through a warm-front surface, it will experience a sudden clockwise shift in wind. If the aircraft ascends through a cold-front surface, it will experience a sudden counterclockwise wind shift (for example, from northwest to west, or from west to southwest). If it rises through the frontal surface north of the center of the depression, it will experience a shift from an easterly to a westerly direction. To the south of the center the westerly winds reach up to great heights; to the north of the center there is usually a westerly current above the easterly. These features are easily deduced from the diagram shown in Fig. 178. The center of the depression, being at the peak of the warm sector, will have a more northerly position in the upper atmosphere than near the earth's surface.

The wind shift is usually more pronounced along cold fronts than along warm fronts. The passage of a cold front is often accompanied by abrupt changes in wind direction through more than 90° in a clockwise direction. Simultaneously, wind squalls and violent turbulence may occur. This happens particularly at quickly moving cold fronts of considerable intensity. The air along the surface of the earth is retarded by friction, while the air above moves with a velocity corresponding with the pressure gradient. This allows the cold front to become very steep, or even overhanging like a wedge with its point at some distance above the ground. Thus, in extreme cases, the cold air behind the cold front may overrun the warmer air in advance of it. In this way great instability is created in the lower layer, and strong convectional currents and squalls result. The convectional cloud that develops in this way is the well-known *line-squall cloud* (see Fig. 155). The line squall is accompanied by heavy rains, hail, and thunderstorms, and the turbulence and the wind squalls that accompany it may be of destructive intensity, particularly to lighter-than-air craft. In extreme cases, tornadoes or waterspouts may occur along the squall line.

Airships and Fronts.—Pronounced fronts are of importance in airship operations because of (1) the forces that a sudden wind shift imposes on the side of the ship and (2) the sudden change in air density, which may cause the ship to change its altitude in an undesirable way.

The ship is designed to withstand large forces from dead ahead. When the wind shift occurs in a period too brief for the ship to turn against the newly arrived wind, the forces on the sides of the ship may damage the fabric and the structure. Wind shifts are of equal importance for ships in the air and for moored ships.

A second danger to airship operation is the sudden change in air density that occurs along pronounced fronts. An airship that flies through a front from the cold toward the warm side will arrive in the warm air with a negative superheat. The ship is then too heavy for the warm air, and so sinks along the frontal surface, where it is exposed to the effects of the wind shift along the front. The change in air density along pronounced fronts may be sufficient to drop the ship 1,000 ft. or more. *A strong front should, therefore, not be crossed too close to the earth's surface.*

An airship that flies through a front from the warm to the cold side will arrive in the cold air with a positive superheat. The ship will ascend along the frontal surface and be under the influence of the undesirable effects of the wind shift along the front.

The difference in density along weak fronts is usually so small that the ascent or descent of the ship can be checked; but at strong fronts the changes may be too rapid to allow adequate precautions.

The Tropical Cyclone.—The tropical cyclones are small cyclonic depressions, having nearly circular isobars and very strong winds circulating in a counterclockwise direction in the northern hemisphere and in a clockwise direction in the southern hemisphere. The tropical cyclones are called *cyclones* in India, *hurricanes* in the West Indies, and *typhoons* in China. They originate in the doldrums over the oceans between Lats. 6°N. and 20°N., or 6°S. and 20°S:, and they travel in the direction of the trade wind along the isobars on the equatorial side of the subtropical anticyclones. At the western end of the subtropical anticyclones they recurve poleward.

The wind is light and variable in the center (the "eye") of the tropical cyclone around which there is a whirl of hurricane winds and torrential rainfall. The horizontal diameter of the cyclone varies from a few miles up to several hundred miles. The diameter increases when the cyclone recurves poleward. The wind velocity often exceeds 100 m.p.h. The cyclone travels with a moderate speed of about 10 to 20 m.p.h.

There is some evidence that the tropical cyclones in the Atlantic and Pacific Oceans originate along the equatorial front as a disturbance like the depressions along the polar front. The cyclones in the Indian Ocean, however, do not form along any such front, and it is plausible that thermal instability plays an important part in the formation of all tropical cyclones.

Tornadoes and Waterspouts.—A tornado is a circular whirl of great intensity and small horizontal extent, in which the wind velocity is usually of superhurricane force. The horizontal diameter of the tornado varies from a few feet up to 1 mile. The wind velocities sometimes exceed 200 m.p.h. The pressure in the center of the tornado is much lower than in the immediate surroundings, and this, together with the high winds, produces destructive effects. The air in the center is rising rapidly, and the whirl is accompanied by heavy rain or hail, thunder, and lightning. The decrease in pressure in the center of the tornado cools the air below its dew point, and, as a result, a funnel-shaped cloud marks the core of the whirl.

The tornadoes are short-lived and usually do not last more than an hour or two. They usually occur in connection with a strong cold front of the type that gives line squalls. They often form in series and travel in almost parallel paths following the squall line. Tornadoes occur quite frequently in the Mississippi Valley. In Europe they are rather rare and not so violent as those that occur in the United States.

Waterspouts are tornadoes that form at sea.

The Anticyclone.—An anticyclone is an area of high pressure surrounded by closed isobars. The wind blows around the anticyclone in a clockwise direction in the northern hemisphere and in a counterclockwise direction in the southern hemisphere. In the center the winds are light and variable. On the whole the wind velocity is moderate in anticyclones. The wind has an outward drift from the central part of the anticyclone, which usually is compensated by descending air in higher levels. The descending motion dissolves the high and medium clouds. The anticyclone is therefore often a region of stable and fair weather. In winter, radiation fog frequently forms over land in the stagnant parts of the anticyclones.

The anticyclones do not always give fair weather. Fronts may often penetrate far into the anticyclonic region, and, over oceans and warm continents, convection may give showers. The weather is usually fair in anticyclones with increasing pressure, while the dissipating anticyclones may have unsettled or bad weather in places.

A *wedge* is a region of relatively high pressure that extends from the center of the anticyclone toward the col. The traveling wedges usually bring clearing weather, except in the vicinity of the col, where the weather may be very variable, as has been explained in an earlier paragraph.

WEATHER REPORTS

Most countries of the world have adopted the international code for meteorological reports, which is a numeral code consisting of several groups, each containing five numerals that give the meteorological observations in a definite order. These reports are collected in great distributing centers and transmitted by radio according to an international timetable. We shall describe the salient features of the international system in a later section.

In the United States, the weather reports are distributed over the teletype and radio circuits of the Bureau of Commerce. In order to shorten the transmission as much as possible, a system of symbols and abbreviations is used, which in several respects differ from the international system.

Form of the U. S. Weather Reports.—The report from each station has the following form:

> STATION DESIGNATOR, one space; CEILING, SKY, VISIBILITY, WEATHER, OBSTRUCTIONS TO VISION, one space; TEMPERATURE, DEW POINT, WIND, one space; BAROMETRIC PRESSURE, one space; FIELD CONDITIONS, one space; REMARKS.

Figure 180 is a condensed explanation of symbol weather reports.

The United States Symbols.—The following symbols are used in the United States for drawing manuscript charts:

Sky Covering (Symbols inside the Station Circle)

Clear	○	Broken clouds	◕
Scattered clouds	◔	Overcast, cloudy	●
Partly cloudy	◑	Sky obscured by obstruction	⊘

Wind

Symbols for wind direction and force are as in the international system.

State of Weather (Symbols outside the Station Circle)

Rain (light, moderate, or heavy)	•	:	⁝
Shower (light, moderate, or heavy)	◦	▽:	▽⁝
Mist (light, heavy)		,	;
Hail (light, moderate, or heavy)	▲	▲▲	▲▲▲
Snow (light, moderate, or heavy)	*	**	***
Snow showers (light, moderate, or heavy)	▽*	▽**	▽***
Sleet (light, moderate, or heavy)	△	△△	△△△
Fog (light, moderate, or dense)	–	=	≡
Ground fog (light, moderate, or heavy)		≡	
Haze (light, thick)		∞	<u>∞</u>
Thunderstorm (mild, moderate, or severe)			
Thunderstorm in progress at time of observation	|O		
Thunderstorm in progress at time of previous observation	O|		
Thunderstorm in progress at previous observation, and another at current observation	|O|		
Thunderstorm in progress at previous observation, another (others) later, but none at time of current observation	<u>O|</u>		
Thunderstorm(s) between observation	<u>O</u>		
(Entry of time of occurrence of a thunderstorm is optional)			
Distant lightning	↳		
Squalls or squally weather			
Heavy (or severe) squall(s)			
Sign that tropical storm is forming	ϩ		
Signs that tropical storm has formed			
Blowing snow (light, thick)	BS	<u>BS</u>	
Blowing sand (light, thick)	BSA	<u>BSA</u>	
Blowing dust (light, thick)	BD	<u>BD</u>	
Smoke (light, thick)	K	<u>K</u>	
Dust (light, thick)	D	<u>D</u>	
Line squall	in plain language		

Surface Condition

Light frost	*Light*
Heavy frost	*Heavy*
Light or moderate freezing rain (glaze)	∽
Heavy freezing rain (heavy glaze)	<u>∽</u>

Pressure Changes

Symbols are the same as in the international system, except that the amount of changes is indicated in hundredths of inch; for example, 4 equals 0.04 in.

Air-mass Abbreviations

Symbols for fronts as in the international system. Abbreviations for air masses of polar origin, for example, Pc = Polar continental; Npp = Transitional polar pacific, etc., are indicated in blue pencil, and air masses of tropical origin are indicated in red pencil, for example, Tg = Tropical air from the Gulf; Ntp = transitional tropical air from the Pacific.

Clouds

Cloud forms are entered in abbreviations as shown on page 200. The direction of the movement of the clouds is indicated by an arrow.

Weather Analysis.—When the observations have been entered on the chart, the analysis of the weather situation begins. The theory and the principles of weather analysis and forecasting form one of the most intricate branches of meteorology and are far beyond the purview of this brief outline. We can here indicate only roughly how the analysis is performed.

It should be emphasized at the outset that a satisfactory analysis can be based only on a succession of charts, and not on a single one. Only a succession of charts will show the life history of the air masses and the life history of the traveling fronts and depressions, etc.

The first step in the analysis is to sketch in the isobars, or the lines through points of equal pressure, in order to obtain an idea of the positions of anticyclones and depressions, and also the main air currents. It is most convenient to draw isobars for each 0.1 in. or for each fifth millibar. The exact position of the isobars should be interpolated between the stations. In doing this we should also take into consideration the direction and the velocity of the wind. The wind should blow along the isobars but with a slight drift toward lower pressure. The stronger the wind, the more crowded are the isobars. This is important to note, particularly where the distance between the stations is large.

Having sketched the isobars lightly, the next step is to examine the cloud systems and the weather phenomena in order to find the fronts and the air masses, using the characteristics described in previous sections. When the fronts have been drawn, the isobars should be adjusted and drawn as heavy lines. It is particularly important to draw the isobars accurately in the vicinity of the fronts, because the velocity of the front may be obtained from the distance between the isobars crossing the front.

Next, areas of fog and precipitation should be indicated, and visibility and cloud heights should be studied. Finally, lines should be drawn through the points of equal barometric tendency, in order to find the areas of rising and falling barometer. These lines will indicate in which way the pressure distribution is going to change. The chart thus analyzed must agree with the previous chart in such a way that the fronts and air masses have moved with a velocity that corresponds with that of the air current.

Figure 181 shows a section of a simplified weather chart for the United States, Jan. 8, 1937, at 8 A.M. The Bermuda High in conjunction with the frontal trough of low pressure along the Mississippi Valley causes a flow of tropical air from the Gulf of Mexico (*Tg*) toward the north and northeast. This air arrives in the United States with a temperature that is considerably higher than that of the surface over which it travels. It is essentially a *warm air mass.* In Florida and the West Indies the sky is partly clouded. When the air moves northward, it

FIG. 180.—EXPLANATION OF SYMBOL

To illustrate the method used in transmission and deciphering of symbol element of the report appears directly above a description of all symbols Elements of observations are always transmitted in the same order; therefore,

WA	N	SPL	1624E	E30	⊕15⦶	2V
Station	**Classification of report**	**Type of report**	**Time of report**	**Ceiling**	**Sky**	**Visibility**
Lists of station names and their representative call letters are posted on Weather Bureau airport station bulletin boards for the information of all concerned	The symbols C, N, or X are used immediately following, after one space, the station letters to classify weather conditions at airports specifically designated as controlled airports If no classification letter is used, the station is not located at a controlled airport C: satisfactory for contact flight N requiring observance of instrument-flight rules X take-off and landing suspended	SPL, meaning "special report," appears when crucial changes have occurred in weather conditions since the last report The absence of the observation-type letter group SPL indicates an observation where no crucial changes have occurred since the last transmitted observation LCL, meaning "local extra observation," appears only on reports sent over local circuits. Such reports are made every 15 minutes during periods of low ceiling and/or visibility	Time groups are in figures based on the 24-hr clock with following letters showing the standard of time used, *e.g.*, 1440E means 2.40 P.M., Eastern Standard Time; 0030C means 12:30 P.M., Central Standard Time; 2359M means 11:59 P.M., Mountain Standard Time; 2015P means 8:15 P.M., Pacific Standard Time, etc. SPL reports that are sent alone, and all LCL reports bear the time of observation immediately following, after one space, the observation-type letter group SPL or LCL SPL reports appearing in sequences do not show the time-of-report group, and the time of observation is considered as the time of all other reports in the sequence as indicated in the sequence heading	The absence of a "ceiling" group indicates an "unlimited" ceiling (above 9,750 ft) Figures representing the number of *hundreds* of feet that apply are used to indicate the height of the ceiling between 51 and 9,750 ft above the station, *e.g.*, 35 indicates 3,500 ft, 3 indicates 300 ft, etc The figure naught (0) is used when the ceiling is zero (below 51 ft) When the height is estimated the letter E precedes the ceiling figures. A plus sign (+) is used preceding the ceiling figures to indicate the ceiling balloon was blown from sight at the height represented by the figures and before reaching the clouds The letter V is used, immediately following the figures for ceiling, if the height of the ceiling is changeable and below 2,000 ft	The absence of a symbol for sky indicates that precipitation or obstructions to vision are present and reduce the ceiling to zero and/or the visibility to 1/16 mile or less and make the sky unobservable The sky condition is indicated by the following symbols unless the condition given above is present: O Clear ⦶ Scattered clouds ◐ Broken clouds ⊕ Overcast ⦶/ High scattered ◐/ High broken ⊕/ High overcast ⊕◐ Overcast, lower broken ⊕⦶ Overcast, lower scattered ◐◐ Broken, lower broken ◐⦶ Broken, lower scattered ⦶◐ Scattered, lower broken ⦶⦶ Scattered, lower scattered ⊕/◐ High overcast, lower broken ⊕/⦶ High overcast, lower scattered ◐/◐ High broken, lower broken ◐/⦶ High broken, lower scattered ⦶/◐ High scattered, lower broken ⦶/⦶ High scattered, lower scattered The plus (+) or minus (−) sign *preceding* the cloudiness symbol indicates "dark" and "thin," respectively. Height of lower scattered clouds is indicated by the entry of a figure representing the hundreds of feet applying, immediately preceding the scattered-cloud symbol	The absence of a figure for visibility indicates that the visibility is 10 miles or more The value of the visibility below 10 miles is indicated by figures representing the number of miles and/or fractions of miles The letter V is used immediately following the figure for visibility if the visibility is fluctuating rapidly and is 2 miles or less

* SOURCE: U.S. Department of Commerce Weather Bureau; based amendments thereto.

† The report given above would be deciphered as follows: Washington Standard Time; ceiling estimated at 3,000 ft; overcast, lower scattered rain; light blowing dust; barometric pressure, 1015.2 millibars; temperature, wind shift from the south at 4:18 P.M.. Eastern Standard Time: altimeter

WEATHER REPORTS*

weather reports, an example of such a report is given above the boxheads. Each and conditions that might be used in that particular phase of the report. all symbol weather reports may be deciphered by reference to this chart.†

<table>
<tr><td>T-R-</td><td>BD-</td><td>152</td><td>68</td><td>60</td><td>→↘18—↑ 1618E</td><td>996</td><td>+ ⊕NW</td></tr>
<tr><th>Weather</th><th>Obstructions to vision</th><th>Barometric pressure</th><th>Temperature</th><th>Dew point</th><th>Wind</th><th>Altimeter setting</th><th>Remarks</th></tr>
<tr>
<td rowspan="5">The "weather" element of the report is indicated, when appropriate, by the following symbols.
R– Light rain
R Moderate rain
R+ Heavy rain
S– Light snow
S Moderate snow
S+ Heavy snow
ZR– Light freezing rain
ZR Moderate freezing rain
ZR+ Heavy freezing rain
L– Light drizzle
L Moderate drizzle
L+ Heavy drizzle
ZL– Light freezing drizzle
ZL Moderate freezing drizzle
ZL+ Heavy freezing drizzle
E– Light sleet
E Moderate sleet
E+ Heavy sleet
A– Light hail
A Moderate hail
A+ Heavy hail
AP– Light small hail
AP Moderate small hail
AP+ Heavy small hail
OP– Light snow pellets
OP Moderate snow pellets
OP+ Heavy snow pellets
SQ– Mild snow squall
SQ Moderate snow squall
SQ+ Severe snow squall
RQ– Mild rain squall
RQ Moderate rain squall
RQ+ Severe rain squall
T– Mild thunderstorm
T Moderate thunderstorm
T+ Severe thunderstorm
SW– Light snow showers
SW Moderate snow showers
SW+ Heavy snow showers
RW– Light rain showers
RW Moderate rain showers
RW+ Heavy rain showers
TORNADO (always written out in full)</td>
<td rowspan="5">The "obstructions to vision" element of the report is indicated, when appropriate, by by the following symbols:
F– – Damp haze
F– Light fog
F Moderate fog
F+ Thick fog
FF Dense fog
GF– Light ground fog
GF Moderate ground fog
GF+ Thick ground fog
GFF Dense ground fog
IF– Light ice fog
IF Moderate ice fog
IF+ Thick ice fog
IFF Dense ice fog
H Hazy
K– Light smoke
K Moderate smoke
K+ Thick smoke
D– Light dust
D Moderate dust
D+ Thick dust
BS– Light blowing snow
BS Moderate blowing snow
BS+ Thick blowing snow
GS– Light drifting snow
GS Moderate drifting snow
GS+ Thick drifting snow
BD– Light blowing dust
BD Moderate blowing dust
BD+ Thick blowing dust
BN– Light blowing sand
BN Moderate blowing sand
BN+ Thick blowing sand</td>
<td rowspan="5">The barometric pressure is indicated by a group of three figures; tens, units, and tenths of millibars involved. Thus, a pressure of 1015.2 millibars would be written as 152; 999 9 as 999; 1,025 7 as 257; etc Sent only by stations equipped with mercurial barometers</td>
<td rowspan="5">Temperature is indicated by figures giving its value to the nearest degree Fahrenheit.

Values below 0°F are indicated by the entry of a minus sign (–) immediately preceding the figures for temperature Zero is entered as 0</td>
<td rowspan="5">Dew point is indicated by figures giving its value to the nearest degree Fahrenheit.

Values below °F are indicated by the entry of a minus sign (–) immediately preceding the figures for dew point</td>
<td rowspan="5">The wind direction is indicated by arrows, as follows:
↓ North
↓↙ North-northeast
↙ Northeast
←↙ East-northeast
← East
←↖ East-southeast
↖ Southeast
↑↖ South-southeast
↑ South
↑↗ South-southwest
↗ Southwest
→↗ West-southwest
→ West
→↘ West-northwest
↘ Northwest
↓↘ North-northwest

The velocity is indicated by figures representing its value in miles per hour, calm being indicated by the letter C. If estimated, this is indicated by the entry of the letter E immediately following the velocity figures

The character of the wind is indicated, when appropriate, by entry, immediately following the velocity, of a minus sign (–) for "fresh gusts" and a plus sign (+) for "strong gusts." No indication of character means the wind is steady

Wind shifts that have occurred at the reporting station are indicated, immediately following the other wind data, by an arrow, showing the direction (to eight points only) from which the wind was blowing prior to the shift, followed by the local time, on the 24-hr clock, at which the shift occurred, with following letter showing the standard of time used The intensity of the shift is indicated by the minus sign (–) for "mild," the absence of a sign for "moderate," and the plus sign (+) for "severe," the signs being entered immediately following the standard-of-time letter</td>
<td>Indicated by a group of three figures representing the inch and hundredths of an inch of pressure involved. Thus, 30.00 would be written as 000; 29.98 as 998, etc. Sent only by designated stations equipped with mercurial barometers</td>
<td>Remarks are transmitted in authorized English abbreviations and teletype symbols. Lists of the abbreviations are available for inspection at all Weather Bureau airport stations. The teletype symbols used are shown on this chart</td>
</tr>
<tr><th colspan="2">Special data</th></tr>
<tr><td colspan="2">Special data comprising pressure change and characteristic, 5,000-ft pressure at selected stations, cloud, thunderstorm, and snow depth data, Great Lakes water temperature, etc., data from selected stations, etc., are entered in code at certain times by the stations designated to do this, as separate groups, immediately following the report proper. These data are intended primarily for the preparation of maps for forecasting</td></tr>
<tr><th colspan="2">Missing data</th></tr>
<tr><td colspan="2">Elements normally sent, but for some reason missing from the transmission, will be indicated by the letter M, entered in the report in place of the missing data</td></tr>
</table>

on instructions contained in *Weather Bureau Circular N*, 1939, and

—satisfactory for instrument flight only; special report at 4:24 P.M., Eastern clouds at 1,500 ft; visibility, 2 miles, variable; mild thunderstorm; light 68°F; dew point, 60°F; wind west-northwest 18 mph., fresh gusts; moderate setting 29.96 in.; dark to the northwest.

cools from below, and a more or less continuous layer of stratus and fog develops. Note that fog forms preferably where the wind speed is slight, whereas stratus or strato-cumulus forms where the wind is strong enough to produce sufficient mixing. It is of interest to note that the fog and stratus are of the *advection* type, and, since they occur together with appreciable wind forces, it is likely that the cloud layer in the tropical air is very deep.

The polar front separates the tropical air from the polar air to the north and northwest. The source of the polar continental air (*Pc*) is the anticyclonic region over Canada. In its source the air is cold and specifically dry. When streaming southward to the west of the center of LOW it is heated from below, whereby it develops toward instability. It is essentially a *cold air mass.* However, since not much moisture is supplied to the air from the surface of the earth, showers are not numerous.

Along the warm front, the warm air overruns the cold air, and a continuous cloud system develops along the front from which continuous precipitation falls.

To the southwest of the center of LOW the cold air cuts under the warm air. The Gulf air, being conditionally unstable, is here lifted so much that the instability is released and thunderstorms form.

Three wave cyclones have formed along the polar front. The first wave is situated somewhere to the south of Newfoundland. The second one, which is only slight, is situated to the east of Toronto. The third wave is situated to the southwest of Chicago. The center (*L*) will move parallel to the direction of the warm sector isobars (*i.e.*, toward the northeast). The barometric tendencies in the warm sector are negative, which indicate that the wave is *unstable, i.e.*, it will deepen and develop into a cyclone while it occludes.

The center of HIGH between Bismarck and Sioux Lookout is moving rapidly southward; this is evident from the fact that the barometric tendency south of the HIGH is plus and north of HIGH is minus. The HIGH near Moosonee will move toward the southeast with considerable speed; this also follows from the distribution of the barometric tendencies. Thus at God's Lake the barometer is falling 0.10 in. in 3 hr., while at Father Point it is rising 0.18 in 3 hr.

It is readily understood that a center of HIGH must move in the direction from the maximum of falling toward the maximum of rising barometer; in the same manner, a center of LOW must move in the direction from the maximum of rising toward the maximum of falling barometer.

Figure 182 shows the weather chart 12 hr. after the previous one. The reader is advised to try to analyze it in order to gain experience in interpreting the observations. The analysis of this chart should be based on that of the previous one, and the principal fronts, air masses,

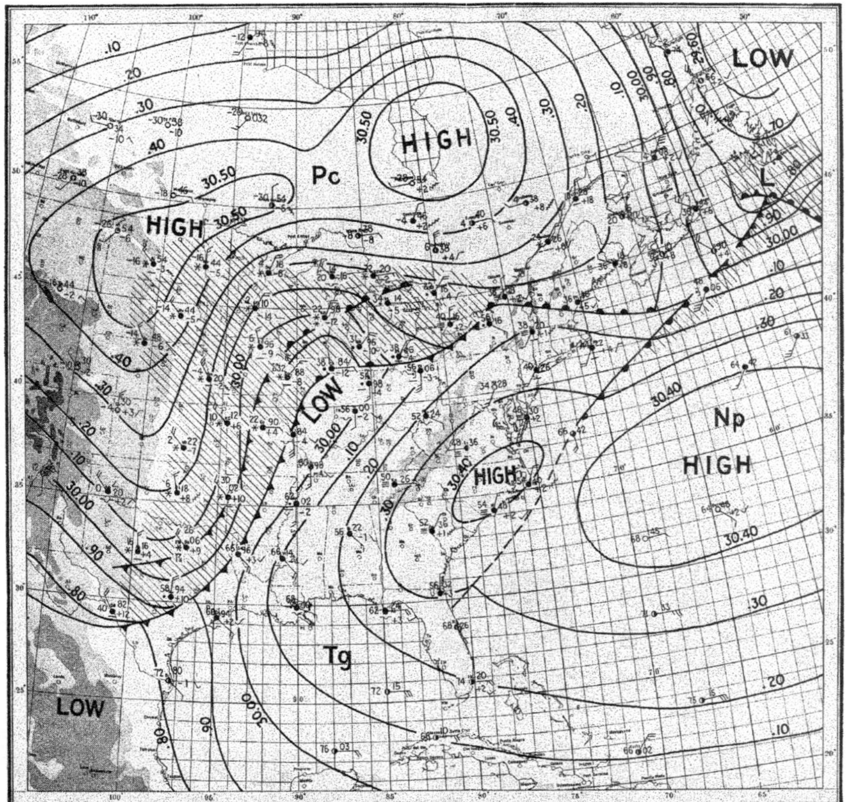

Fig. 181.—The weather map Jan. 8, 1937, 8 A.M. Lines with filled semicircles indicate warm fronts; lines with filled triangles indicate cold fronts; hatched areas indicate areas of continuous precipitation.

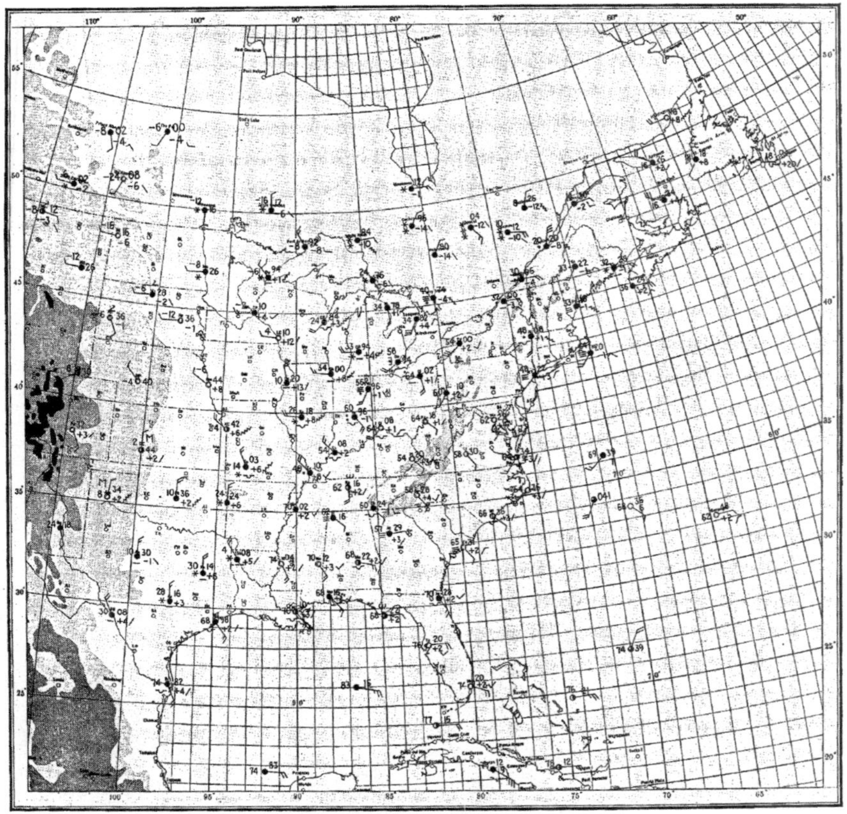

FIG. 182.—The weather map Jan. 8, 1937, 8 P.M., with isobars omitted.

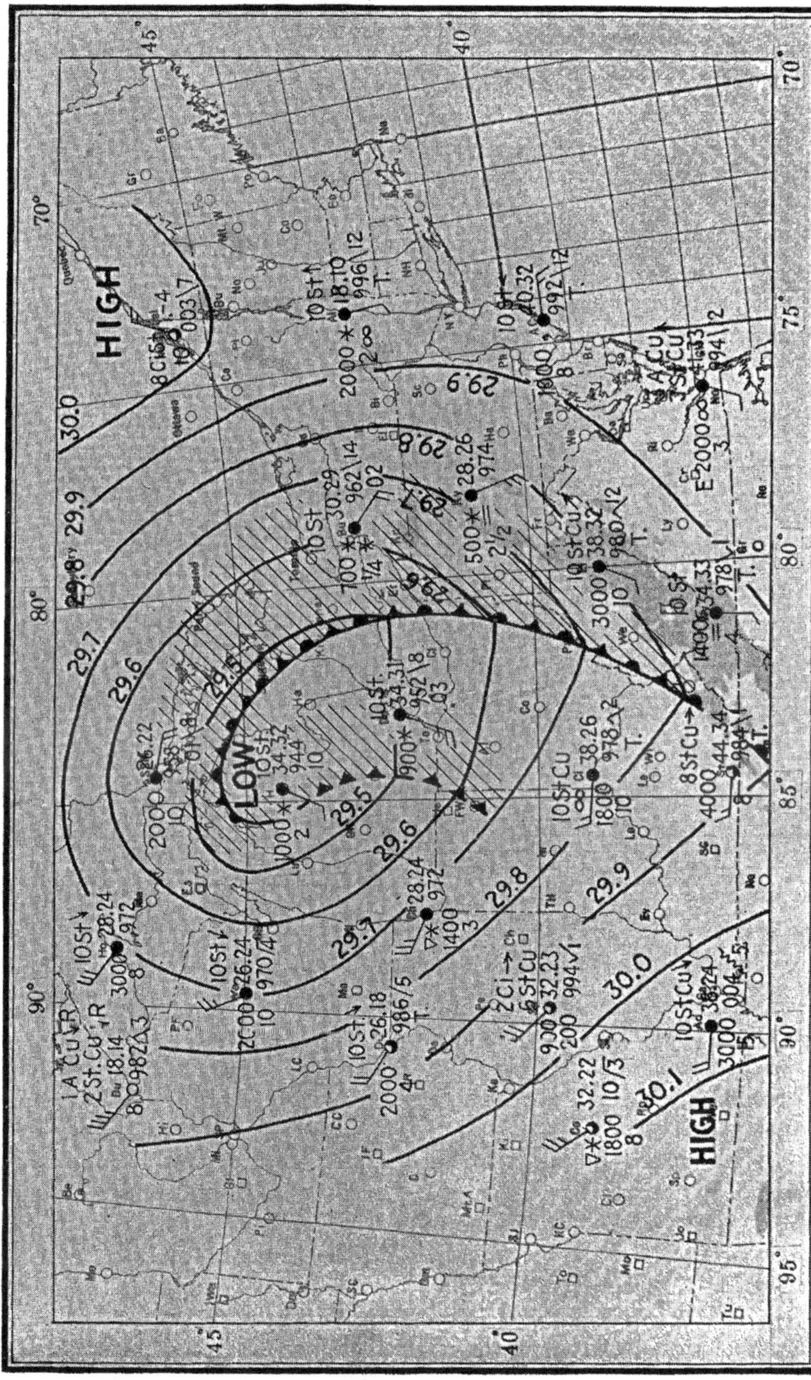

FIG. 183.—Jan. 12, 1938, 1:30 P.M. In this map all observed elements have been plotted. The elements are plotted around the station circles in the following manner: Sky covering in the station circle; cloud forms and amounts above the station circle; temperature and dew point to the right of the station circle; barometric pressure, tendency and amount of precipitation under temperature and dew point; weather, ceiling (in feet) and visibility (in miles and fractions thereof) to the left of the station circle.

etc., should be identified. The weather conditions in the tropical air mass have improved on the whole during the day on account of the diurnal heating, which has a tendency to dissolve fog and stratus.

Figure 183 shows a small section of a weather chart plotted according to the U. S. Weather Bureau regulations. In it all elements have been plotted. The reader is advised to use the list of symbols in order to interpret the observations. Only those who are thoroughly familiar with the symbols will be able to make adequate use of the weather charts that are prepared at most principal airports.

In some aviation weather services, cross sections are prepared in order to facilitate the interpretation of the upper air data. Figure 184

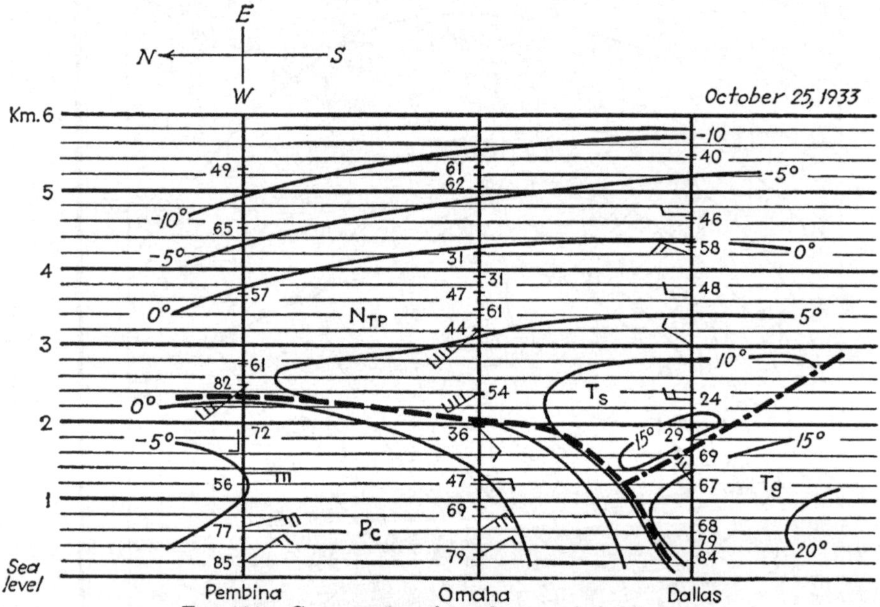

FIG. 184.—Cross section through an occluded front.

shows such a cross section through an occluded front. The full lines are isotherms (centigrade); the values of relative humidity are plotted for significant points along the vertical; the direction of the wind arrows refers to the direction diagram above; the broken lines represent the intersection between the frontal surfaces and the cross section.

The cross-section diagram furnishes the best basis for choosing the most convenient flight altitude. The position of the freezing-point isotherm is of importance for the estimation of the possibilities for icing.

Forecasting.—The analysis of the weather situation shows (in a rough way) the state of the atmosphere at a given moment. The weather that will occur in a certain locality within the forecasting period will then depend mainly on the following factors:

1. The travel of the air masses, fronts, etc., across the chart.

2. The changes that will occur within the air masses and in the structure of the fronts and depressions, etc., while they travel across the chart.

In order to be able to forecast the weather with reasonable accuracy, it is necessary to have a thorough knowledge of the physics of the atmosphere, several years' experience in general forecasting, and also a thorough knowledge of the numerous local influences due to the terrain.

The problem of forecasting becomes most intricate when the forecasting period is large. For forecasts of short period (6 hr. or so) it usually suffices to extrapolate the positions of the depressions, fronts, and anticyclones by means of their previous displacements, assuming that they will continue to move with the speed they had during the last 6 hr. Two or more consecutive charts at 3- or 6-hr. intervals will then furnish the basis for the forecasting of the displacements of the pressure systems. (The movement of the pressure systems can also be computed from the distribution of pressure and pressure changes; but the procedure is too elaborate to be described in this brief outline.) The movement of the air masses during 6 hr. can be obtained from the geostrophic wind (see table below). The movement of the fronts can also be obtained with fair accuracy from the geostrophic wind, using the following rules:

a. A warm front (or a warm-front-type occlusion) moves in the direction of the warm-sector isobars (or the isobars in the warmer air) with a velocity of about 60 to 85 per cent of the geostrophic wind in the rear of the front.

b. A cold front (or a cold-front-type occlusion) moves with a velocity of about 80 to 100 per cent of the geostrophic wind in the rear of the front.

The following table gives the distance between the isobars, the corresponding geostrophic wind velocity, and the distance traveled in 6 hr. The distances are given in miles, and the velocities in miles per hour.

Distance between 0.10-in. isobars	60° N. or S.		50° N. or S.		40° N. or S.	
	Geostrophic wind	Travel in 6 hr.	Geostrophic wind	Travel in 6 hr.	Geostrophic wind	Travel in 6 hr.
25	121	728	137	821	164	983
50	61	364	68	410	82	491
75	40	242	46	274	55	328
100	30	182	34	205	41	245
125	24	146	27	164	33	199
150	20	121	23	137	27	164
175	17	104	20	117	23	139
200	15	91	17	103	21	123

When the distance between the isobars is larger than 200 miles, the computed displacements are inaccurate, but the displacements are then less than 125 miles in 6 hr. The geostrophic wind is not suitable for extrapolation purposes between Lat. 35°N. and 35°S., because the representative wind in the equatorial regions may differ considerably from the geostrophic wind.

When the approximate displacements of the pressure systems and the fronts and air masses have been determined, the next point to consider is the changes that will occur while they travel. The changes in the air masses are, as we have seen in previous sections, mainly of the following three kinds:

a. Changes due to the travel of the air toward warmer or colder regions.

b. Changes due to diurnal heating and cooling.

c. Changes due to local influences of the terrain.

The changes *a* may be considerable during a large forecasting period but are usually insignificant as far as aviation forecasts (for the first 6 hr. or so) are concerned. We can, therefore, assume that the air masses will not change their properties essentially during 6 hr., except when they travel from water to land, or from bare ground to snow-covered ground, or from a warm sea current to a cold one. In such cases the changes may be considerable, and one must apply the principles of air-mass analysis in order to determine in which direction and to what extent the masses will change their properties in these cases. This has been explained in the section on air masses.

The changes *b* depend on the season and the time of day. The reader is referred to earlier sections where these phenomena have been explained.

The changes *c* depend mainly on whether the air current blows uphill or downhill and on whether the air is stable or not. This, too, has been explained in earlier sections.

The forecasting procedure, therefore, is the following:

1. To find the fronts, pressure systems, and air masses, and their properties.
2. To determine their travel during the forecasting period.
3. To determine what changes will occur while they travel.

In this way it is possible, *with experience*, to forecast the weather phenomena with fair accuracy.

Glossary

Adiabatic changes in temperature.—If a body of air is moved upward or downward in the atmosphere and no heat is supplied to or withdrawn from it, its temperature will change adiabatically. This change is caused by the expansion or contraction of the air caused by the variation in pressure. A body of non-saturated air will cool

1°C. for each 100 meters it is lifted; this temperature lapse rate is called the dry-adiabatic lapse rate. If the air is saturated it cools less—approximately 1°C. per 200 meters of elevation; this lapse rate is called the moist-adiabatic lapse rate.

Advection.—The process of transfer by horizontal motion.

Aerology.—The branch of meteorology that deals with the free atmosphere.

Air Pockets.—Regions of descending air, upon entering which an aircraft experiences a sudden decrease in lift. Air pockets that occur in connection with irregularities in the wind caused by obstacles may be experienced in the upper air to a height equal to four times that of the obstacle. Most frequently air pockets occur in squally weather, notably in connection with thunderstorms (see also *Bumpiness*).

Altimeter.—An aneroid barometer graduated to show height instead of pressure. Assuming normal distribution of temperature along the vertical, pressure becomes a simple function of height. It is, therefore, possible to graduate the dial to show height instead of pressure (see also *Aneroid Barometer*).

Anemograph.—An instrument for recording the velocity and direction of the wind.

Anemometer.—An instrument for measuring the velocity of the wind.

Aneroid Barometer.—An instrument for measuring the air pressure. It consists of a shallow airtight metal box that expands or contracts in proportion to the variations in pressure. The instrument should be compared frequently with a mercury barometer as its zero is subject to change.

Anticyclone.—A region in which the pressure is high relative to that in its surroundings. The wind circulation in an anticyclone is clockwise, and the weather is usually of a settled type.

Arctic Air.—Air originating over the arctic (or antarctic) fields of ice and snow.

Arctic Smoke.—A fog that occurs, mostly in arctic or antarctic regions, when very cold air streams over open water.

Atmosphere Circulation.—The wind system of the earth as a whole, also called the "general circulation."

Backing.—Counterclockwise change in wind direction. The opposite to *veering*.

Barograph.—A self-recording barometer, usually of the aneroid type.

Barometer.—An instrument for measuring the pressure of the atmosphere.

Blizzard.—High winds accompanied by great cold and drifting or falling snow.

Bumpiness.—A flying sensation usually caused by turbulence in the air. Bumpiness may be caused by (1) vertical currents set up as a result of irregular heating of the earth's surface, (2) irregularities due to roughness of the ground, and (3) change in lift when the aircraft passes a discontinuity in the wind.

Buys-Ballott's Law.—This law states that if, in the northern hemisphere, you stand with your back to the wind, the pressure is lower on your left hand than on your right hand. In the southern hemisphere the reverse is true.

Centigrade.—A thermometric scale where zero is the melting point of ice, and 100° represents the boiling point of pure water at standard atmospheric pressure. A centigrade degree is 9⁄5 of a Fahrenheit degree. To convert centigrade degrees to Fahrenheit degrees, multiply by 9⁄5 and add 32.

Cloudburst.—A sudden and heavy downpour of rain usually associated with thunderstorms.

Col.—The saddle-backed region between two anticyclones and two depressions.

Cold Front.—A line on the weather chart along which cold air replaces warmer air.

Condensation.—The process of formation of water from water vapor.

Convection.—Vertical currents in the air caused by thermal instability.

Convectional Rain, etc.—Rain, etc., caused by convectional currents.

Convergence.—There is convergence into an area or a volume if more air flows into the area or volume than out of it. This is the opposite to *divergence*.

Cyclone.—An area of low pressure. In middle and high latitudes areas of low pressure are most frequently called *depressions*, the word cyclone being reserved for tropical cyclones.

Cyclonic Rain.—Rainfall associated with cyclones.

Density.—The mass of unit volume of a substance.

Depression.—(See *Cyclone.*)

Dew.—Water condensed on grass, leaves, etc.

Dew Point.—The temperature to which air can be cooled without causing condensation.

Divergence.—See *Convergence.*

Doldrums.—The equatorial regions of calms or light variable winds with heavy rains, thunderstorms, and squalls.

Drizzle.—Precipitation of numerous and very small drops.

Dust.—Finely divided earth whirled up by the wind (blowing dust).

Dynamic Cooling.—The fall of temperature caused by expansion due to diminished pressure (see *Adiabatic*).

Eddy.—The deviation from steady flow in any fluid that streams past obstacles, or in streams that flow in contact with one another.

Equatorial Air.—Air originating in the doldrums.

Fahrenheit.—A thermometric scale where the freezing point of water is 32°, and the boiling point of water is 212°. One degree Fahrenheit is 5⁄9 of a degree centigrade. To convert Fahrenheit degrees to centigrade degrees, subtract 32° and multiply by 5⁄9.

Fohn.—A warm, dry wind that blows down the leeward slopes of mountains.

Friction Layer.—The lower part of the atmosphere (usually 1,500 to 3,000 ft. deep) in which the friction along the earth's surface influences the flow of air.

Gale.—A wind of force 8 on the Beaufort scale.

Glazed Frost.—Frozen rain forming a layer of smooth ice upon objects.

Gradient.—The decrease in an element per unit distance. The gradient of pressure is particularly important. The horizontal pressure gradient is directly proportional to the wind velocity. The closer the isobars are to one another the larger is the gradient, and the stronger is the wind. The vertical pressure gradient varies only within narrow limits, being on an average 1⁄8 millibar per meter near the earth's surface. The vertical temperature gradient is expressive of the stability conditions in the air. In recent years the expression *lapse rate* has been given to gradient in the vertical direction.

Gradient Wind.—The wind velocity necessary to balance the pressure gradient. The true wind above the friction layer is approximately equal to the gradient wind.

Gust.—A sudden increase in the velocity of the wind of short duration.

Hail.—Hard pellets of ice precipitated from clouds. Soft hail is a variety that is small, white, opaque, and soft.

High.—A high-pressure area.

Hoarfrost.—Deposit of ice formed in the same way as dew.

Horse Latitudes.—Regions of calms or light variable winds coinciding with the subtropical belts of high pressure.

Humidity.—(See *Relative Humidity.*)

Hurricane.—Wind of force 12 on the Beaufort scale. The name hurricane is also given to violent windstorms originating in tropical regions.

Hygrograph.—Self-recording hygrometer.

Hygrometer.—An instrument for determining the humidity of the air.

Instability.—An air mass is unstable when, if non-saturated, its lapse rate exceeds the dry adiabatic, and, if saturated, its lapse rate exceeds the moist adiabatic. Convection, cumulus clouds, showers, squalls, thunderstorms, gustiness, etc., are phenomena favored or caused by instability.

Inversion.—Layer in which the temperature increases with increasing altitude instead of the normal decrease.

Isallobars.—Lines of equal barometric tendency.

Isobars.—Lines of equal barometric pressure.

Isotherms.—Lines of equal temperature.

Lapse Rate.—Vertical temperature gradient. Usually the decrease in temperature per 100 meters.

Lightning.—The flash of an electrical discharge between two clouds or between a cloud and the earth.

Line Squall.—Heavy squalls occurring simultaneously along a line (cold front). The phenomena characteristic of the passage of a line squall are an arch or line of low black cloud; a rapid rise in wind velocity; a sudden change in wind direction; extreme turbulence and gusts; heavy rain or hail; thunder and lightning; a sudden drop in temperature; a rapid rise in pressure. Line squalls occur along pronounced cold fronts. The clouds are produced by the lifting of the warm air along the cold-front surface, while the drop in temperature and the rise in pressure are the results of the invading cold air. Only very strong cold fronts produce line squalls. Line squalls are dangerous to air navigation chiefly because of the extreme turbulence and electric discharges that accompany them. The clouds usually reach up to great altitudes, and, since they are arranged in a continuous line (often 250 to 400 miles in length), it is often difficult for aircraft to fly over or around them.

Low.—A region of low pressure or a depression.

Mackerel Sky.—A sky covered with cirro-cumulus or alto-cumulus clouds.

Meteorograph.—A self-recording instrument for recording pressure, temperature, and humidity in the free air.

Meteorology.—The science of the atmosphere.

Mist.—A thin fog whose visibility is greater than 1 km. In the United States mist means drizzle or rain of small drops.

Monsoon.—Winds that blow with great persistence in opposite directions at different seasons. They are caused by the annual variation in temperature between oceans and continents. They are analogous to land and sea breezes, but their period is a year instead of a day, and they blow over vast regions.

Nephoscope.—An instrument for determining the direction and velocity of the motion of clouds.

Occlusion.—A front that occurs when the cold front overtakes the warm front in a depression.

Orographic Rain.—Rain caused by air blowing up the slopes of a mountain range.

Ozone.—An allotropic form of oxygen. It is present in considerable quantities in the upper atmosphere above 20 miles above the ground.

Pilot Balloon.—A small free balloon that is set adrift to enable, through observation of its movements, determination of the direction and velocity of the wind aloft.

Polar Air.—Air originating in polar regions.

Polar Front.—The line of separation between air of polar origin and air of tropical origin.

Potential Temperature.—The temperature a specimen of air would attain if it were brought adiabatically to standard pressure (*i.e.*, 1,000 millibars).

Precipitation.—Deposition of solid or liquid water on the surface of the earth.

Rain.—Precipitation of water drops.

Relative Humidity.—The ratio (expressed as a percentage) of the actual amount of water vapor in the air to the amount the air could hold if it were saturated at the same temperature.

Ridge.—An area of relatively high pressure extending from an anticyclone.

Sandstorm.—A strong wind carrying dust or sand.

Scud.—Ragged fragments of low clouds drifting rapidly.

Secondary Cold Front.—A cold front in the polar air, following the primary front.

Secondary Depression.—A small depression on the outskirts of a larger or primary depression.

Shower.—Precipitation falling from convective clouds.

Sleet.—Rain and snow falling together, or melting snow. In the United States sleet means ice pellets or frozen rain.

Smoke.—Particles of foreign matter (resulting from combustion) in the air.

Snow.—Precipitation in the form of minute ice crystals usually falling in irregular masses or flakes.

Specific Humidity.—The number of grams of water vapor contained in 1 kg. of air.

Squall.—A strong wind that rises suddenly and lasts for some minutes. Squalls are of longer durations than gusts.

Stability.—An air mass is stable when its lapse rate is less than the adiabatic rate (see *Instability*).

Storm.—A wind of force 11 on the Beaufort scale. The word is also used in such connections as thunderstorm, snowstorm, rainstorm, dust storm, etc.

Stratosphere.—The upper layer of the atmosphere in which there is no convection. The height of the base of the stratosphere is about 10 miles in the tropics, and about 6 miles in intermediate latitudes. It is higher in anticyclones than in depressions.

Subsidence.—Slow downward motion.

Synoptic Chart.—A weather chart showing the meteorological conditions over a large area at a given moment.

Thermograph.—A self-recording thermometer.

Thermometer.—An instrument for measuring temperature.

Thunder.—The noise accompanying a lightning discharge.

Tornado.—A violent counterclockwise whirl attended by a more or less funnel-shaped cloud.

Trade Winds.—The winds that blow from the subtropical high-pressure belts toward the equatorial region of low pressure.

Tropical Air.—Air originating in low latitudes, notably in the regions of the subtropical anticyclones.

Tropical Cyclone.—Small and violent depression originating in tropical regions, usually of destructive intensity.

Tropopause.—The base of the stratosphere.

Troposphere.—The layer of the atmosphere under the stratosphere. In the troposphere the temperature decreases with height; in the stratosphere the temperature is sensibly constant.

Trough.—A region of relatively low pressure extending from the center of a depression.

V-shaped Depression.—A trough of low pressure bounded by V-shaped isobars.

Vapor Pressure.—The partial pressure of the water vapor in the air.

Veering.—Clockwise change in wind direction. The opposite to *backing.*

Visibility.—The maximum distance at which an object can be discerned.

Warm front.—A line on the weather chart along which warmer air replaces colder air.

Warm Sector.—The part of a depression that is occupied by warm (usually tropical) air.

Waterspout.—Funnel-shaped tornado cloud at sea.

Wedge.—An area of relatively high pressure extending from an anticyclone.

Wind.—Motion of the air.

CHAPTER XI

CELESTIAL NAVIGATION—THEORY

> The radio beam's best effort cannot, to me, compare with the stark independence and simplicity of the star navigator's apparatus and methods. He depends on no fallible mass of "electrickery" and its human crew. In "this majestical roof, fretted with golden fire" (Shakespeare haunts this page today!) there are a million leading lamps trimmed by a steadier than mortal hand and located in eternity with divine exactitude.

Introductory Remarks.—Having covered the methods of navigation by pilotage, dead reckoning, and radio, we now arrive at that of *celestial navigation*, the determination of position by means of celestial bodies.

Before we can explain the theory of the subject to the beginner in a clear manner, we must use known terms. Like learning any strange machine, game, or process, the beginner must first learn new definitions and the principal "rules of the game." Once the definitions and rules are mastered, celestial navigation becomes easy. In Chap. I we discussed position, direction, and distance on the earth's surface. We shall now discuss briefly the universe outside the earth.

Celestial Sphere.—When we step off the earth into space, there is no theoretical limit to this space; but for practical purposes we may consider that it has limits and we may indicate it in a sketch, as in Fig. 185. With the earth and observers on its surface at the assumed center of the universe, the eye of an observer projects all heavenly bodies against the huge dome overhead which we call the *celestial sphere*.

A navigator in a plane flying over the earth's surface sees the various heavenly bodies as if they were situated on the interior surface of this enormous hollow sphere. By imagining this spherical surface to be of infinite radius, the position of objects in the heavens can be projected on its inner surface. With the sphere of infinite dimensions, the radius of the earth is so small as to be negligible, so the further assumption is made that the eye of the observer is actually at the center of the earth. With the eye in this position, not only the heavenly bodies, but points on the earth's surface, parallels of latitude, and meridians of longitude on the earth may be projected on this celestial sphere.

Definitions and Abbreviations.—The *zenith* (*Z*) of an observer is the point of the celestial sphere vertically overhead.

The *nadir* (*Na*) is the point of the celestial sphere directly beneath an observer.

The *celestial horizon* is the great circle of the celestial sphere formed by passing a plane through the center of the earth perpendicular to the straight line joining the zenith and nadir. The celestial horizon differs somewhat from the visible horizon, which is the line appearing to an observer as marking the intersection of earth and sky. This difference arises from two causes:

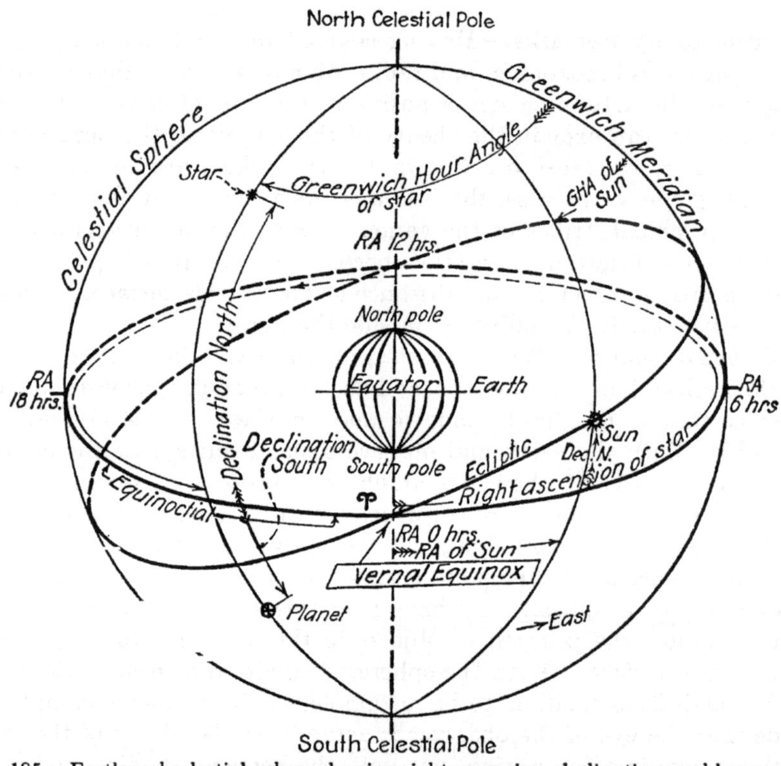

FIG. 185.—Earth and celestial sphere showing right ascension, declination, and hour angle.

1. The eye of the observer is always elevated above the sea level, thus giving him a range of vision exceeding 90° from the zenith.
2. The observer's actual position is on the surface of the earth instead of at its center.

These two causes give rise, respectively, to dip of the horizon and parallax, which will be explained in Chap. XIV.

When flying over land the visible horizon is of no practical value for navigation because of its varying elevations. When flying over the sea at low altitudes, the visible horizon may be used for observations.

An *artificial horizon* is the true celestial horizon determined by some mechanical device. In air navigation the usual practice is to determine the horizon by some optical means, such as the bubble built into the sextant.

Vertical circles or *circles of altitude* are great circles of the celestial sphere which pass through the zenith and the nadir. These circles are perpendicular to the horizon.

The *prime vertical* is the vertical circle whose plane is at right angles to the plane of the meridian and intersects the horizon at its east and west points.

The *altitude* (*h* or alt.) of any point on the celestial sphere is its angular distance from the horizon measured upon the vertical circle passing through the point.

The *zenith distance* (*z*) of any point is its angular distance from the zenith measured upon the vertical circle passing through the point. The zenith distance of any point that is above the horizon is therefore equal to 90° minus the altitude.

The *azimuth* of any point on the celestial sphere is the angle at the zenith between the meridian of the observer and the vertical circle passing through the point. It is usually measured from north to the right from 0° to 360°. Also, it is sometimes measured from the north or south point of the horizon to the east or west through either 90° or 180° and is named accordingly; as N. 30°E., S. 110°W.

The *equinoctial* or *celestial equator* is the great circle of the celestial sphere formed by extending the plane of the earth's equator until it intersects the celestial sphere. The celestial equator intersects the horizon at its east and west points.

The *celestial poles* are the projections of the north and south poles of the earth upon the celestial sphere. To an observer on the earth in north latitude the north celestial pole is above the horizon; the south celestial pole is below the horizon. Hence we often refer to the "elevated pole," which is the pole of the same name as the observer's latitude upon the earth.

Hour circles (sometimes called *declination circles*) are great circles of the celestial sphere passing through the poles. They are, therefore, formed by projecting the meridians of longitude on the earth to the celestial sphere. The hour circle containing the zenith is the *celestial meridian* of the observer.

The *declination* (*d* or dec.) of any point on the celestial sphere is its angular distance from the celestial equator measured on the hour circle that passes through that point; it is designated as north or south according to the direction of the point from the celestial equator. North declinations are regarded as positive and are written with the plus sign

(+); south declinations are considered as negative and are written with the minus sign (−). It should be remembered that declination upon the celestial sphere corresponds to latitude on the earth.

The *polar distance* (p) of any point is its angular distance from the pole, measured on the hour circle passing through the point. It must therefore be equal to 90° minus the declination, if measured from the

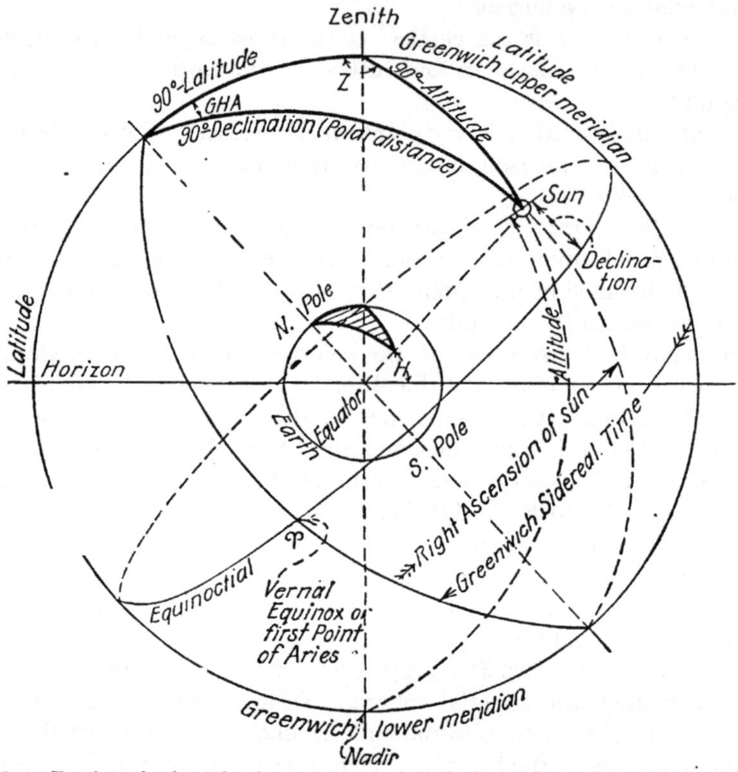

FIG. 186.—Earth and celestial sphere showing altitude, latitude, and the celestial triangle.

pole of the same name as the declination, or 90° plus the declination, if measured from the pole of opposite name (see Fig. 186).

The *hour angle* (t or HA) of any point on the celestial sphere is the angle at the pole between the meridian of the observer and the hour circle passing through the point; it is measured by the arc of the celestial equator intercepted between those circles. Hour angle is measured toward the west as a positive direction through 24^h or 360°, or it may be measured from 0^h to 12^h or 0° to 180° east or west. It is sometimes called *local hour angle* (LHA), to distinguish it from Greenwich hour angle.

The *Greenwich hour angle* (GHA) of a heavenly body is the angle at the pole between the meridian of Greenwich and the hour circle of the body. It is measured along the celestial equator from the meridian of Greenwich to the west from 0° to 360° (see Fig. 185).

The *ecliptic* is the great circle representing the path on the celestial sphere in which the sun appears to move by reason of the annual revolution of the earth. The plane of the ecliptic is inclined to the plane of the celestial equator at an angle of 23°27½′.

The Vernal Equinox.—Since there is no fixed point in the heavens to correspond exactly with Greenwich, which is the origin of time and longitude for the earth, a point has been chosen, called the *First Point of Aries* or the *vernal equinox* and designated by the sign of the ram's horn, ♈. This point is the intersection of the ecliptic (the path of the sun) with the celestial equator in the spring when the sun is traveling north.

The *autumnal equinox* is the point of intersection, as the sun passes from north to south declination on or about September 23.

The *right ascension* of any point on the celestial sphere is the angle at the pole between the hour circle passing through the point and the hour circle passing through the vernal equinox. It is measured from the vernal equinox positive to the eastward from 0^h to 24^h or 360° (as shown in Figs. 185 and 186). *Right ascension* is of great importance to astronomers but of little interest to navigators who use *hour angle* instead.

Sidereal hour angle (SHA) is 360° minus *right ascension*, expressed in arc. This is an important new term for navigators. The positions of stars are given in terms of SHA and declination. The SHA of the observed star *added* to the GHA♈ gives the GHA of the observed star. See details about this in Appendix B.

Systems of Coordinates.—To define the position of a point on the surface of a sphere, some great circle must be selected as the primary, and some particular point of it as the origin. Then a series of great circles that intersect the primary at right angles are chosen as secondaries. Then the position of the point is defined by two coordinates. First, the distance along the primary measured from the origin to the secondary circle passing through the point; second, the distance along the secondary measured from the primary to the point.

In the case of the earth, it was explained in Chap. I that the primary selected is the equator and the origin is the point at which the meridian of Greenwich cuts the equator. The secondary circles are the meridians of longitude. The longitude is measured along the primary, which is the equator from the meridian of Greenwich to the secondary passing through the point. The latitude is then measured along the secondary from the equator to the point.

In the case of the celestial sphere there are two systems of coordinates in use in air navigation for defining the position of any point, as follows:

1. Altitude and azimuth.
2. Declination and hour angle.

In the system of *altitude and azimuth*, the primary circle is the celestial horizon, the secondaries to which are the vertical circles or circles of altitude. The horizon is intersected by the celestial meridian in its northern and southern points. One of these two points—usually that adjacent to the elevated pole—is selected as an origin for reckoning coordinates. The azimuth indicates in which vertical circle the point to be defined is found, and the altitude gives the position of the point in that circle measured from the horizon.

In the system of *declination and hour angle*, the primary circle is the celestial equator, the secondaries to which are the hour circles. The origin is that point of intersection of the celestial equator and the observer's celestial meridian which is above the horizon. The hour angle indicates in which hour circle the point to be defined is found, and the declination gives the position of the point in that circle, measured from the equator.

Geographical Position (G.P.).—The *geographical position* of a heavenly body is the point on the earth's surface that has the body in its zenith; in other words it is the *substellar, subsolar*, or *sublunar* point. Its latitude is defined by the declination and its longitude by the Greenwich hour angle (reckoned westward from 0° to 360°) of the body as taken from the Almanac. This is an important definition, and it is unfortunate that we do not have a less awkward name for it, although we shall frequently refer simply to the *position* of the body concerned.

With the modern use of the *terrestrial* triangle instead of the *celestial* triangle in Fig. 186, most of the terms pertaining to the celestial sphere are replaced by terms applying to the earth

Celestial Triangle	Terrestrial Triangle
Celestial horizon	Terrestrial horizon
Celestial equator	Earth's equator
Celestial poles	Terrestrial poles
Hour circles	Meridians of longitude
Declination	Latitude of the geographical position of the body
Polar distance	Distance from the pole to the geographical position of the body
Greenwich hour angle	Longitude of the geographical position of the body
Local hour angle	Difference of longitude of the observer and the geographical position of the body

Relation of the Solar System to the Universe.—The following facts should be kept in mind:

1. The moon is the only heavenly body near the earth (mean distance 239,000 miles).

2. The entire solar system, consisting of the sun, planets, and the moons or satellites of the planets, may be considered an infinitesimally small fraction of the universe.

3. For practical purposes the earth may be considered a point at the center of the universe.

These facts make it clear why the stars, which are nearly stationary in the heavens relative to each other, are more convenient than bodies of the solar system for determining position, since the latter are continually moving, thus necessitating increased tabulation and interpolation.

Causes of the Apparent Motions of the Heavenly Bodies.—An observer on the earth's surface is constantly changing his position relative to the heavenly bodies projected on the celestial sphere. This gives these bodies an apparent motion, due to four causes:

1. The diurnal (daily) motion of the earth due to its rotation on its axis.

2. The annual motion of the earth arising from its revolution about the sun in its orbit.

3. The orbital motion of bodies in the solar system, and the actual motion in space of the stars, known as "proper motion."

4. The motion of the observer on the earth's surface.

The changes in position produced by the diurnal motion are different for observers at different points on the earth and depend on the latitude and longitude of the observer. But the changes arising from causes 2 and 3 are independent of the observer's position and would be the same for an observer at the center of the earth. Hence, *Air Almanac* tabulations are for the center of the earth and are based on the gravitational laws that have been found by long years of observation to govern the actual and apparent motions of the various heavenly bodies.

An observer traveling over the surface of the earth at, say, 180 knots directly toward a celestial body would superpose an increase of altitude at the rate of 3′ per minute on the changes of altitude due to other causes.

Methods of Reckoning Time.—The instant at which any point of the celestial sphere or any heavenly body is on the meridian of the observer is known as the *transit, culmination,* or *meridian passage* of that point; when on that half of the meridian which contains the zenith, it is designated as *superior* or *upper* transit; when on the half of the meridian containing the nadir it is known as *inferior* or *lower* transit; the two parts of the meridian are known as the upper and lower branches, respectively.

The rotation of the earth on its axis from *west* to *east* causes the heavenly bodies to *appear* to revolve round the earth from *east* to *west*. The period of this apparent revolution is conveniently measured by the time elapsing between two successive transits of a heavenly body over the same meridian, and is called the day, deriving its name from the body whose apparent revolution is considered; thus, in the case of the sun, it is called a *solar* day, and in the case of the moon a *lunar* day, and in that of a star a *sidereal* day.

Units of Time.—The basic unit of time is the *day,* which is the period between two successive transits of the point of reference over the same branch of the meridian. The day is divided into 24 equal parts, called *hours;* each hour is divided into 60 equal parts called *minutes;* and each minute is divided into 60 equal parts called *seconds.* These are called units of time. Since the reference point in a day travels in a complete circle from the meridian and back to it again, it follows that a day, or 24 hr., is equal to 360°. It is sometimes convenient to express time in units of arc or in degrees, minutes, and seconds.

As 24 hr. equal 360°, it follows that 1 hr. is equal to 15°. Time may therefore be converted into arc by multiplying hours, minutes, and seconds of time by 15, to get degrees, minutes, and seconds of arc. Thus $8^h00^m00^s$ may be written 120°00′00″ and $4^h15^m22^s$ may be written 63°50′30″, or better, 63°50′.5.

Civil Time.—The time which we measure by watches, clocks, and chronometers, and which we use to regulate our daily lives, is *mean* or *civil time.* A *civil day* is the interval between two successive transits of the mean sun over the lower branch of the meridian.

The real sun, as explained in the preceding chapter, appears to move in a path known as the ecliptic, which is inclined to the equinoctial at an angle of 23°27½′. The apparent movement of the sun in the ecliptic is variable because of the varying speed of the earth in its orbit. This variable movement of the sun becomes still more variable when projected on the equinoctial on which all hour angles are measured.

Since the real sun's rate of change of hour angle is not uniform, it would not furnish a practical measure for time, since the units of time determined by it would be constantly changing in length. To avoid this disadvantage, time measurement is based on the motion of an imaginary sun called the *mean sun,* which moves to the eastward in the equinoctial at a uniform rate equal to the average rate of the true sun in the ecliptic.

Sidereal Time.—*Sidereal* or *Star time* is the hour angle of the vernal equinox. Since the position of the vernal equinox is fixed in the celestial sphere, except for precession (which is here disregarded), and does not, like the sun, moon, and planets, have actual or apparent motion, it is similar in this respect to the fixed stars.

A *sidereal day* is the interval between two successive transits of the vernal equinox across the upper branch of the same meridian. There are 366.24 sidereal days in a year and only 365.24 solar days. Thus, since 366.24 sidereal days equal 365.24 solar days, a sidereal day of 24 hr. is about 3 min. 56.6 sec. less than a solar day. This is due to the passage of the earth once round the sun during the year.

The Time Diagram.—The simplest way for a beginner to grasp the definitions and problems in time and hour angle is to study a sketch called the *time diagram.* If it is assumed that the observer's eye is at the south celestial pole and that he is looking at the celestial sphere projected on the plane of the celestial equator, the simple sketch of the actual conditions which he sees constitutes the *time diagram.*

The polar axis of the earth appears as a point at the center of a circle formed by the celestial equator, as shown in Fig. 187. The earth

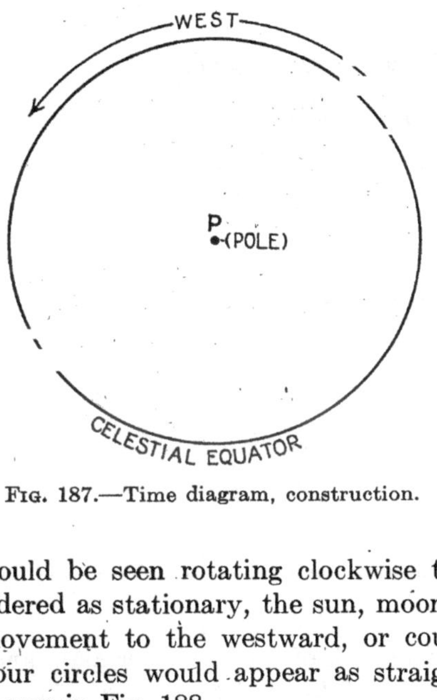

Fig. 187.—Time diagram, construction.

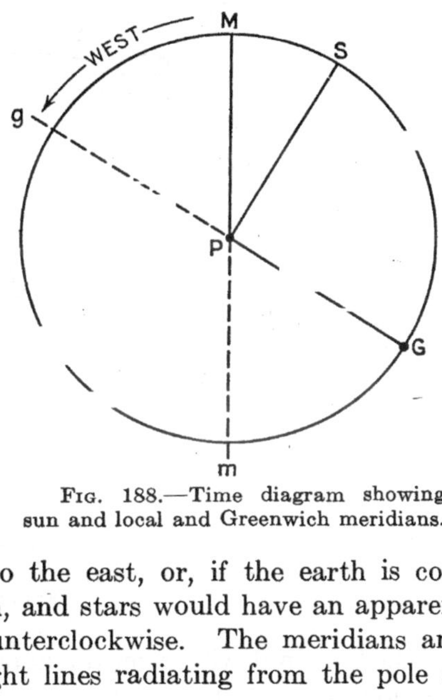

Fig. 188.—Time diagram showing sun and local and Greenwich meridians.

would be seen rotating clockwise to the east, or, if the earth is considered as stationary, the sun, moon, and stars would have an apparent movement to the westward, or counterclockwise. The meridians and hour circles would appear as straight lines radiating from the pole as shown in Fig. 188.

The upper vertical radius of the circle represents the meridian of the observer as shown by *PM* in Fig. 188. The required angles are set off from this line to get the position of the various other hour circles and meridians.

The lower branch of the observer's meridian, which is located 12 hr. from the upper branch and from which civil and apparent time are reckoned, is shown by the dotted line *Pm* in Fig. 188. The Greenwich meridian may be located by making the angle *MPG* equal to the longitude. In the case shown in Fig. 189 the longitude is 5 hr., or 75°W. In the same way the meridian of any other place with a known longitude may be shown as a straight line. The position of a heavenly body may be shown by drawing its hour circle. The straight line *PS* represents the hour circle of the sun in Fig. 188.

In the case of the sun, we can readily find its position relative to any given meridian by reading a watch set to the time of that given meridian. At the start of the civil day at midnight the watch will read 12 o'clock or 0^h and we know that the sun is on the lower branch of the local meridian or at the point *m*. An hour later by watch the sun will be 1 hr. or 15° west of *m*. At 6^h, it will be 90° west of *m*, and at

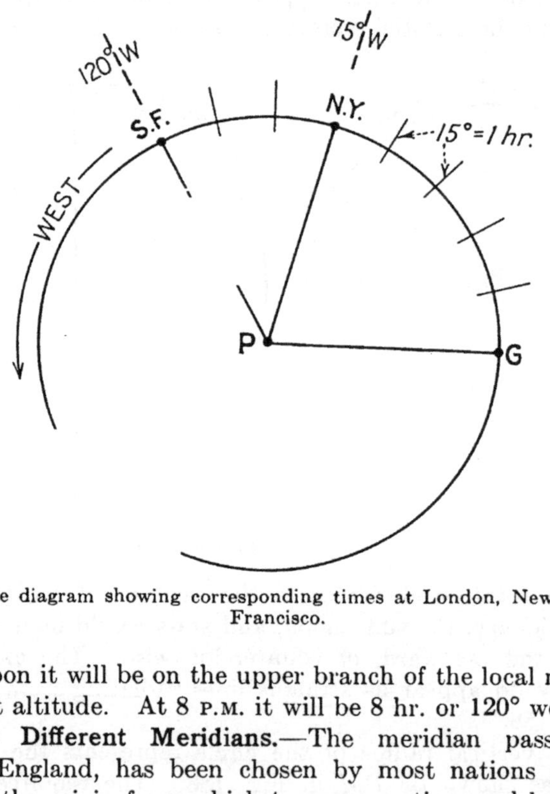

FIG. 189.—Time diagram showing corresponding times at London, New York, and San Francisco.

12 o'clock noon it will be on the upper branch of the local meridian, and at its highest altitude. At 8 P.M. it will be 8 hr. or 120° west of *M*.

Time at Different Meridians.—The meridian passing through Greenwich, England, has been chosen by most nations as the *prime meridian*, or the origin from which to measure time and longitude. By remembering that a day is 24 hr. and that in a day the three reference points for measuring time start from a given meridian, cover the complete 360° of longitude, and are back again on the same meridian, it is evident that 1 hr. must be equal to 360/24 = 15° of longitude. That is, in 1 hr. the sun apparently moves to the westward 15° of longitude. Therefore, if the observer is 15° in longitude from Greenwich, there must be a difference of 1 hr. between his local time and the Greenwich time. For instance, if it is 12 o'clock noon at Greenwich and the observer is 15° west of Greenwich, it is only 11 o'clock at the observer's meridian.

The time diagram given in Fig. 189 will make clear the different times that would be shown at any given moment on clocks in London, New York, and San Francisco. Suppose the moment is selected when the sun is on the local meridian of New York, approximate Long. 75°W. Then from the diagram it is seen that the sun has passed over 75° since it was over Greenwich, and, since the sun travels at the rate of 15° per hour, 5 hr. have elapsed since Greenwich noon and it must be 5 P.M. by Greenwich civil time. San Francisco, on the other hand, is in approxi-

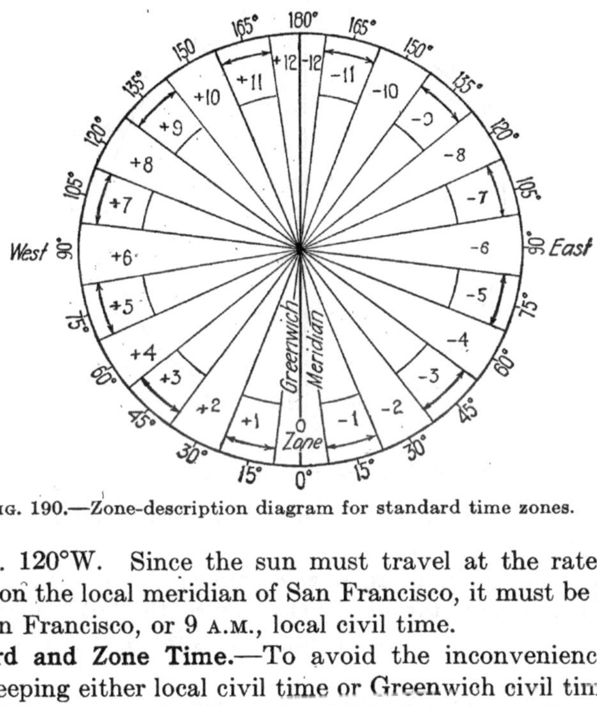

FIG. 190.—Zone-description diagram for standard time zones.

mate Long. 120°W. Since the sun must travel at the rate of 15° per hour to be on the local meridian of San Francisco, it must be 3 hr. before noon at San Francisco, or 9 A.M., local civil time.

Standard and Zone Time.—To avoid the inconvenience and confusion of keeping either local civil time or Greenwich civil time, a system of standard time zones has been established to cover the earth. The surface of the earth is conceived to be divided into 24 zones, each bounded by meridians 15° of arc or 1 hr. of time apart in longitude. The initial zone is the one that has the meridian of Greenwich running through the middle of it, and the meridians 7½° east and 7½° west of Greenwich marking its eastern and western limits. It is called the *zero zone*, because the difference between the standard time of this zone and the Greenwich civil time is zero. Each of the zones in turn is designated by a number representing the number of hours by which the standard time of the zone differs from Greenwich civil time.

The number of a zone prefixed by the word "plus" or the plus sign (+) or by the word "minus" or the minus sign (−) constitutes the *zone description* of the time of that zone. The zone-description diagram for the standard time zones is given in Fig. 190.

Greenwich Date.—The elements are tabulated in the "Nautical Almanac" for certain intervals of Greenwich civil time for each day of the year. Therefore, to use the tabulated data it is necessary to know the Greenwich date as well as the Greenwich civil time. Since the Greenwich civil date is not always the same as the local civil date, it becomes

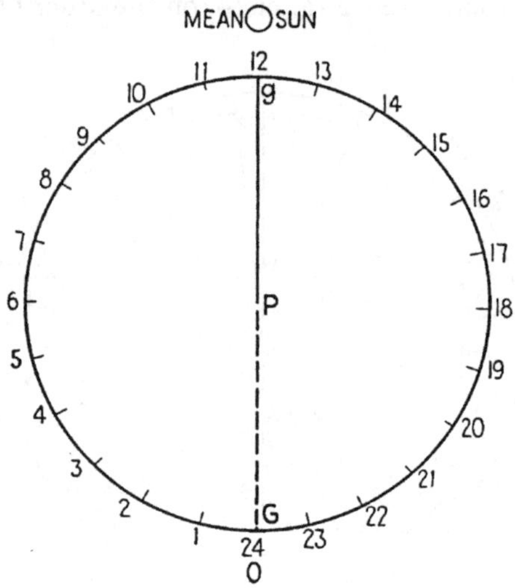

FIG. 191.—Time and date over the world at Greenwich noon.

necessary for the navigator to know how to turn local civil time and date into Greenwich civil time and date.

To avoid confusion in dates, an *international date line* has been established near the middle of the Pacific Ocean near the 180th meridian, and the date is changed 1 day when moving across this line to either the east or west.

Since the watch is set *ahead* for each change of 15° of longitude to the east, the date correction is made by setting the date back one full day when crossing the date line moving to the east. In other words, *going east* the watch is set ahead by the hour for each 15° change of longitude, and the date is set back a day on crossing the international date line. Going west, the watch is set back by the hour for each change of 15° of longitude, and the date is set ahead a day on crossing the international date line.

The instant of Greenwich civil noon is worth noting for it is at this instant, and this instant only, that the same date prevails throughout the earth. This may be seen by referring to Fig. 191.

An important use of the time diagram (Fig. 192) is to settle the Greenwich date. For this purpose *M*, *G*, and *g* are drawn; now mark *m* opposite *M*. Then the date changes at Greenwich as the sun *S* moves from west to east through *g*, and it changes at the observer's locality as *S* passes *m*. Now mark off the rough position of *S*. Then *S* will lie to one side or other of the line *Gg*, and it will be morning or afternoon at Greenwich according to whether *S* lies to the west or east of *g*. Similarly, it will be morning or afternoon at *M* according to whether *S* lies to the west or east of *m*.

A little thought will show that it is the same day at *G* and *M* when *S* does not lie on the shorter arc between *m* and *g*.

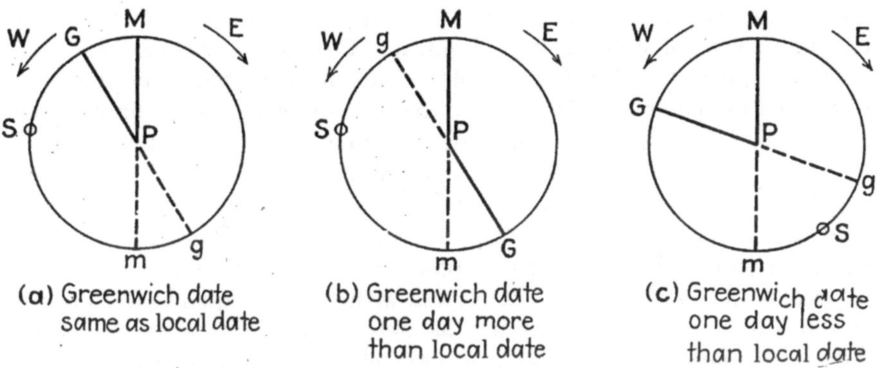

FIG. 192.—Using time diagram to determine Greenwich date.

If *S* lies between *m* and *g*, there will be one day difference in date between *M* and *G*; if the position of *S* is such that it has passed *g* but not reached *m* in its journey in the westerly direction, the Greenwich date is one day more than the local date; if it has passed *m* but not reached *g*, the Greenwich date is one day less. In other words, if we go around the outer circle in a counterclockwise direction and find *S* on the shorter arc between *m* and *g*, the more advanced date belongs to *m* or *g*, whichever the sun has passed.

Local Hour Angle, LHA.—The navigator who has to determine his position by celestial observation is concerned primarily with his local hour angle, and the relationship between this, his longitude, and the GHA is given by the equation:

$$\text{LHA} = \text{GHA} - \text{Long. W.}$$

which can be written in the alternative form:

$$\text{LHA} = \text{GHA} + \text{Long. E.}$$

The Theory of Celestial Navigation.—An observer at any point on the earth's surface, except the poles, sees the altitude of any heavenly body change with time. The altitude can be calculated for any instant of time by methods to be described later.

The altitude of a heavenly body also changes with a change in the position of the observer. The amount of this change depends on the direction in which the observer moves relative to the geographical position (G.P.) of the observed body. If he moves 1 mile directly toward the observed body, the altitude is thereby increased by 1′; if he moves away, the altitude is decreased by 1′; if he moves in any other direction, the altitude is changed by a predictable amount.

The converse is also true, namely, that, if the altitude of a heavenly body is known, the distance from the geographical position of the body for the instant of observation is known. Celestial navigation is based on the application of this principle. Therefore we may say that the work of the navigator in celestial navigation consists in finding his geographical coordinates of latitude and longitude by locating himself with reference to the geographical position of one or more celestial bodies.

The exact position of a body, in terms of declination and Greenwich hour angle, may be taken from the Almanac for any instant of time, and this position may be plotted on a globe or chart as latitude and longitude. A circle with this position as center, and with a radius equal to the observer's distance from this position as determined by the sextant (*i.e.*, 90° − altitude), would be the observer's circle or position; the observer must lie somewhere on this circle. If two bodies were observed, the intersections of the circles of position would determine the position (or two possible positions) of the observer.

Since the circle of position as determined above would be very large, a small arc of the circle may be considered as a straight line, at some point on which is the observer's position. If it were possible to lay off on a chart with sufficient accuracy the distance of the observer from the geographical position of the body, celestial navigation would be easy. The observer would doubtless use a sextant graduated to read zenith distance (instead of altitude) in minutes, or nautical miles, and this distance would simply be used as the radius of an arc to be laid off on a chart from the plotted position of the body. Unfortunately, the great distances—up to 5,400 miles—to be plotted make this simple and obvious method impracticable, but the desired effect is secured by the means about to be described.

The Navigation Triangle.—In Figs. 193, 194, and 195, which are representations of part of the earth's surface, *P* is the elevated pole, *Z* the position of the observer, and *S* the geographical position of a star. These three points form the vertices of a spherical triangle *ZPS*, which is known as the *astronomical* or *navigation triangle.* The side PS is the polar distance (90° − declination) of the star, ZS is the zenith distance (90° − altitude) of the star, and PZ is the colatitude (90° − latitude) of the observer. The angle ZPS at the pole is the local hour angle, and the angle PZS at the zenith is the azimuth of the star. The angle PSZ at S, which is seldom used, is called the parallactic angle.

Solving the Astronomical Triangle.—The quantities from which the astronomical triangle is usually solved are the sides PZ and PS, and the included angle at P. The assumed latitude subtracted from 90° gives the side PZ; the declination subtracted from 90° gives the side PS. The angle at P is the difference between the longitude of the body (geo-

FIG. 193.—The astronomical triangle.

FIG. 194.—Circles of equal altitude.

graphical position) and the assumed longitude of the observer, which is the local hour angle.

By specially arranged tables, the latitude, declination, and local hour angle are used directly in solving the triangle for the computed altitude, which is 90° minus the side ZS, and the azimuth. The computed altitude is then compared with the measured altitude to determine the position of the observer relative to the assumed position.

Circles of Equal Altitude.—Since the altitude of a heavenly body would be the same for every position on the earth's surface equidistant from the body's geographical position, it follows that one altitude will not give definite fix; nevertheless, it will give a circle on which the observer's position must be.

In Figs. 194 and 195, S is the geographical position of a star at a given time. If the observer were at S at this instant, he would see the star in his zenith, *i.e.,* the observed altitude would be 90°. But if he were, say, 2,700 nautical miles away from S, his zenith would be over some point Z on the altitude circle instead of at S. Now we know that 1 nautical mile represents an angle of 1′ of arc, so 2,700 nautical miles represent an angle of 45°. Thus it is clear that the star *S* would be 45° away from the observer's new zenith. Trials would soon indicate that there were a great number of positions on the earth from all of which this heavenly body would have the same zenith distance of 45°; and all these positions would lie on a small circle having the geographical position of the star as its center. Such a circle is known as a *circle of equal altitude*. Having measured the altitude of a star and knowing its geographical position, the navigator can find the circle of equal altitude upon whose circumference he must be situated. By immediately observing a second star he can determine a second circle of equal altitude, which also passes through his position. The intersection of the two

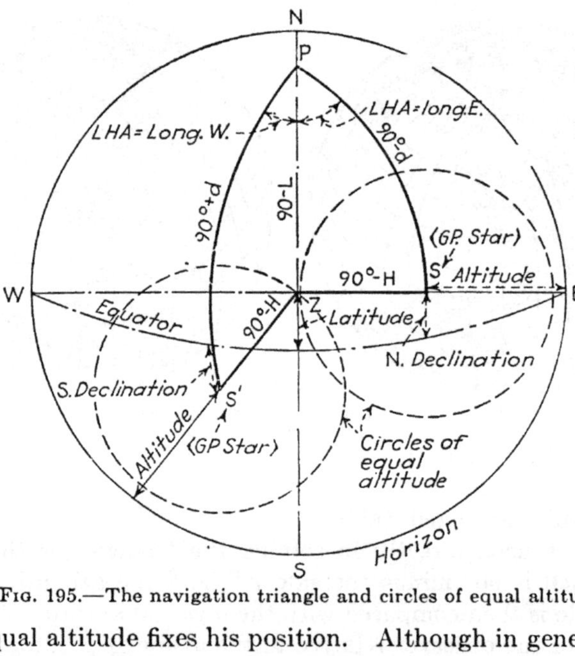

Fig. 195.—The navigation triangle and circles of equal altitude.

circles of equal altitude fixes his position. Although in general there are two intersections of these circles, the navigator always knows his position closely enough to eliminate one of them.

The Line of Position.—It is not usual to include the whole earth on any one chart; only that part of immediate interest to the navigator is shown. Hence the whole circumference of circles of equal altitude would not appear; as the radius of the circle is usually large, the small, slightly curved arc in which the navigator is interested may be replaced by a straight line without appreciable error. This line is called a *line of position* or *position line* or *Sumner line* after Captain Thomas H. Sumner, an American shipmaster, who discovered it in 1837. The direction of this line of position on the map must clearly be at right angles to the radius joining the navigator's position to the geographical position of the heavenly body; in other words, at right angles to the true bearing or azimuth of the heavenly body at the instant at which the observation was made. The various methods of finding the line of position are discussed in Chap. XIII. All the methods of interest to the navigator are based on the method of Marcq St. Hilaire, which makes use of the difference between the observed altitude and the altitude computed with the aid of an assumed position.

Plotting a Position Line on a Chart.—To lay down a line of position on a chart by the method of Marcq St. Hilaire, it is necessary to take the dead-reckoning position, or some convenient point close to it, and to compute for this assumed position the altitude and azimuth of the observed body at the time at which the observation was made. If the altitude observed with the sextant agrees exactly with the computed altitude, the altitude difference is zero. Then it is only necessary to draw from the assumed position A (Fig. 196) a line of bearing AB along the computed azimuth of the body and a line CAD at right angles to this; then CAD is the required line of position.

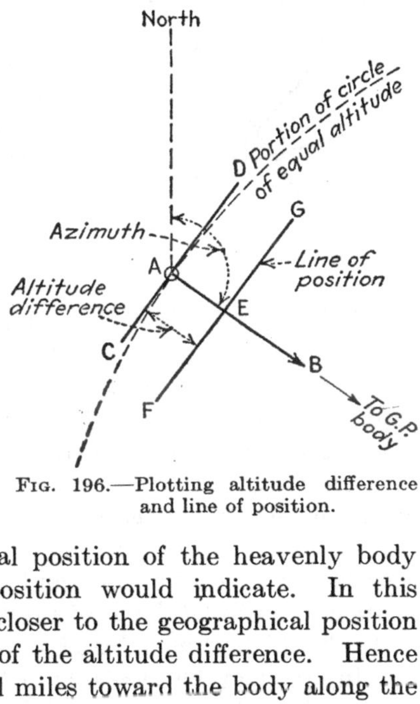

Fig. 196.—Plotting altitude difference and line of position.

If, however, the observed altitude is greater than the computed altitude, say by 15′, which equals 15 nautical miles on the earth's surface, the observer is nearer to the geographical position of the heavenly body than the assumed dead-reckoning position would indicate. In this case the position line must be moved closer to the geographical position of the heavenly body by the amount of the altitude difference. Hence the intercept AE is laid off 15 nautical miles toward the body along the computed azimuth, and the position line FEG is drawn, as before, at right angles to azimuth line AB.

The Fix.—If two position lines similar to FG in Fig. 196 are found, the actual position or fix, assuming that there are no errors in the work, will be at their intersection. If three position lines such as AB, CD, and EF in Fig. 197 are determined, the true position may be reckoned to be somewhere within the shaded triangle or "cocked hat."

To determine a number of lines of position, it is necessary to take nearly simultaneous observations of different heavenly bodies or to get radio bearings of stations of known position. It should be remembered that the nearer lines of position are at right angles to each other, the sharper and more clear-cut their intersection will be. As the angle between two lines of position diminishes, the intersection becomes less clearly defined, until, at 15° or less, little faith can be placed in the fix obtained.

Whenever stars are visible, it is an easy matter to select a pair of stars whose position lines will give a good "cut." During the daytime, however, the sun is usually the only heavenly body available. In marine navigation it is the practice to derive a position line from two observations of the sun several hours apart; by moving the first line parallel to itself for a distance equal to the estimated run between the observations, it is made to cut the second line and thus give a *running fix*.

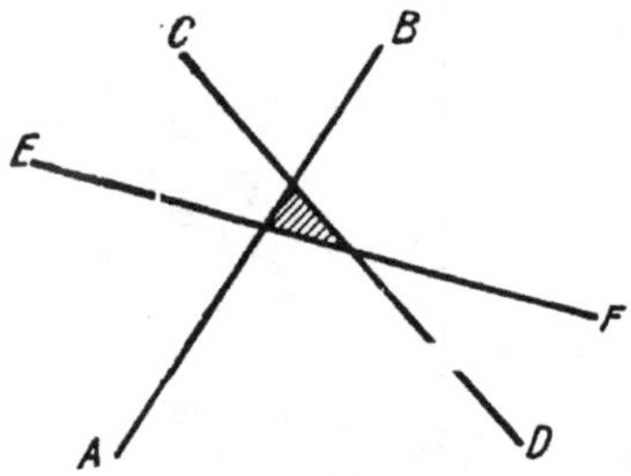

FIG. 197.—Obtaining a fix.

In the air, however, this is of doubtful value because the high speed of the aircraft, and possibly of the wind, makes it impossible to estimate accurately the run between observations. It will usually be found more satisfactory to consider these lines separately in celestial navigation. A single position line will almost always give valuable information and often all that the navigator requires.

Consider the case where the position line is parallel to the course of the aircraft. Then its distance from the course, as laid down by dead reckoning, will give the navigator an indication of the accuracy of his dead reckoning and perhaps warn him of a change in the direction or intensity of the wind. Similarly, if the position line is perpendicular to the course, the navigator can determine the approximate distance he has traveled. In this case the course plotted by dead reckoning may be regarded as a position line, and its intersection with the astronomical position line will be the most probable position of the aircraft.

It should be remembered that at certain times the moon, Venus, and Jupiter may be available in the daytime and may be used to give additional position lines that will cut the sun line, or cut each other.

Position Lines by Azimuths.—The careful reader may wonder why the navigator does not get a fix by using the observed altitude of a heavenly body in conjunction with its observed azimuth. The answer is that he can do so, but it is not worth while, because the accuracy so obtainable is very poor. Also owing to convergence of meridians, it would be difficult to plot the azimuth line accurately.

The altitude can be measured with sufficient accuracy, but the azimuth cannot be obtained with a comparable degree of accuracy. On an airplane the azimuth determination of a body at altitudes up to about 30° is accurate to about 1°; this would give a range of intersection with a position line at right angles to the azimuth of 50 to 60 miles each way.

This inaccuracy arises from the great distance of the geographical position of the body from the observer's position. As soon as this distance is decreased by using another body having a nearer geographical position, the altitude of the body is increased so that, although the distance is then small enough to reduce the error due to distance to a reasonable amount, the altitude is so great that the observation of azimuth becomes very inaccurate owing to the effect of errors of the level of the observing instrument.

Limit of Accuracy.—To be of practical use to the navigator, the results obtained by celestial navigation must fall within the limits of speed and accuracy required in the air. We may approach the problem by saying that if a position within 5 miles may be obtained in 5 min. or less, then celestial navigation is of real value to aviation.

Further experience may change these figures somewhat, but the principle is correct, namely, that methods that are to prove useful in navigation must be within certain limits of speed and accuracy. Actually, a line of position may now be worked in the air in less than 3 min. during the daytime; if working with stars during the night, a definite fix may be determined in about 2 min., with an average accuracy of about 5 miles.

Summary.—Briefly, celestial navigation consists in finding and plotting on a chart one or more lines of position. One line of position gives the navigator certain useful information. Two or more lines of position determine by their intersection a definite position or fix. To determine a line of position, the navigator must take the following principal steps:

1. Observe the sextant altitude of a known heavenly body.
2. Note the exact time of the observation.
3. Take the declination and Greenwich hour angle of the observed body from the Almanac for the GCT of observation.
4. Apply the assumed longitude to the Greenwich hour angle to find the local hour angle.
5. Compute the altitude and azimuth, using as arguments, the assumed latitude, hour angle, and declination.
6. Compare the computed and observed altitudes to obtain the altitude difference.
7. Set off the altitude difference along the computed azimuth and draw the line of position at right angles to it.

CHAPTER XII

CELESTIAL NAVIGATION—EQUIPMENT

All the equipment discussed in Chap. VII for dead reckoning is required by the celestial navigator. In addition, four principal items that are applicable solely to celestial navigation are needed:

1. The sextant.
2. The accurate timepiece.
3. The Almanac.
4. Methods of converting observations into positions

Figure 198 shows a sample celestial-navigation outfit.

THE SEXTANT

The sextant is an instrument for measuring the angle between two objects by bringing into coincidence at the eye of the observer rays of light received directly from the one object and by reflection from the other object, the measure being afforded by the inclination of the reflecting surfaces relative to each other.

Optical Principle.—One of the fundamental principles of optics is that "when a ray of light is reflected from a plane surface, the angle of incidence is equal to the angle of reflection." From this it may be proved that when a ray of light undergoes two reflections in the same plane the angle between its first and its last direction is equal to twice the inclination of the reflecting surfaces. Upon this fact the construction of the sextant is based.

Since, in direct observations, altitudes may vary from 0° to 90°, it follows from the above principle that an angle of 45° between the two reflecting surfaces would be sufficient to measure the altitude of any heavenly body. Since an arc of 45° is one-eighth of a circle, instruments of this general type, having a maximum angle of 45° between the two reflectors, are known as *octants.* Instruments having an arc or limb of 60°, or one-sixth of a circle, for measuring the angle between reflectors are called *sextants,* and those having arcs of 90°, or one-quarter of a circle are called *quadrants.* Either octants or sextants are suitable for aerial use, but the quadrant is too bulky. For a great many years practically all instruments used by marine navigators for measuring altitudes have been sextants, and the term sextant has become so gen-

Fig. 198.—Celestial-navigation equipment.

erally used that it is now applied to all instruments for measuring altitudes of heavenly bodies whether they are actually quadrants, sextants, or octants.

The principle of the sextant may readily be understood by a glance at Fig. 199.

The sextant consists essentially of the following parts:

1. An arc, or limb, MM of a little more than 60° with a scale usually of silver graduated to read degrees and fractions. This scale is graduated to read 2° for each degree of arc through which the index arm moves, owing to the optical principle of the sextant.

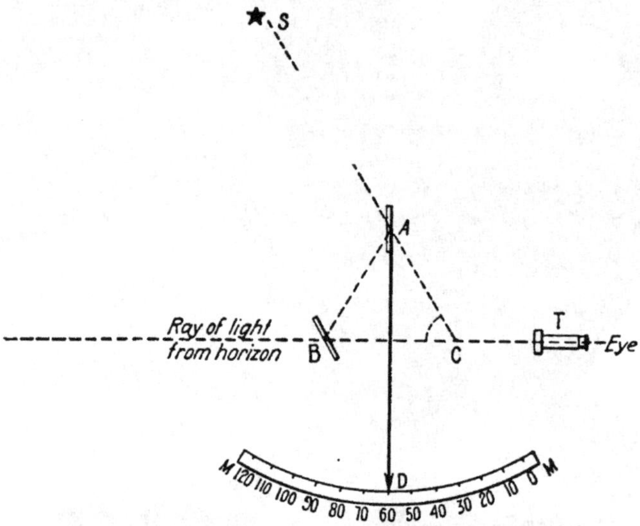

FIG. 199.—Principle of the sextant.

2. An index arm AD arranged to pivot about the point A, which is the center of curvature of the limb MM. At the lower end D of the index arm is a vernier, in old-style sextants, and a tangent screw and micrometer drum on modern sextants, for reading the scale more accurately. On the upper end of the index arm is a mirror A called the index glass, which is mounted perpendicular to the plane of the limb.

3. A horizon glass B fixed to the frame of the sextant. For use at sea the half of the glass next to the frame is a mirror and the other half is clear glass.

4. A telescope T, to direct the light of sight of the observer and to magnify objects that are observed.

If, now, the observer's eye is placed at the telescope T, the sea horizon is seen through the unsilvered half of the horizon glass B. The mirror A is then turned on its pivot until the heavenly body S appears to coincide with the sea horizon, when the angle *SCB* gives the angular elevation of *S* above the horizon, which is known as the altitude of the heavenly body.

The Aircraft Sextant.—The marine navigator has for a great many years measured altitudes of heavenly bodies from the sea horizon. With moderately clear weather this sea horizon forms an accurate reference line all through the day and during morning and evening twilight.

The air navigator, unfortunately, cannot depend upon the sea horizon. Most of his flying is done overland where no sea horizon is available. Even when he is flying over the sea, if the plane is at any considerable altitude, the sky and sea tend to blend into each other and there is seldom a clear-cut horizon line. Even in case the horizon can be seen, the altimeter does not always indicate the height of the plane with sufficient reliability to determine the height-of-eye correction accurately. One alternative to using the true horizon is to use a cloud or haze horizon. This has the disadvantage that the height of a cloud or haze horizon at a distance has to be guessed and may frequently be guessed wrong.

For these reasons the aircraft sextant must embody its own artificial horizon within itself. Three lines of approach to this desideratum have been attempted:

1. Gyroscopic horizons.
2. Pendulous horizons.
3. Bubble horizons.

The gyroscopic sextants have been either mechanically or operationally complicated. The pendulous sextants have found little favor. Development has been mainly concentrated upon bubble sextants, because of their simplicity.

All three of these types of sextants suffer from the disadvantage that the horizon element is acted upon, not by the gravitational acceleration alone, but by the resultant of the gravitational acceleration and whatever other accelerations may exist upon the airplane at the instant of taking the sight. No device suitable for incorporation in an aircraft sextant is known at the present time that will react to the gravitational acceleration alone, ignoring the other accelerations.

Since these accelerational errors, which frequently are of the order of 2°, or 120 miles, are inherent in the design of the sextant, it follows that success in its use is very largely dependent upon the skill of the pilot in flying the plane steadily. Since bubble sextant errors may be either too large or too small and follow the law of probability and chance, a large portion of the acceleration error may be eliminated by averaging a series of observatives, as described below.

Although, for the above reasons, the bubble sextant is not so accurate as the mariner's sextant using the sea horizon, this loss of accuracy is compensated for by having the horizon available for observations 24 hr of the day, so that observations can be made whenever a heavenly body can be seen.

A modern airplane may easily cover 1,500 miles in the interval between evening and morning twilight. Moreover, it is during this period of darkness that piloting is most difficult, that dead reckoning becomes least reliable because of the difficulty of ascertaining the direction and force of the wind with any degree of accuracy, and that celestial navigation by means of stellar observations becomes the most handy and the most accurate.

Some of the sextants in use are the Bausch and Lomb, the Pioneer, the Fairchild, the RAF, and the Link. All late models of these sextants include means for averaging a series of observations, and most of them are similar in many respects. Since the new Link averaging bubble sextant is the only aircraft sextant available to the public at this writing, it will be described in some detail.

Link Averaging Bubble Sextant.—The Link bubble sextant (Fig. 200) was designed to fill the urgent need for a low-priced bubble sextant that is both accurate and easy to use. It weighs less than 3 lb. and is mechanically simple and rugged. Since particular attention has been paid to weight distribution, the instrument is well balanced when held in the hand. This combination of light weight and correct balance eliminates muscular exertion and makes accurate observations easy to obtain. The sextant is so engineered as to permit rapid quantity production.

The optical principle of the Link sextant is similar to that in the Bureau of Standards type of sextant, which permits the object to be observed without an astigmatizer either direct or reflected with the image appearing in front of the bubble. Structurally, the sextant is similar to the marine sextant with resulting accuracy. Averaging is accomplished as follows:

1. With the instrument held in the position shown in Fig. 200, turn the recording drum until the image appears centered in the sextant bubble and press the recording trigger.
2. Repeat this operation 10 or 20 times, noting the average time.
3. Turn the drum until the recording pencil is at the average of the observed series of altitudes and read the average altitude with the index vernier.

The vernier is more difficult to read than the micrometer drum used in some sextants, but this apparent disadvantage is counterbalanced by the fact that only one reading is required for a series and by the important fact that the cost is about half that of other types of sextants.

An illustrated booklet giving the details of construction and use is furnished with each Link sextant.

Bubble-sextant Errors.—This type of sextant is subject to two errors:

1. Bubble error.
2. Index error.

The *bubble error* is caused by failure of the bubble to indicate the true horizontal position. There are four practical means for determining the bubble error:

FIG. 200.—The Link bubble sextant.

1. With the eye within 2 or 3 ft. of sea level, to eliminate dip, observe the position of the bubble relative to the sea horizon. The center of the bubble should coincide with the horizon when there is no bubble error.

2. Observe a distant object at the same level as the sextant as determined by a theodolite or other accurate means. If there is no bubble error, the center of the bubble should coincide with the distant object.

3. With the index error reduced to zero, as explained below, work several lines of position from a position. If the lines pass through the position of the observer, there is no bubble error; if the lines are

bunched, say 8 nautical miles toward the observed body, the bubble is in error by 8′ of arc, the bubble in this case being below the horizontal position.

4. By far the most satisfactory method for determining sextant errors is by means of the Link sextant collimator as shown in Fig. 201. It

Fig. 201.—The collimator.

was especially designed for the purpose and permits a check for sextant error for all altitudes. An incidental use of the collimator is student classroom practice in taking observations.

The *index error* is caused by the zero position of the index glass not being properly recorded on the counter.

Bubble-error Adjustment.—With the bubble error determined as already described, this error is then adjusted by raising or lowering the bubble by adjusting screws usually provided with each sextant. If adjusted by the sea horizon, by the collimator, or by a distant object level with the sextant, the bubble center is simply brought into coincidence with the horizon or with the object. When adjusting by lines of position, the adjustment is made by trial and error until the lines pass through the position of the observer.

Index-error Adjustment.—The index error, or error caused by the position of the index glass not being properly recorded on the counter, may easily be corrected by bringing the reflected and the projected images of a distant body into coincidence, and then resetting the counter to zero.

Every navigator should be able to make quick and accurate practical counter-reading adjustments in the air as well as on the ground. This

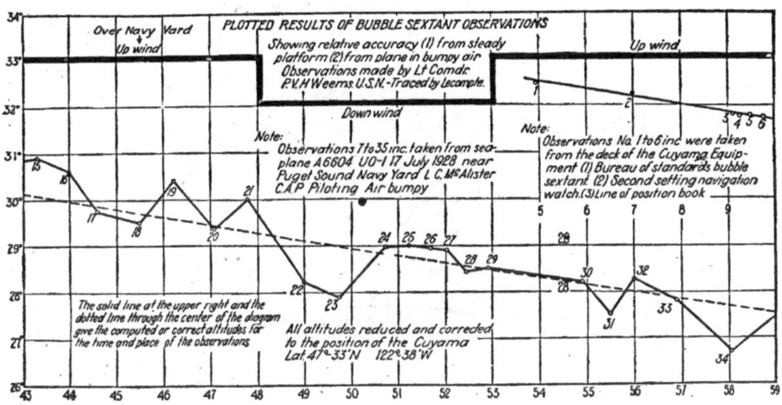

FIG. 202.—Plotted results of bubble-sextant observations.

is most important, since the safety of the plane and observer may be dependent on the ability of the observer to make these adjustments in the air. This may be done by setting the counter near zero, and then by bringing the direct and the reflected images of the sun, moon, star, or distant object into coincidence. When in coincidence, the counter should read zero or, if there is a known error in the bubble, the counter reading should be set to allow for this error.

Bubble-sextant Acceleration Error.—In an effort to demonstrate the effect of the bubble acceleration when taking observations in the air, the author observed the data shown in the accompanying graph plotted on Fig. 202. The straight diagonal line is the computed altitude for the time and place where the data were observed. This line is the correct altitude, and the difference between the values picked from the plotted line and the measured altitudes is the error of the observations. It will be noted that observations 1 to 6 taken from the deck of the *U.S.S. Cuyama*, when practically stationary, are very close to the true altitudes as shown by the plotted line of true altitudes. On the other hand, the observations taken in the air a few minutes later, under the same conditions except that there was motion on the plane and hence an acceleration error, showed much larger errors.

The observations were made from an open two-seater seaplane under average conditions, except that frequent turns were made, which perhaps caused slightly larger errors than would be incurred when on a steady flight.

Although the errors of the observations taken in the air were in some cases very large, the *average* or *mean* of the observations followed fairly closely the plotted true altitudes. It will be seen that, although an individual sight cannot be depended on for accuracy, the average of a large

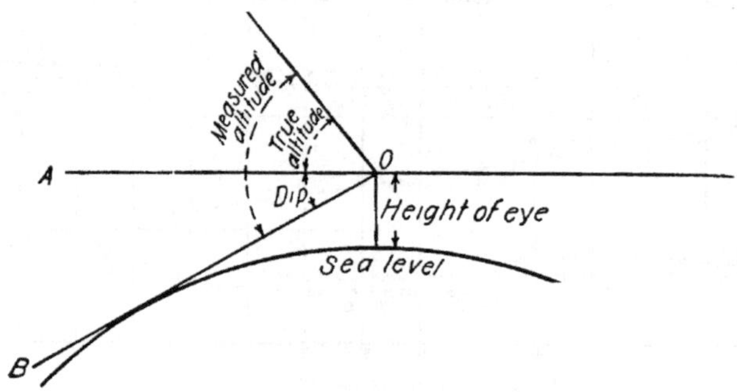

FIG. 203.—The effect of dip.

number of observations gives results that are sufficiently close for practical purposes.

Altitude Corrections.—The following corrections to sextant altitudes are required:

1. Index error. Eliminate by adjusting sextant.
2. Dip. Required for sea-horizon sights only.
3. Refraction. Principal correction to be made.
4. Parallax. Required for moon sights only.
5. Semidiameter. Required for sea-horizon sights of sun and moon only.
6. Earth's rotation. Required for bubble-sextant sights only.

In practice all these corrections, except the first and last, are combined into one table.

Index Correction.—Index correction should be applied to the sextant reading when there is a known index or instrumental error in the sextant. The skilled navigator will keep his sextant correctly adjusted and not be bothered with this correction.

Dip.—Dip is the angle of depression of the visible horizon below the true horizon due to the elevation of the observer above the surface of the earth. Owing to the spherical shape of the earth, a line joining an observer in an aircraft to the visible horizon will be sloped below the horizontal. Thus, in Fig. 203, the true horizontal for an observer at O is OA, whereas the straight line to the visible horizon will have the slope OB. The angle AOB is the dip. Dip error is not incurred when using a bubble sextant, because the bubble sextant uses an artificial horizon incorporated in the sextant and not the true or visible horizon.

Refraction.—It is a well-known principle of optics that a ray of light in passing from one medium into another medium of greater density is

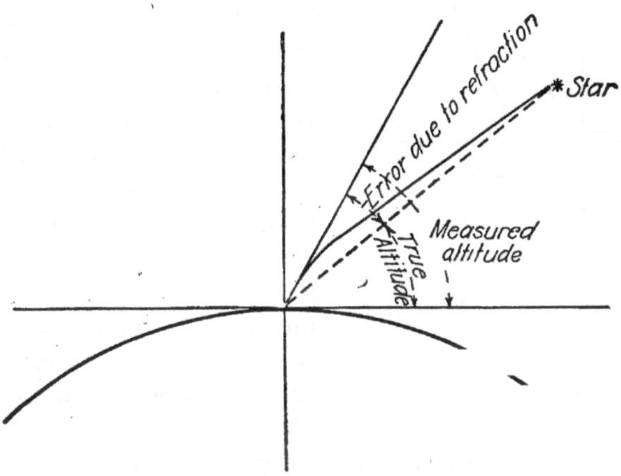

FIG. 204.—The effect of refraction.

deflected toward a perpendicular to the surface separating these mediums. This is known in optics as refraction.

Similarly, in astronomy a ray of light from a heavenly body as it enters the denser layers of atmosphere near the earth is deflected toward a perpendicular to the surfaces of these layers, which are parallel to the surface of the earth. This change of direction of the ray, as shown in Fig. 204, causes the altitude measured with the sextant to appear greater than it really is.

Parallax.—*Parallax* is the error in the altitude due to the fact that the observer is actually at some position on the earth's *surface* instead of at the *center* of the earth. Navigational tables are computed on the assumption that the observer's eye is at the center of the earth. The difference between the altitude of a heavenly body measured from the center of the earth and from a point on its surface is clearly shown in Fig. 205. Thus, an observer at A would find the zenith distance of the body S to be the angle ZAS. If he were at the center of the earth, he would find the zenith distance to be ZES, which is less than ZAS by the angle ASE. The angle ASE is the correction for parallax. The correction for parallax must be added to the sextant altitude to get the true observed altitude. For practical navigation the moon is the only heavenly body for which a correction for parallax need be applied.

Semidiameter.—The magnifying telescope of a sextant will not show any disk for stars or planets. For the sun and the moon there is a decided disk so that, when a marine sextant is used, the point brought into coincidence with the horizon may be the top of the disk or "upper limb," the center of the disk, or the bottom of the disk or "lower limb." The navigational tables are calculated for the center of heavenly bodies so that, if the "upper limb" is observed, the semidiameter must be subtracted and, if the "lower limb" is observed, the semidiameter must be added to make the final result apply to the center of the body. Since the angular diameter of both sun and moon is always about 30′ of arc, the correction for semidiameter of either sun or moon will be approximately 15′ of arc.

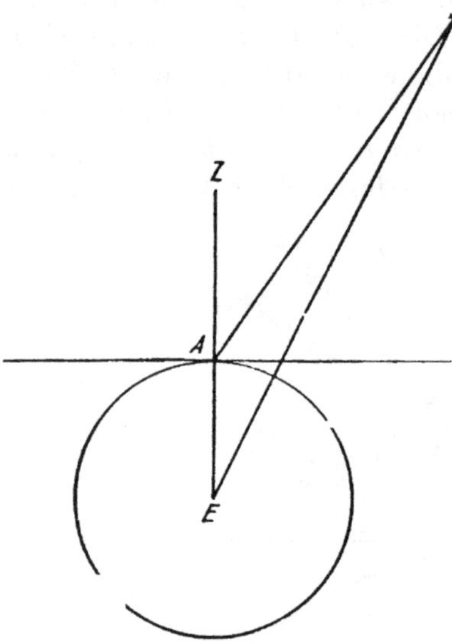

FIG. 205.—The effect of parallax.

If, as is generally the case, the navigator is using the modern bubble sextant, it is more convenient to observe the center of the body and thus avoid the necessity of correcting for semidiameter.

Lat.	Ground speed, m.p.h.						
	100	150	200	250	300	350	400
0	0	0	0	0	0	0	0
10	0.4	0.7	0.9	1.1	1.4	1.6	1.8
20	0.9	1.3	1.8	2.2	2.7	3.2	3.6
30	1.3	2.0	2.6	3.3	3.9	4.6	5.3
40	1.7	2.5	3.4	4.2	5.1	5.9	6.8
50	2.0	3.0	4.0	5.0	6.0	7.1	8.1
60	2.3	3.4	4.6	5.7	6.8	8.0	9.1
70	2.5	3.7	4.9	6.2	7.4	8.7	9.9
75	2.5	3.8	5.1	6.4	7.6	8.9	10.2

Translate all lines $\frac{R}{L}$ in $\frac{N}{S}$ hemisphere, *perpendicular to track.*

Correction in same units as GS.

FIG. 206.—Correction for rotation of the earth.

Bubble Sextant Correction for Earth's Rotational Effect.—Apart from the well-known transient acceleration errors of the bubble sextant in flight, there exists a persistent and predictable error caused by the rotation of the earth. This error varies with latitude and ground speed, as shown in Fig. 206, and is corrected as illustrated in Fig. 207, *i.e.*, all celestial position lines are translated: (*a*) 90° to the right of track in the northern hemisphere, (*b*) 90° to the left of track in the southern hemisphere.

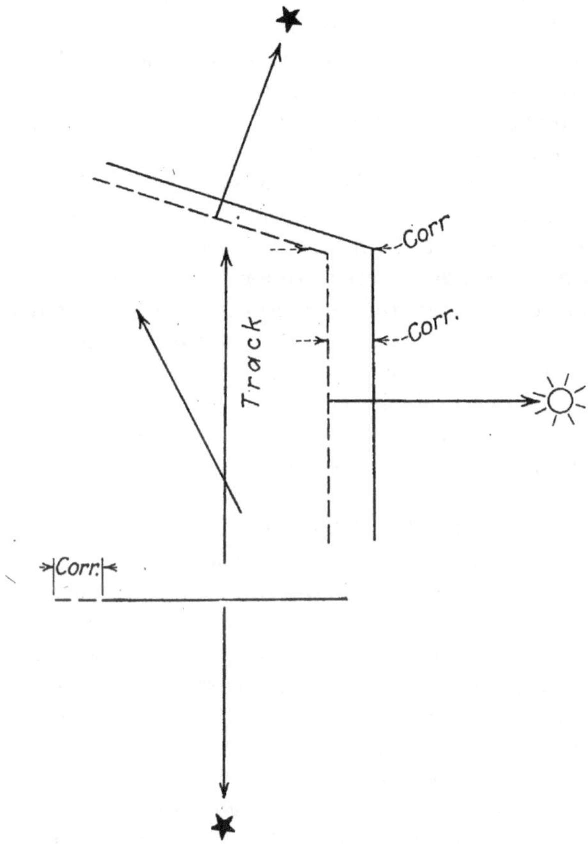

FIG. 207.—Applying correction in northern hemisphere.

THE TIMEPIECE

History.—The first crude portable watches date from the year 1500. The first of these were thick clocklike timepieces called "Nuremberg eggs," from their shape and place of manufacture.

There were three principal mechanical difficulties to overcome to produce the modern accurate timepiece:

1. Errors in rate due to the changing tension on the main spring as it unwound.

2. Need of a suitable escapement to replace the pendulum used with clocks.

3. Erratic rates due to temperature changes.

The problem of variable spring tension was corrected by a series of inventions between 1515 and 1540, whereby the leverage of the spring was increased in proportion to the decrease in tension. The first "dead-beat escapement" was invented by Thomas Tompion (1639–1713), to whom watchmaking probably owes most. The first "temperature correction" was made in 1726 by Harrison.

It is difficult to realize that as late as the end of the eighteenth century, there was no practical means for determining longitude at sea. This was due to the fact that at sea it was necessary to have the exact time of the observations, and no practical timepiece had been developed that would withstand the motion and temperature changes aboard ship. In 1714, the British government appointed "Commissioners for the Discovery of Longitude at Sea" with authority to grant the following prizes for determining the longitude at sea:

1. Within 80 nautical miles, $50,000 (near shores of greatest danger).
2. Within 60 nautical miles, $50,000.
3. Within 40 nautical miles, $75,000.
4. Within 30 nautical miles, $100,000.

Since 30 miles represents an error of at least 2 min. in time, it will be seen that a dollar watch, with the daily radio ticks now available, would have won the biggest prize.

In 1735 the first chronometer was made. In 1731 the other important item of navigation equipment, the marine sextant, was improved to the point where it was fairly reliable. Modern celestial navigation may be fairly dated from either of these inventions.

In 1761–1762 a chronometer made by Harrison was tested on a voyage from England to Jamaica. The voyage out lasted 147 days, during which the chronometer lost 1 min. 55 sec. (equal to 18 miles in longitude); on the return, it lost 1 min. 49 sec. (equal to 16 miles in longitude). Finally, in 1765, Harrison was paid $50,000 and, in 1773, the remaining $50,000. In 1828, the "Commission for the Discovery of Longitude at Sea" was disbanded, having in 114 years expended a total of $505,000, largely toward the development of the modern chronometer.

The modern watch is one of the mechanical marvels of the age. A moderately priced watch is expected to count the 86,400 sec. of each day without missing more than one or two of them, or to an accuracy within 1:40,000.

Although it is not essential to have a navigation watch run to within a fraction of a second per day, some watch owners insist on getting the best possible rate on their watches. One record which has come to the attention of the writer is most impressive. This watch (Fig. 208) belonging to E. A. Link, Jr., inventor of the Link Trainer and other navigation devices, was carried by airplane, train, automobile, and at sea during the period covered, and gave the following record:

LINK WATCH RECORD

	Error + (fast) − (slow)	Gain +, or loss −	Number of days	Rate + gaining − losing
	sec.	sec.		sec.
Nov. 25, 1939	− 25			
Dec. 21, 1939	− 21	+ 4	26	+.15
Mar. 23, 1940	− 65	−44	93	−.5
Apr. 21, 1940	− 68	− 3	29	−.1
June 28, 1940	− 92	−24	68	−.35
July 22, 1940	− 98	− 6	24	−.43
Nov. 17, 1940	−121	−23	118	−.21
Jan. 3, 1941	−104	+17	47	+.36
		Net −79	405	Net −.19

Not only is the record of this particular watch impressive as regards accuracy, but also the fact that it was not permitted to run down over a period of 407 days is in itself a remarkable achievement, especially for a person like Link, who must travel all over the country and is conducting an active business.

Second-setting Watch.—Owing to the low gear ratio between the hour and minute hands as compared with the second hand, and to other causes, no provision is made for setting the second hand of an ordinary watch. This means that a watch set to the correct hour and minute may still be in error as much as 30 sec., which, near the equator, represents an error in longitude of 7.5 miles. The second-setting watch was devised to permit the exact second to be set, thus avoiding the correction ordinarily required by the navigation watch. This is done in the standard model by rotating an inner seconds dial, as shown in Fig. 209, or by rotating a special bezel on the wrist model as shown in Fig. 210. The hour and minute hands are set in the usual way. When the minute hand is properly set, it should of course be exactly at a minute division when the second hand is at 60.

Rating of Watches.—Strange to say, keeping the watches running correctly is one of the most difficult matters in navigation. Even a

Fig. 208.—Link navigation watches.

Fig. 209.—Standard second-setting navigation watch.

Fig. 210.—Wrist-model second-setting watch. (*Courtesy Longines-Wittnauer Company.*)

perfect watch will stop unless it is wound up, and this is one cause of trouble. Lincoln Ellsworth, member of the 1925 Amundsen-Ellsworth Polar Expedition, told the writer that it was Amundsen's custom to wind his watch both evening and morning, so that he would not forget it.

Watches used for navigation should be handled as carefully as possible. The writer has used continuously for about 15 years the first pair of second-setting watches made, and right now they are running within 3 sec. of the correct time per day. Both are losing 2 or 3 sec. per day but this does not mean that they are *changing rate that much.* So long as we know just how much they gain or lose we are all right, for we can then allow for the small gain or loss. This pair of watches are large patrol-boat watches belonging to the Navy. They have been used in the air, on the ground, in submarines, and at sea. Although they have been given rather rough treatment, they are in practically perfect condition.

Radio Time Signals.—Accurate radio time signals are broadcast all over the world several times daily. For instance, Washington (NAA)

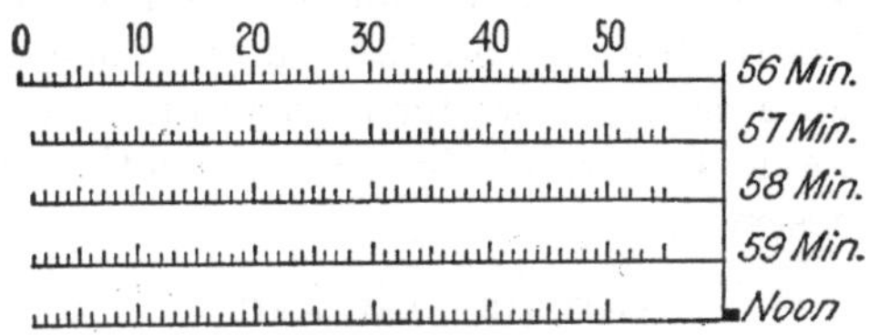

FIG. 211.—U. S. Naval Observatory time signals.

broadcasts the time signals on frequency 113 kc., from 5 min. before each hour to the hour, except at 0200, 0400, 1400, and 1600, GCT. Other stations broadcast the time tick at various times and on various frequencies.

The time signal begins 5 min. before the hour to be marked and consists of a dot for each second. The dot for the 29th, 56th, 57th, 58th, and 59th sec. of each minute is omitted and also the following: 51st of the first minute; 52d of the second; 53d of the third; 54th of the fourth; and the 51st, 52d, 53d, 54th, and 55th of the fifth. The silence after the 50th sec. of the fifth minute is followed by a 1-sec. dash, the beginning of which marks the time signal. In this code, the number of dots between the two omissions at the end of each minute indicates at once the number of minutes of signal yet to be sent. It is also believed that the shortened interval before the 1-sec. dash will make possible more accurate comparisons of the final 1-sec. dash. This scheme is shown graphically in Fig. 211.

In addition, there is a time service furnished by the Bureau of Standards, little known to the navigation public, which sends out every second of time throughout the day. This service, known as the "Standard

Frequency Broadcast," is sent out on 5 megacycles as impulses each second, like a grandfather's clock, throughout the day. In order to designate which instant of time is being sent out, a standard musical pitch, on frequency 440 cycles per second, is superposed on the 5-megacycle impulse for each second. The standard musical pitch is sent out continuously except for the first minute of each 5-min. period. This permits an accurate check of two instants in each 5-min. period throughout the day. This musical note corresponds to A above middle C, and may therefore be used not only for scientific purposes but also by musicians, as well as by navigators. This combination time signal is sent out from station WWV and is announced in telegraphic code while the musical tone is off for 1 min. in the 5-min. period.

This excellent universal time service should prove a great boon to navigators who find it inconvenient to wait for the even hour, or as formerly, to wait for the noontime signal or other principal signals during the day.

Of course, radio time signals should be backed up by reasonably accurate navigation timepieces and their proper operation. It is not necessary to pay an excessive price for navigation timepieces—not more than $100—but a navigation watch should be given the attention that any good mechanism deserves.

THE AIR ALMANAC

To fix a position on the earth by observations of celestial bodies, it is necessary to determine the position of the observer relative to the position of the observed body or bodies. To find the position of heavenly bodies we use the *Air Almanac,* in which are tabulated the positions of all bodies used in navigation for certain intervals of time.

In 1928, the author suggested the direct tabulation of Greenwich hour angle, and as a result the "Lunar Ephemeris for Aviators for September 1 to December 31, 1929" was published by the U.S. Naval Observatory. In 1933, the U.S. Naval Observatory published the *Air Almanac,* designed by the author, which omitted the equation of time, right ascension, and other data not required by the air navigator.

Unfortunately, in 1934 the *Air Almanac* was combined with the *Nautical Almanac* instead of being published separately, and the former publication was discontinued. In 1936, the author suggested to the superintendent of H.M. Nautical Almanac Office that the British issue a separate *Air Almanac.* This was done, and the splendid British *Air Almanac* was published in the United States under license by the author until a similar American *Air Almanac* was issued in 1941.

The *Air Almanac* represents a great advance in navigation, which explains the author's pride in being associated with others in its development. Appendix B gives extracts from it, together with auxiliary tables, sample problems, and instruction for use. Although the almanac includes a star chart, supplementary data are given below.

Identification of Stars and Planets.—The frequency with which the principal navigational stars are observed makes it essential for a navigator to recognize them in clear weather at sight, by their relative positions.

When the weather is unfavorable, the navigator should be able to identify a star observed through a rift in the clouds by the use of star finders or star maps such as the Brownell-Weems Star Finder, or Illyne's Star Chart. Instructions for use are furnished with these devices.

Stars and Planets Used for Navigation.—Only about 40 of the stars are used to any extent in practical navigation. It is possible to distin-

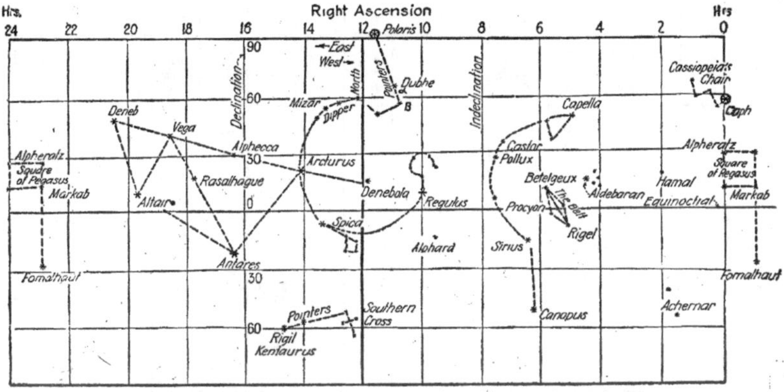

FIG. 212.—Navigational stars projected on celestial sphere.

guish between fixed stars and planets by the fact that the latter change their positions in the heavens relative to the former and to each other. Venus and Jupiter are brighter than any fixed stars. Mars is usually distinguished by its reddish color.

The angle between the sun and Venus as viewed from the earth is never more than 47°, which is equivalent to about 3 hr. of time, so that, allowing for difference in declination, Venus is never visible more than about 3 hr. before sunrise or 3 hr. after sunset, except in high latitudes. When Venus is to the eastward of or following the sun, it is visible as soon as the sun has set (or even before) and is then called the "evening star." When it is to the westward or ahead of the sun, it is visible during morning twilight and is then known as the "morning star." When at or near its maximum brilliancy, it is easily seen in full daylight.

The Constellations.—By referring to a star chart or to Fig. 212, or by observing the heavens, it will be noted that the principal stars group themselves into certain conspicuous figures. Since the early ages these groups or constellations have borne fanciful names assigned to them

Direction	Constellations	Stars	July	Aug.	Sept.	Oct.	Nov.	Dec.	Jan.	Feb.	Mar.	Apr.	May	June
South (passing over meridian)	ORION	Betelgeuse, Bellatrix, Rigel, Sirius, Aldebaran, Procyon			6 A.M.	4 A.M.	2 A.M.	Midnight	10 P.M.	8 P.M.	6 P.M.			
	LEO CORVUS	Regulus, Denebola, Spica						6 A.M.	4 A.M.	2 A.M.	Midnight	10 P.M.	8 P.M.	6 P.M.
	SCORPIO	Antares, Shaula	10 P.M.	8 P.M.	6 P.M.						6 A.M.	4 A.M.	2 A.M.	Midnight
	PEGASUS	Alpheratz, Markab, Deneb Kaitos, Fomalhaut	4 A.M.	2 A.M.	Midnight	10 P.M.	8 P.M.	6 P.M.						
East (rising)	ORION				Midnight	10 P.M.	8 P.M.	6 P.M.						
	LEO CORVUS							Midnight	10 P.M.	8 P.M	6 P.M.			
	SCORPIO										Midnight	10 P.M.	8 P.M.	6 P.M.
	PEGASUS		10 P.M.	8 P.M.	6 P.M.									Midnight
West (setting)	ORION							6 A.M.	4 A.M.	2 A.M.	Midnight			
	LEO CORVUS										6 A.M.	4 A.M.	2 A.M.	Midnight
	SCORPIO		4 A.M.	2 A.M.	Midnight									
	PEGASUS				6 A.M.	4 A.M.	2 A.M.	Midnight						
Zenith (or vicinity)	AURIGA GEMINI	Capella, Castor, and Pollux			6 A.M.	4 A.M.	2 A.M.	Midnight	10 P.M.	8 P.M.	6 P.M.			
	URSA MAJOR ("THE DIPPER")	Dubhe, Alioth, Mizar, Arcturus (BOÖTES)						6 A.M.	4 A.M.	2 A.M.	Midnight	10 P.M.	8 P.M.	6 P.M.
	LYRAE (VEGA) CYGNUS	Altair (Aquilae), Deneb	10 P.M.	8 P.M.	6 P.M.						6 A.M.	4 A.M.	2 A.M.	Midnight
	CASSIOPEIA	Caph, Ruchbah	4 A.M.	2 A.M.	Midnight	10 P.M.	8 P.M.	6 P.M.						6 A.M.

FIG. 213*a*.—Times when constellations of Fig. 213*b* are visible. North latitude.

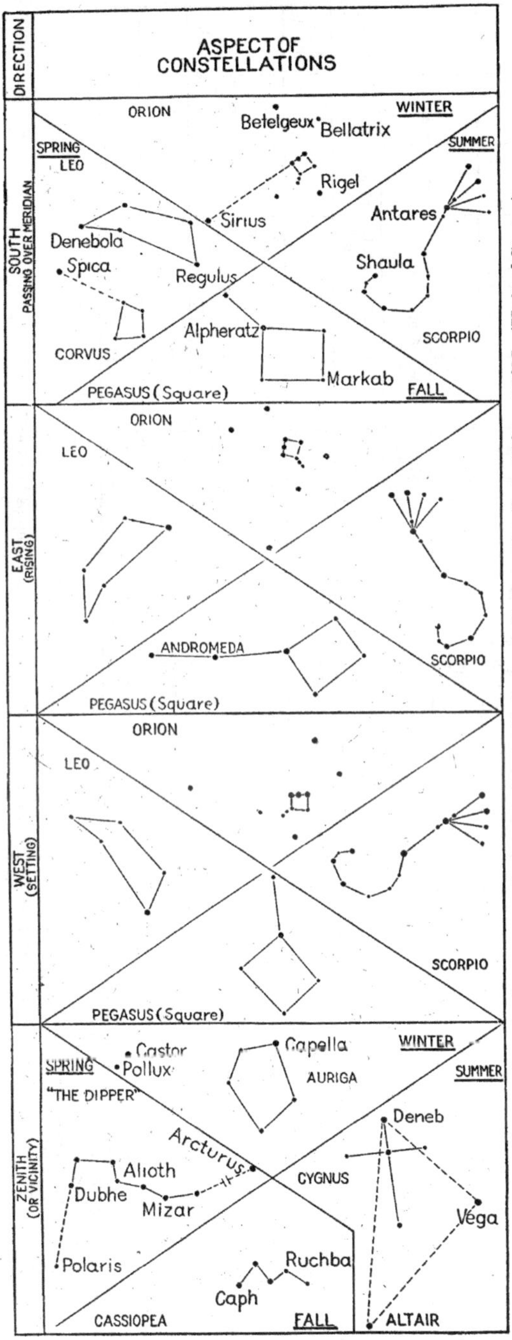

FIG. 213b.—Appearance and names of principal constellations for Lat. 30° to 50°N. (United States).

In the use of the table, the following facts should be kept in mind:

1. Directions mentioned are general.
2. Time is approximate (within 1 hr.).
3. Date of the month is around the 21st.
4. Stars of other constellations close to principals and mentioned in the table may vary up to 2 hr.
5. In northern latitudes the Dipper and Cassiopeia climb higher toward the zenith.

MONTHLY STAR IDENTIFICATION TABLE
Mean Lat. 35°S.

Direction	Constellations	Stars	July	Aug.	Sept.	Oct.	Nov.	Dec.	Jan.	Feb.	Mar.	Apr.	May	June
North (Passing over meridian)	ORION	Betelgeuse, Bellatrix, Rigel, Sirius, Aldebaran, Procyon, Pollux, Capella			6 A.M.	4 A.M.	2 A.M.	Midnight	10 P.M.	8 P.M.	6 P.M.			
	LEO CORVUS	Regulus, Denebola, Spica, Arcturus						6 A.M.	4 A.M.	2 A.M.	Midnight	10 P.M.	8 P.M.	6 P.M.
	SCORPIO SAGITTARIUS	Antares, Shaula, Nunki, Kaus Australis, Vega, Deneb, Altair	10 P.M.	8 P.M.	6 P.M.						6 A.M.	4 A.M.	2 A.M.	Midnight
	PEGASUS	Alpheratz Markab, Deneb Kaitos Fomalhaut	4 A.M.	2 A.M.	Midnight	10 P.M.	8 P.M.	6 P.M.						
East (Rising)	ORION				Midnight	10 P.M.	8 P.M.	6 P.M.						
	LEO CORVUS							Midnight	10 P.M.	8 P.M.	6 P.M.			
	SCORPIO SAGITTARIUS										Midnight	10 P.M.	8 P.M.	6 P.M.
	PEGASUS		10 P.M.	8 P.M.	6 P.M.									Midnight
West (Setting)	ORION							6 A.M.	4 A.M.	2 A.M.	Midnight			
	LEO CORVUS										6 A.M.	4 A.M.	2 A.M.	Midnight
	SCORPIO SAGITTARIUS		4 A.M.	2 A.M.	Midnight									
	PEGASUS				6 A.M.	4 A.M.	2 A.M.	Midnight						

In the use of this table, the following facts should be kept in mind:
1. Directions mentioned are general.
2. Time is approximate (within 1 hr.), local.
3. Date of the month is around the 21st.
4. Stars of other constellations close to principals and mentioned in the table may vary in time up to 2 or 3 hr.
5. If close to celestial equator, stars rise due east, climb to meridian in 6 hr., and set due west 6 hr. later.
If south declination, they rise in S.E. quadrant (with increasing declination, closer to south), climb to meridian in more than 6 hr., and set in S.W. quadrant.
Stars with declination greater than 90° minus latitude never set, and revolve around the south pole, clockwise.
If north declination, they rise in N.E. quadrant (with increasing declination, closer to north), climb to meridian in less than 6 hr., and set in N.W. quadrant.
Stars with declination greater than 90° minus latitude are not seen.
Stars with south declination equal to latitude pass meridian at zenith.

FIG. 214*a*.—Times when constellations of Fig. 214*b* are visible. South latitude.

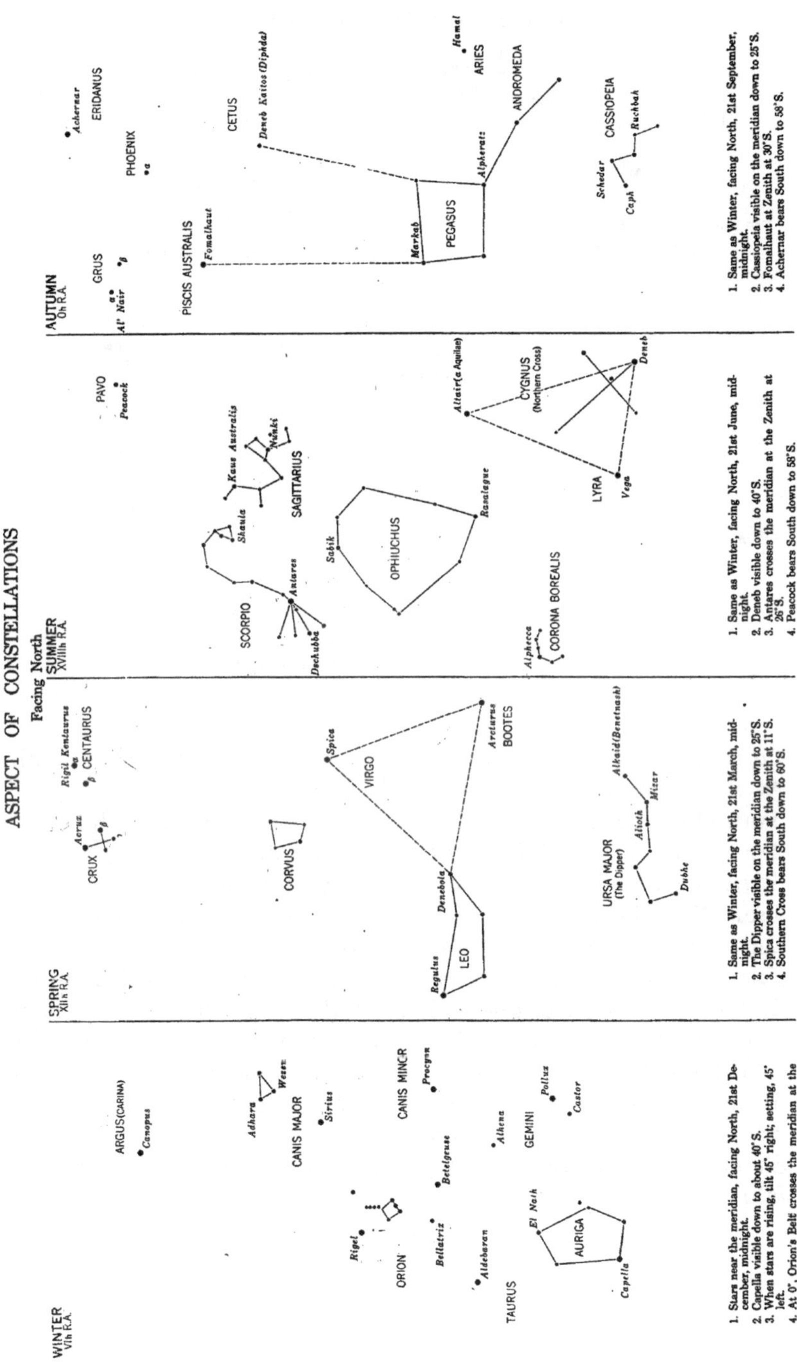

FIG. 214*b*.—Appearance and names of principal constellations for mean Lat. 35°S.

PRINCIPAL STARS AROUND THE SOUTH POLE

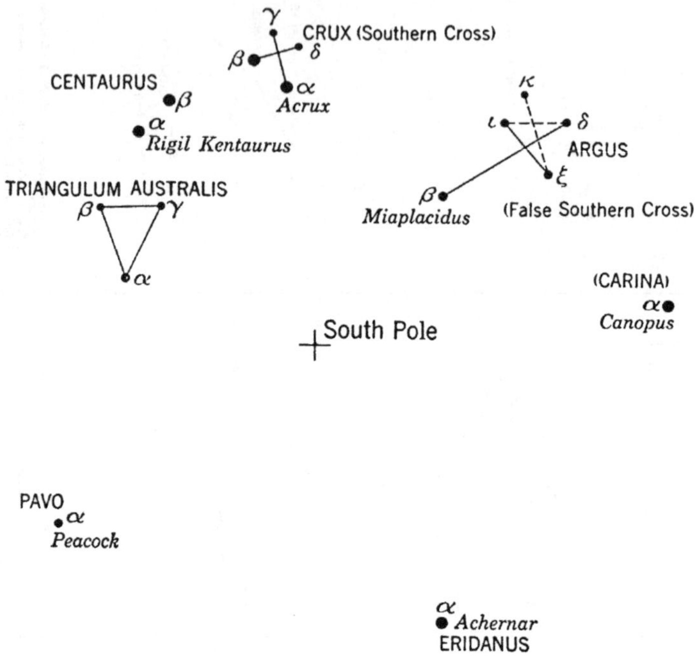

1. Position of Southern Cross, 21st March, midnight, facing south. At this time, Southern Cross is about zenith at 60°S.
2. Turn chart 90° left for position for 21st December, midnight.
3. Turn chart 90° right for position for 21st June, midnight.
4. Turn chart 180° for position for 21st September, midnight.
5. From 35°S. down to south pole, the Southern Cross is always visible, and revolves clockwise around the south pole.
6. The False Southern Cross is shown in two ways: solid lines and dotted lines, however one looking at the sky wishes to imagine it. The dotted lines show the False Cross as commonly adopted. The solid lines show a false Southern Cross as imagined by the author in order to show more navigation stars (Σ and β Miaplacidus) in it; also because its main direction is closer to the North-South direction of the actual Crux and further because K Argus is fainter.

Fig. 214c.

by the ancients, and these names, in their later form, are still used. The individual stars of each constellation are designated by letters of the Greek alphabet prefixed to the Latin name of the constellation in the genitive. For instance, the bright star in the constellation Ursa Major (the Great Bear, but often known as the Dipper) nearest to Polaris is α Ursae Majoris. In addition to these designations, most of the bright stars have individual names, usually of Arabic origin. Thus α Ursae Majoris mentioned above bears the name Dubhe; α Canis Majoris, the brightest star in the heavens, is better known as Sirius.

Star Magnitudes.—The brightness of stars is expressed in the form of magnitudes. The 20 brightest stars are of the first magnitude, and those just visible to the naked eye are of the sixth. Typical second magnitude stars are those in the belt of Orion. Two stars, Sirius and Canopus, are so bright that their magnitudes have to be expressed as negative numbers, as have also those of Venus and Jupiter.

Star Configurations.—The simplest way to learn the stars is to select (1) a conspicuous constellation in the northern heavens about which to group stars of high northern declination; (2) one or more of several in the region of the celestial equator for fixing stars in this region and others not too far north or south; (3) one in the southern heavens for stars of high southern declination. The constellations best suited for this plan are

1. Ursa Major, or the Dipper.
2. Orion in winter, Leo in spring, Scorpio in summer, and Pegasus in the autumn.
3. Crux Australis, or the Southern Cross.

After becoming familiar with the brightest stars of these constellations, one should learn the brightest stars close to them, and finally pick out others by

1. Prolonging a straight or curved line through two or more known stars until it passes through the star in question.
2. Noting the geometrical figure formed by three or more bright stars.

As an aid to the beginner, a star identification table is provided in Fig. 213*b*, which lists and indicates the appearance of the principal constellations for the latitude of the United States at various hours for each month of the year.

Ursa Major.—By referring to Fig. 212 and starting at the Dipper, or Ursa Major, it is noted that two bright stars (α and β) point to Polaris, or the Pole Star; for this reason they are often known as the Pointers. By continuing the curved sweep of the handle we come to Arcturus, a very bright star with a reddish tint, and later to Spica, which is further identified by the "sail." By following the sweep of this curve round and back toward the north we come to Regulus, which may also be identified by the "sickle." Regulus is a bright white star, which forms with Spica and Arcturus a triangle right-angled at Spica. To the east of Arcturus are three bright stars forming a large triangle; these are Vega, Deneb, and Altair. Antares, to the southwest, also forms another large triangle with Vega and Altair.

Across the pole from the Dipper, and at about the same distance from Polaris, is a group of stars forming a W, or an M, according to the position of the constellation in its diurnal path; this group is popularly called Cassiopeia's Chair. The westerly star of this group, Caph, is important in that it is near the meridian of zero right ascension; when Caph transits the upper branch of the local meridian the local sidereal time is approximately 0^h. Directly south of Caph is the Great Square of Pegasus, the western side of which points to Fomalhaut, one of the few bright stars in the southern hemisphere available for use in northern latitudes.

Orion and Stars Identified from It.—Orion, which is the most brilliant and beautiful constellation in the heavens, is outlined by a quadrilateral of three bright stars and one of lesser magnitude. The bright yellowish northeast star is Betelgeuse, the northwest star is Bellatrix, and the bright bluish-white star to the southwest is Rigel.[1] Inside the quadrilateral are three second-magnitude stars nearly equidistant, forming the well-known Belt of Orion. The three stars of the Belt point southeastward to Sirius, the brightest star of the heavens, which shines with a scintillating white light. To the southward of Sirius, and next to it in brilliancy, is Canopus, which is not seen in high northern latitudes. Forming a gentle curve to the northward of Sirius are Procyon, Pollux, Castor, and Capella, in the order named. Sirius, Procyon, and Betelgeuse form an almost perfect equilateral triangle.

The Southern Cross.—This is the most conspicuous constellation of the southern hemisphere, and is outlined by five bright stars. When this group is above the pole α Crucis is the southernmost, β Crucis the easternmost γ Crucis the northernmost, and δ Crucis the western star. A line from α Centauri through β Centauri points directly to the Cross, and these two stars are therefore known as the Pointers.

The position of the south celestial pole is not marked by any conspicuous star. It may be found by taking the intersection of a line through α and γ Crucis (which have nearly the same right ascension) and a line perpendicular to α and β Centauri (which have nearly the same declination). Or again, it may be taken as halfway between β Centauri and Achernar.

METHODS OF CONVERTING OBSERVATIONS INTO POSITIONS

There are any number of methods, or variations of methods, for converting the observed altitudes and times of celestial bodies into posi-

[1] As seen from northern latitudes.

tions. With the publication of "H.O. 214," the "Star Altitude Curves," the "Line of Position Book," and other short methods, practical navigators have largely settled on several of the best so that solutions of questionable value described in earlier editions of this book may be omitted in this one. There is little choice among several of the new methods and they are easy to learn. We therefore restrict ourselves here to "H.O.

FIG. 215.—Selected methods for celestial navigation.

214," "Line of Position Book," and "Star Altitude Curves" as illustrated in Fig. 215.

The "Air Navigation Tables (RAF)," published as "H.O. 218," is similar to "H.O. 214," and the "Astrograph" is a development of the "Star Altitude Curves." Because of their restricted nature, they cannot be described in detail here.

Solving the Navigation Triangle.—Let the spherical triangle used in navigation and projected on the plane of the horizon be represented by Fig. 216. Let the parts of the triangle be designated as follows:

P = pole.
Z = zenith of the observer. The azimuth (angle PZM) is also called Z.
M = body observed.
L = latitude.
d = declination.

t = hour angle of body M.
H = altitude.
EQ = equinoctial.
N = perpendicular let fall from Z' on PM. This is an auxiliary part.
O = intersection of N with PM.
K = distance from O to the equinoctial. This is an auxiliary part introduced to facilitate the solution of the triangle. *K always takes the sign of L.*

In the solution of this triangle, modern fast methods use an assumed position as shown at Z' in Fig. 216, which provides integral values of hour angle and latitude and thereby simplifies the work. Both "H.O. 214" and the "Line of Position Book" use this artifice. In the case of "H.O. 214," three arguments—latitude, declination, and hour angle—are used in the solution of the triangle direct. In the case of the "Line of Position Book," the triangle is divided as shown in Fig. 216, into two right triangles, which are solved in turn by using two arguments as described below. "H.O. 214" is a shorter solution, but requires a longer book in several volumes. The "Star Altitude Curves" is a graphical solution for position direct from observations of two or three stars. Although restricted to the selected stars, the latter method is much faster than either of the other two.

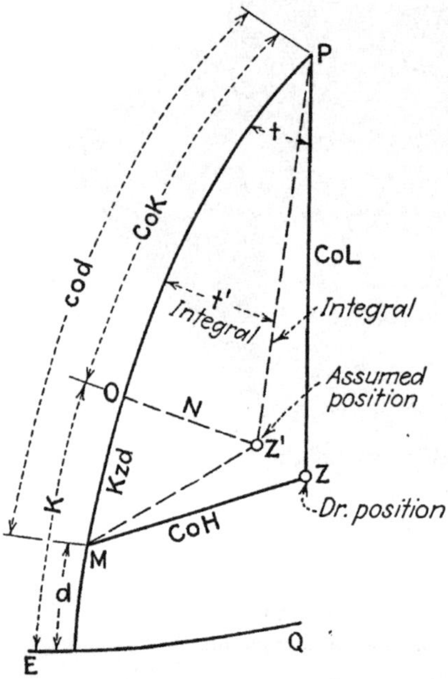

FIG. 216.—The astronomical triangle, showing method of solution, assuming a position to give hour angle and latitude as integral values. (*From Weems, Line of Position Book.*)

"H.O. 214."—Many will prefer "H.O. 214" owing to its short solutions when working from an assumed position and the absence of any arithmetical work except for the one simple linear interpolation for declination. "H.O. 214" is bound for each 10-degree band of latitude in a volume 9.5 by 11.75 in., containing about 270 pages. The cost, weight, and size are disadvantages. Values of H_c and Z are tabulated for integral values of t and L and for every half degree of declination. Figure 221 shows an extract from "H.O. 214." With assumed integral degrees of LHA and latitude, and the exact value of declination, the values of H_c and Z are found with one interpolation for the minutes of declination.

Figure 219 shows the solution for six lines of position worked first by "H.O. 214," and then by the "Line of Position Book." Similar results are plotted in Fig. 218. Detailed instructions for use are given in the book.

"Line of Position Book."—This book of 44 pages includes Ogura's altitude tables (Fig. 217) (by permission), consisting of two parts and 27 pages, Rust's azimuth diagram (by permission), and necessary auxiliary tables.

In the right triangle $PZ'O$ (Fig. 216) we know the values of t and L. Table A gives for all integral values of t and L the corresponding values of $A(= \log \sec N)$ and of K. The angle K is combined with d algebraically to get $K \sim d$, and with $K \sim d$ as an argument, we find in Table B the value of $B\,[= \log \sec\,(K \sim d)]$. The sum of A and B is log cosec H, from which H is found by entering Table B at the bottom. In order to avoid interpolation in Table A, an assumed position, Z', is used such that both t and L will be integral degrees. Figures 217*a* and *b* show extracts from Table A and Table B. Figure 217*c* shows a portion of Rust's azimuth diagram.

Rust's azimuth diagram is based on the formula,

$$\sin t \cos d = \sin Z \cos H$$

or its reciprocal,

$$\operatorname{cosec} Z = \frac{\operatorname{cosec} t \sec d}{\sec H}$$

The values of t are shown in the left margin, of Z in the right margin. Both H and d are plotted on the same curves and numbered alike.

Since this formula uses only secants and cosecants, and since Table B in the "Line of Position Book" is merely a convenient table of secants when entered from the top and cosecants when entered from the bottom, this table of only nine pages may be used for the accurate computation of the azimuth, leaving 27 pages of altitude tables in the briefest and most efficient form.

Figure 220 shows the solutions for azimuth from given values of t, d, and H_c. In practice, Rust's azimuth diagram will be found most satisfactory. Where possible, the azimuth should be observed at the time of sight to save computation.

The first sight in Fig. 235 is discussed here to show the steps taken. In Table A (see Fig. 217*a*), with hour angle 18° at the top of the page and Lat. 33° in the vertical column at the left side of the page, take out the values $A = 1509.8$, or to the nearest integer, 1510, and $K = 34°19'.6$ (use 34°20′). Combine 34°20′ with the declination 21°23′ S., adding if

TABLE A

H.A. / Lat.	16° (1h 4m)		17° (1h 8m)		18° (1h 12m)		19° (1h 16m)		20° (1h 20m)	
	A	K	A	K	A	K	A	K	A	K
°		° ′		° ′		° ′		° ′		° ′
0	1715.8	0 0.0	1940.4	0 0.0	2179.4	0 0.0	2433.0	0 0.0	2701.4	0 0.0
1	1715.3	1 2.4	1939.7	1 2.7	2178.6	1 3.1	2432.2	1 3.5	2700.6	1 3.8
2	1713.7	2 4.8	1937.9	2 5.5	2176.6	2 6.2	2429.8	2 6.9	2697.9	2 7.7
3	1711.0	3 7.2	1934.8	3 8.2	2173.1	3 9.2	2426.0	3 10.4	2693.5	3 11.5
4	1707.2	4 9.6	1930.5	4 10.9	2168.2	4 12.3	2420.5	4 13.8	2687.4	4 15.3
5	1702.3	5 12.0	1925.0	5 13.6	2162.0	5 15.4	2413.4	5 17.2	2679.6	5 19.1
6	1696.4	6 14.4	1918.2	6 16.3	2154.3	6 18.4	2404.9	6 20.6	2670.0	6 22.9
7	1689.4	7 16.7	1910.3	7 19.0	2145.4	7 21.4	2394.8	7 23.9	2658.7	7 26.7
8	1681.3	8 19.1	1901.1	8 21.6	2135.0	8 24.4	2383.2	8 27.3	2645.8	8 30.4
9	1672.2	9 21.4	1890.8	9 24.2	2123.3	9 27.3	2370.1	9 30.6	2631.1	9 34.0
10	1662.1	10 23.7	1879.3	10 26.8	2110.4	10 30.2	2355.5	10 33.8	2614.8	10 37.7
11	1650.9	11 25.9	1866.6	11 29.4	2096.1	11 33.1	2339.5	11 37.0	2596.9	11 41.2
12	1638.8	12 28.1	1852.8	12 31.9	2080.5	12 35.9	2322.0	12 40.2	2577.4	12 44.7
13	1625.7	13 30.3	1837.9	13 34.3	2063.7	13 38.7	2303.1	13 43.3	2556.3	13 48.2
14	1611.6	14 32.4	1821.9	14 36.8	2045.6	14 41.4	2282.8	14 46.3	2533.7	14 51.6
15	1596.6	15 34.5	1804.8	15 39.1	2026.4	15 44.1	2261.2	15 49.3	2509.6	15 54.9
16	1580.6	16 36.6	1786.7	16 41.5	2005.9	16 46.7	2238.3	16 52.3	2484.0	16 58.2
17	1563.8	17 38.6	1767.6	17 43.7	1984.3	17 49.2	2214.0	17 55.1	2456.9	18 1.3
18	1546.0	18 40.6	1747.4	18 46.0	1961.5	18 51.7	2188.5	18 57.9	2428.4	19 4.4
19	1527.4	19 42.5	1726.3	19 48.1	1937.7	19 54.2	2161.8	20 0.6	2398.6	20 7.4
20	1508.0	20 44.3	1704.2	20 50.2	1912.8	20 56.5	2133.9	21 3.2	2367.5	21 10.4
21	1487.7	21 46.1	1681.3	21 52.2	1886.9	21 58.8	2104.8	22 5.8	2335.1	22 13.2
22	1466.7	22 47.8	1657.4	22 54.2	1860.0	23 1.0	2074.7	23 8.2	2301.5	23 15.9
23	1445.0	23 49.5	1632.7	23 56.1	1832.2	24 3.1	2043.5	24 10.6	2266.6	24 18.6
24	1422.5	24 51.1	1607.2	24 57.9	1803.4	25 5.2	2011.2	25 12.9	2230.7	25 21.1
25	1399.3	25 52.7	1580.8	25 59.7	1773.7	26 7.1	1978.0	26 15.1	2193.6	26 23.5
26	1375.4	26 54.2	1553.8	27 1.3	1743.2	27 9.0	1943.8	27 17.2	2155.5	27 25.9
27	1350.9	27 55.6	1526.0	28 2.9	1711.9	28 10.8	1908.7	28 19.2	2116.5	28 28.1
28	1325.8	28 56.9	1497.6	29 4.5	1679.9	29 12.5	1872.8	29 21.1	2076.5	29 30.2
29	1300.2	29 58.2	1468.5	30 5.9	1647.1	30 14.1	1836.2	30 22.9	2035.6	30 32.1
30	1274.0	30 59.4	1438.8	31 7.2	1613.7	31 15.6	1798.7	31 24.5	1993.9	31 34.0
31	1247.3	32 0.5	1408.5	32 8.5	1579.6	32 17.0	1760.6	32 26.1	1951.5	32 35.7
32	1220.2	33 1.6	1377.8	33 9.7	1545.0	33 18.4	1721.8	33 27.6	1908.3	33 37.4
33	1192.6	34 2.5	1346.5	34 10.8	1509.8	34 19.6	1682.4	34 28.9	1864.5	34 38.9
34	1164.6	35 3.4	1314.8	35 11.8	1474.1	35 20.7	1642.5	35 30.2	1820.0	35 40.2
35	1136.3	36 4.2	1282.7	36 12.7	1438.0	36 21.7	1602.1	36 31.3	1775.1	36 41.5
36	1107.6	37 5.0	1250.2	37 13.5	1401.4	37 22.6	1561.2	37 32.3	1729.6	37 42.6
37	1078.6	38 5.6	1217.4	38 14.3	1364.6	38 23.5	1520.0	38 33.2	1683.8	38 43.6
38	1049.4	39 6.2	1184.3	39 14.9	1327.4	39 24.2	1478.4	39 34.0	1637.6	39 44.5
39	1020.0	40 6.7	1151.0	40 15.4	1289.9	40 24.8	1436.6	40 34.7	1591.0	40 45.2
40	990.4	41 7.1	1117.5	41 15.9	1252.2	41 25.4	1394.5	41 35.2	1544.3	41 45.8
41	960.7	42 7.4	1083.9	42 16.3	1214.4	42 25.7	1352.2	42 35.7	1497.3	42 46.3
42	930.8	43 7.7	1050.1	43 16.5	1176.5	43 26.0	1309.8	43 36.0	1450.2	43 46.6
43	900.9	44 7.8	1016.3	44 16.7	1138.5	44 26.2	1267.4	44 36.2	1403.0	44 46.8
44	870.9	45 7.9	982.4	45 16.8	1100.4	45 26.2	1224.9	45 36.3	1355.9	45 46.9
45	841.0	46 7.9	948.5	46 16.8	1062.4	46 26.2	1182.4	46 36.2	1308.7	46 46.8
46	811.1	47 7.8	914.7	47 16.7	1024.4	47 26.1	1140.1	47 36.1	1261.7	47 46.7
47	781.2	48 7.6	881.0	48 16.5	986.5	48 25.9	1097.8	48 35.8	1214.8	48 46.4
48	751.5	49 7.4	847.4	49 16.2	948.8	49 25.5	1055.8	49 35.4	1168.2	49 45.9
49	722.0	50 7.0	814.0	50 15.8	911.4	50 25.1	1014.0	50 34.9	1121.8	50 45.4
50	692.6	51 6.6	780.8	51 15.3	874.1	51 24.5	972.4	51 34.3	1075.7	51 44.7
51	663.4	52 6.1	747.9	52 14.7	837.2	52 23.9	931.2	52 33.6	1030.1	52 43.8
52	634.5	53 5.6	715.2	53 14.1	800.6	53 23.2	890.4	53 32.8	984.8	53 42.9
53	605.9	54 4.9	682.9	54 13.3	764.3	54 22.3	850.0	54 31.8	940.1	54 41.8
54	577.6	55 4.2	651.0	55 12.5	728.5	55 21.4	810.1	55 30.8	895.8	55 40.7
55	549.7	56 3.4	619.4	56 11.6	693.1	56 20.3	770.7	56 29.6	852.2	56 39.4
56	522.1	57 2.5	588.3	57 10.6	658.3	57 19.2	731.9	57 28.3	809.2	57 37.9
57	495.0	58 1.5	557.7	58 9.5	624.0	58 18.0	693.7	58 26.9	766.9	58 36.4
58	468.3	59 0.5	527.6	59 8.3	590.2	59 16.7	656.2	59 25.5	725.3	59 34.8
59	442.1	59 59.4	498.1	60 7.1	557.1	60 15.2	619.3	60 23.9	684.5	60 33.0
60	416.4	60 58.2	469.1	61 5.8	524.7	61 13.7	583.2	61 22.2	644.5	61 31.1
61	391.3	61 57.0	440.7	62 4.3	492.9	62 12.2	547.8	62 20.4	605.4	62 29.2
62	366.7	62 55.7	413.0	63 2.9	461.9	63 10.5	513.3	63 18.6	567.2	63 27.1
63	342.7	63 54.3	386.0	64 1.3	431.6	64 8.7	479.6	64 16.6	530.0	64 24.9
64	319.4	64 52.9	359.7	64 59.7	402.2	65 6.9	446.9	65 14.6	493.7	65 22.6
65	296.7	65 51.4	334.1	65 58.0	373.6	66 6.0	415.0	66 12.4	458.5	66 20.3
Lat.	A	180° − K	A	180° − K	A	180° − K	A	180° − K	A	180° − K
H.A.	164° (10h 56m)		163° (10h 52m)		162° (10h 48m)		161° (10h 44m)		160° (10h 40m)	

FIG. 217*a*.—Sample page from "Line of Position Book."

TABLE B

K∽d	50°	51°	52°	53°	54°	55°	56°	57°	58°	59°	MIN. ALT.
0′	19193	20113	21066	22054	23078	24141	25244	26389	27579	28816	60
1	19208	20128	21082	22070	23096	24159	25263	26409	27599	28837	59
2	19223	20144	21098	22087	23113	24177	25281	26428	27619	28858	58←
3	19238	20160	21114	22104	23130	24195	25300	26448	27640	28879	57
4	19254	20175	21131	22121	23148	24213	25319	26467	27660	28900	56
5	19269	20191	21147	22138	23165	24231	25338	26487	27680	28921	55
6	19284	20207	21163	22154	23183	24249	25356	26506	27701	28942	54
7	19299	20222	21179	22171	23200	24267	25375	26526	27721	28964	53
8	19314	20238	21195	22188	23218	24286	25394	26545	27741	28985	52
9	19329	20254	21212	22205	23235	24304	25413	26565	27762	29006	51
10	19344	20269	21228	22222	23253	24322	25432	26584	27782	29027	50
11	19359	20285	21244	22239	23270	24340	25451	26604	27802	29048	49
12	19375	20301	21261	22256	23288	24358	25469	26623	27823	29069	48
13	19390	20316	21277	22272	23305	24376	25488	26643	27843	29091	47
14	19405	20332	21293	22289	23323	24395	25507	26663	27863	29112	46
15	19420	20348	21309	22306	23340	24413	25526	26682	27884	29133	45
16	19435	20364	21326	22323	23358	24431	25545	26702	27904	29154	44
17	19450	20379	21342	22340	23375	24449	25564	26722	27925	29176	43
18	19466	20395	21358	22357	23393	24467	25583	26741	27945	29197	42
19	19481	20411	21375	22374	23410	24486	25602	26761	27966	29218	41
20	19496	20427	21391	22391	23428	24504	25621	26781	27986	29239	40
21	19511	20442	21408	22408	23446	24522	25640	26800	28006	29261	39
22	19527	20458	21424	22425	23463	24541	25659	26820	28027	29282	38
23	19542	20474	21440	22442	23481	24559	25678	26840	28048	29303	37
24	19557	20490	21457	22459	23499	24577	25697	26860	28068	29325	36
25	19572	20506	21473	22476	23516	24595	25716	26879	28089	29346	35
26	19588	20522	21490	22493	23534	24614	25735	26899	28109	29367	34
27	19603	20537	21506	22510	23552	24632	25754	26919	28130	29389	33
28	19618	20553	21522	22527	23569	24650	25773	26939	28150	29410	32
29	19634	20569	21539	22544	23587	24669	25792	26959	28171	29432	31
30	19649	20585	21555	22561	23605	24687	25811	26978	28191	29453	30
31	19664	20601	21572	22578	23622	24706	25830	26998	28212	29475	29
32	19680	20617	21588	22595	23640	24724	25849	27018	28233	29496	28
33	19695	20633	21605	22613	23658	24742	25868	27038	28253	29518	27
34	19710	20649	21621	22630	23676	24761	25887	27058	28274	29539	26
35	19726	20665	21638	22647	23693	24779	25907	27078	28295	29561	25
36	19741	20681	21654	22664	23711	24798	25926	27098	28315	29582	24
37	19756	20696	21671	22681	23729	24816	25945	27117	28336	29604	23
38	19772	20712	21687	22698	23747	24835	25964	27137	28357	29625	22
39	19787	20728	21704	22715	23764	24853	25983	27157	28378	29647	21
40	19803	20744	21720	22732	23782	24872	26003	27177	28398	29668	20
41	19818	20760	21737	22750	23800	24890	26022	27197	28419	29690	19
42	19834	20776	21754	22767	23818	24909	26041	27217	28440	29712	18
→43	19849	20792	21770	22784	23836	24927	26060	27237	28461	29733	17
44	19864	20808	21787	22801	23854	24946	26079	27257	28481	29755	16
45	19880	20824	21803	22819	23871	24964	26099	27277	28502	29776	15
46	19895	20840	21820	22836	23889	24983	26118	27297	28523	29798	14
47	19911	20856	21837	22853	23907	25001	26137	27317	28544	29820	13
48	19926	20872	21853	22870	23925	25020	26157	27337	28565	29841	12
49	19942	20889	21870	22888	23943	25039	26176	27357	28586	29863	11
50	19957	20905	21887	22905	23961	25057	26195	27378	28607	29885	10
51	19973	20921	21903	22922	23979	25076	26215	27398	28627	29907	9
52	19988	20937	21920	22939	23997	25094	26234	27418	28648	29928	8
53	20004	20953	21937	22957	24015	25113	26253	27438	28669	29950	7
54	20019	20969	21953	22974	24033	25132	26273	27458	28690	29972	6
55	20035	20985	21970	22991	24051	25150	26292	27478	28711	29994	5
56	20050	21001	21987	23009	24069	25169	26311	27498	28732	30016	4
57	20066	21017	22003	23026	24087	25188	26331	27518	28753	30037	3
58	20082	21033	22020	23043	24105	25206	26350	27539	28774	30059	2
59	20097	21050	22037	23061	24123	25225	26370	27559	28795	30081	1
60	20113	21066	22054	23078	24141	25244	26389	27579	28816	30103	0′
	39°	38°	37°	36°	35°	34°	33°	32°	31°	30°	MIN. ALT.

ALTITUDE Hc

FIG. 217b.—Sample page from "Line of Position Book."

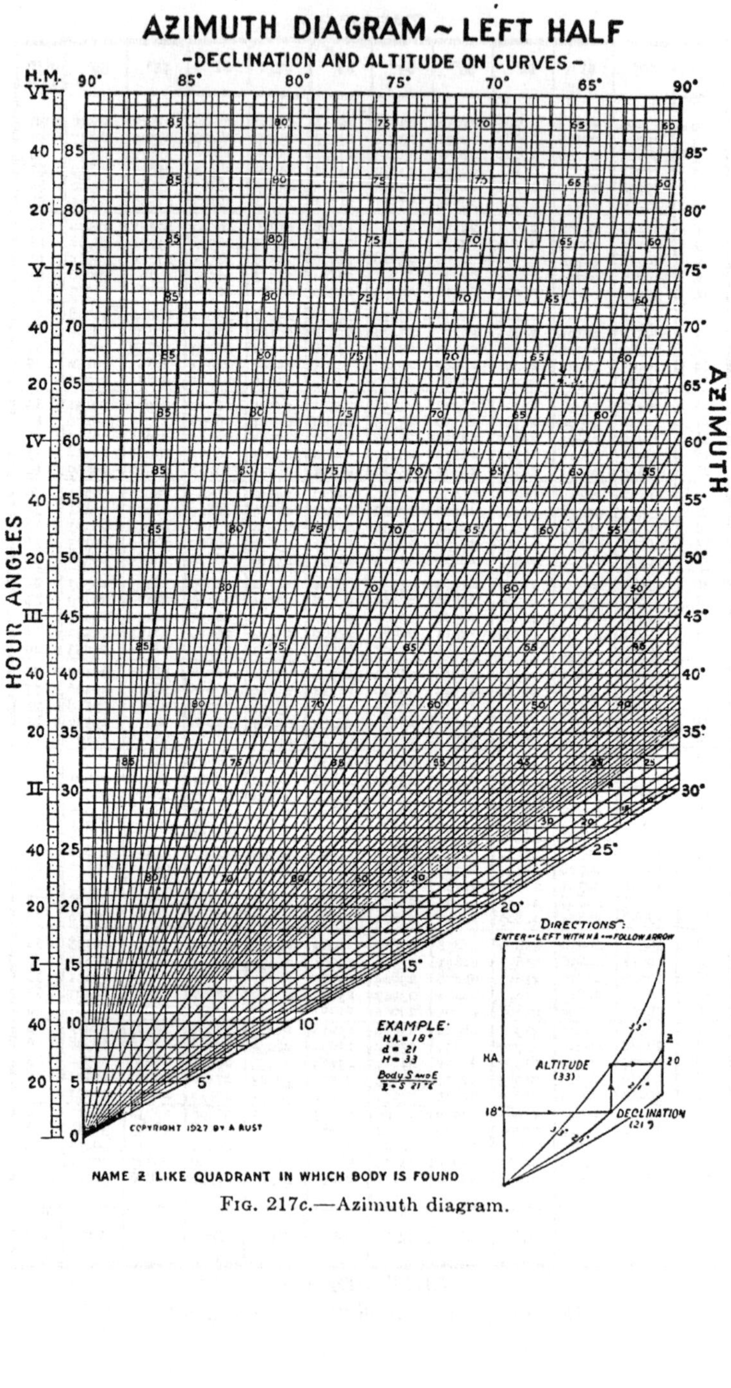

FIG. 217c.—Azimuth diagram.

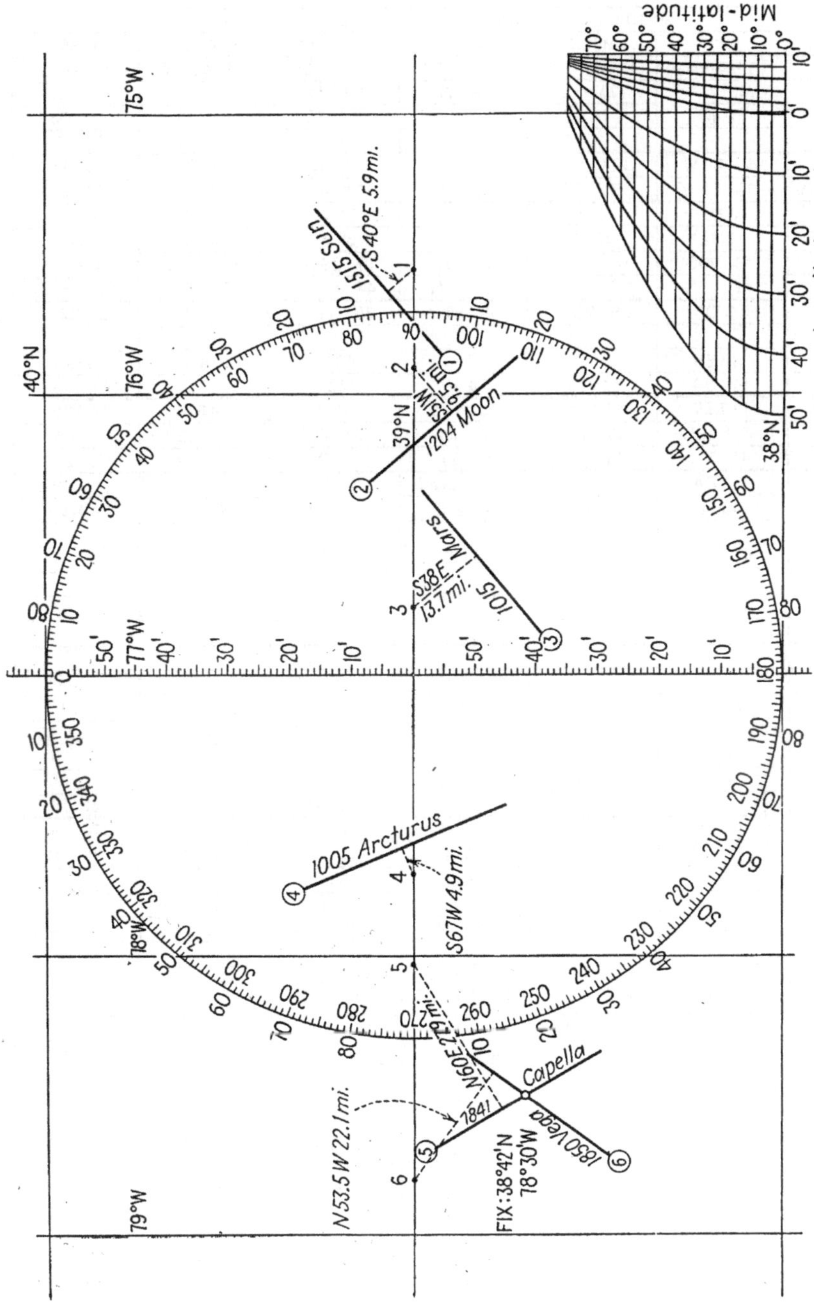

FIG. 218.—Lines of position as plotted.

SOLUTIONS FOR LINES OF POSITION 17 MARCH 1941						
	SUN	MOON	MARS	ARCTURUS	CAPELLA	VEGA
GCT	15h 14m 40s	12 04 08	10 14 52	10 05 28	18 40 40	18 50 20
GHA	45°23'	123 55	34 02	324 35	94 57	97 27
CORR	1 10	1 00	3 43	1 22	10	05
(SHA)				146 45	281 55	81 16
GHA	46 33 W	124 55	37 45	112 42	17 02	178 48
LONG(A)	75 33 W	75 55	76 45	77 42	78 02	78 48
LHA	29 E	49 W	39 E	35 W	61 E	100 W
LAT(A)	39 N	39 N	39 N	39 N	39 N	39 N
DEC	1 20S	14 21S	22 46S	19 29N	45 56N	38 44N
K	42 48N	50 59N	46 11N	44 40N	59 06N	102 06N
K~DEC	44 08	65 20	68 57	25 11	13 10	63 22
A	3325	9155	5937	4810	13461	19137
+B	14404	37951	44468	4337	1157	34845
LOG H_c	17729	47106	50405	9147	14618	53982
H_c	41° 40'	19 45	18 15.5	54 06	45 35	16 46
H_o	41 34	19 54	18 29	54 01	45 07	16 24
a	6A	9T	13.5T	5A	28A	22A
Z	S 40° E	S 51 W	S 38 E	S 67 W	N 60 E	N 53.5 W
H_s	41 35	18 57	18 32	54 02	45 08	16 27

	SUN	MOON	MARS	ARCTURUS	CAPELLA	VEGA
GCT	15h 14m 40s	12 04 08	10 14 52	10 05 28	18 40 40	18 50 20
GHA	45° 23'	123 55	34 02	324 35	94 57	97 27
CORR	1 10	1 00	3 43	1 22	10	05
(SHA)				146 45	281 55	81 16
GHA	46 33	124 55	37 45	112 42	17 02	178 48
LONG(A)	75 33W	75 55	76 45	77 42	78 02	78 48
LHA	29 E	49 W	39 E	35 W	61 E	100 W
LAT(A)	39 N	39 N	39 N	39 N	39 N	39 N
DEC	1 20S	14 21S	22 46S	19 29N	45 56N	38 44N
H	41° 31.7'	19 38.2	18 03.8	54 06.7	45 35.7	16 37.7
CORR	(+85) 8.5	(+78) 7.0	(+86) 12.0	(−64) 0.6	(−24) 1.0	(+60) 8.4
H_c	41 40.2	19 45.2	18 15.8	54 06.1	45 34.7	16 46.1
H_o	41 34	19 54	18 29	54 01.	45 07	16 24
a	6A	9T	13T	5A	28A	22A
Z_N	139.7°	230.9	142.5	247.2	60.2	306.4
H_s	41° 35'	18 57	18 32	54 02	45 08	16 27

FIG. 219.—Solution of six lines of position by "LPB" and "H.O. 214."

AZIMUTHS BY OGURA'S TABLE B

	1	2	3	4	5	6	7	8	9
	° ′	° ′	° ′	° ′	° ′	° ′	° ′	° ′	° ′
t	67 W.	16 W.	25 W.	25 W.	43 W.	56 E.	57 W.	80 E.	33 W.
d	2–21 N.	57–35.7 S.	12–36.8 N.	3–22.6 N	45–01.6 N.	23–23.8 N.	19–31.4 N.	20–55 S.	21–19.8 S.
H_o	18–16.2	30–01.3	40–13	48–31.3	50–07.4	14–14	36–03.5	18–06.3	58–34
cosec t	3597	55966	37405	37405	16622	8143	7641	664.9	26389
sec d	36.5	28309	1061	75	15072	3723	2572	2961.0	3082
cosec t + sec d	3633.5	84275	38466	37480	31694	11866	10213	3625.9	29471
sec h	2247	6255	11713	17632	19305	1354	9236	2205	28274
cosec Z	1386.5	78020	26753	19848	12389	10512	977	1420.9	1197
Z	N. 75–36 W.	S. 9–33 W.	N. 32–41 W.	N. 39–17 W.	N. 48–45 W.	N. 51–43 E.	N. 77–53 W.	S. 75–25 E.	N. 76–37 W.

Formula: $\text{cosec } Z = \dfrac{\text{cosec } t \text{ sec } d}{\text{sec } H}$.

FIG. 220.—Azimuths by Ogura's Table B.

latitude and declination are of contrary names, or subtracting if of the same name, to get the value $K \sim d$.

ᴎATION SAME NAME AS LATITUDE 257

Lat. 39°

H.A.	56° 30′ Alt. ° ′	Δd Δt	Az. °	57° 00′ Alt. ° ′	Δd Δt	Az. °	57° 30′ Alt. ° ′	Δd Δt	Az. °	59° 00′ Alt. ° ′	Δd Δt	Az. °	59° 30′ Alt. ° ′	Δd Δt	Az. °	H.A.
91	31 09.1	42 50	40.1	31 21.6	41 49	39.6	31 33.9	41 49	39.1	32 10.2	40 47	37.5	32 22.1	39 47	36.9	91
2	30 39.1	43 50	39.9	30 51.9	43 49	39.3	31 04.6	42 48	38.8	31 42.0	41 47	37.2	31 54.1	40 46	36.7	2
3	30 09.3	44 49	39.6	30 22.4	44 49	39.1	30 35.5	43 48	38.6	31 13.9	42 47	37.0	31 26.4	41 46	36.5	3
4	29 39.6	45 49	39.3	29 53.1	45 48	38.8	30 06.5	44 48	38.3	30 45.9	43 46	36.7	30 58.8	42 46	36.2	4
95	29 10.2	46 49	39.0	29 24.0	46 48	38.5	29 37.7	46 48	38.0	30 18.1	44 46	36.5	30 31.3	43 45	35.9	95
6	28 40.9	47 48	38.7	28 55.0	47 48	38.2	29 09.1	47 47	37.7	29 50.5	45 46	36.2	30 04.0	45 45	35.7	6
7	28 11.8	49 48	38.4	28 26.3	48 48	37.9	28 40.6	48 47	37.4	29 23.0	47 45	35.9	29 36.9	46 45	35.4	7
8	27 42.9	49 48	38.1	27 57.7	49 47	37.6	28 12.4	49 47	37.1	28 55.7	48 45	35.6	29 10.0	47 45	35.1	8
9	27 14.2	50 47	37.8	27 29.4	50 47	37.3	27 44.3	50 46	36.8	28 28.6	49 45	35.4	28 43.2	48 44	34.9	9
100	26 45.8	52 47	37.5	27 01.1	51 46	37.0	27 16.5	51 46	36.5	28 01.8	50 44	35.1	28 16.7	49 44	34.6	100
1	26 17.5	53 47	37.2	26 33.2	52 46	36.7	26 48.8	52 46	36.2	27 35.1	51 44	34.8	27 50.3	50 44	34.3	1
2	25 49.4	54 46	36.9	26 05.4	53 46	36.4	26 21.4	53 45	35.9	27 08.5	52 44	34.5	27 24.1	51 43	34.0	2
3	25 21.6	55 46	36.5	25 37.9	54 46	36.0	25 54.1	54 45	35.6	26 42.3	53 43	34.2	26 58.2	52 43	33.7	3
4	24 53.9	56 46	36.2	25 10.5	55 45	35.7	25 27.1	55 45	35.3	26 16.2	54 43	33.9	26 32.4	54 43	33.3	4
105	24 26.5	57 45	35.8	24 43.4	56 45	35.4	25 00.3	56 44	34.9	25 50.3	55 43	33.6	26 06.8	55 42	33.1	105
6	23 59.3	58 45	35.5	24 16.5	57 44	35.0	24 33.7	57 44	34.6	25 24.6	56 42	33.2	25 41.5	56 42	32.8	6
7	23 32.3	58 44	35.1	23 49.9	58 44	34.7	24 07.3	58 43	34.3	24 59.2	57 42	32.9	25 16.3	57 41	32.5	7
8	23 05.6	59 44	34.8	23 23.4	59 44	34.3	23 41.2	59 43	33.9	24 33.9	58 41	32.6	24 51.4	58 41	32.1	8
9	22 39.1	60 44	34.4	22 57.2	60 43	34.0	23 15.3	60 43	33.6	24 09.0	59 41	32.2	24 26.7	59 41	31.8	9
110	22 12.9	61 43	34.1	22 31.2	61 43	33.6	22 49.6	61 42	33.2	23 44.2	60 41	31.9	24 02.3	60 40	31.5	110
1	21 46.9	62 43	33.7	22 05.6	62 42	33.3	22 24.2	62 42	32.9	23 19.7	61 40	31.6	23 38.0	61 40	31.1	1
2	21 21.1	63 43	33.3	21 40.1	63 42	32.9	21 59.0	63 42	32.5	22 55.4	62 40	31.2	23 14.0	62 39	30.8	2
3	20 55.6	64 42	32.9	21 14.9	64 42	32.5	21 34.1	64 41	32.1	22 31.3	63 40	30.9	22 50.3	63 39	30.5	3
4	20 30.4	65 42	32.6	20 49.9	65 41	32.1	21 09.4	65 41	31.8	22 07.5	64 39	30.5	22 26.8	64 39	30.1	4
115	20 05.4	66 41	32.2	20 25.2	66 41	31.8	20 45.0	66 40	31.4	21 43.9	65 39	30.2	22 03.5	65 38	29.8	115
6	19 40.7	67 41	31.8	20 00.8	67 40	31.4	20 20.9	67 40	31.0	21 20.6	66 38	29.8	21 40.5	66 38	29.4	6
7	19 16.3	68 40	31.4	19 36.7	68 40	31.0	19 57.0	68 39	30.6	20 57.6	67 38	29.4	21 17.7	67 37	29.0	7
8	18 52.1	69 40	31.0	19 12.8	69 39	30.6	19 33.4	69 39	30.2	20 34.8	68 37	29.1	20 55.2	68 37	28.7	8
9	18 28.3	70 39	30.6	18 49.2	70 39	30.2	19 10.0	70 38	29.8	20 12.3	69 37	28.7	20 33.0	69 37	28.3	9
120	18 04.7	71 39	30.2	18 25.8	71 38	29.8	18 47.0	71 38	29.4	19 50.1	70 37	28.3	20 11.0	70 36	27.9	120
1	17 41.4	72 38	29.8	18 02.8	71 38	29.4	18 24.2	71 37	29.0	19 28.1	71 36	27.9	19 49.3	70 36	27.5	1
2	17 18.4	72 38	29.4	17 40.1	72 37	29.0	18 01.7	72 37	28.6	19 06.4	72 36	27.5	19 27.9	71 35	27.2	2
3	16 55.7	73 37	28.9	17 17.6	73 37	28.6	17 39.5	73 36	28.2	18 45.0	73 35	27.1	19 06.7	72 35	26.8	3
4	16 33.3	74 37	28.5	16 55.5	74 36	28.2	17 17.6	74 36	27.8	18 23.9	73 35	26.7	18 45.9	73 34	26.4	4
125	16 11.2	75 36	28.1	16 33.6	75 36	27.8	16 56.0	75 35	27.4	18 03.0	74 34	26.3	18 25.3	74 34	26.0	125
6	15 49.4	76 36	27.7	16 12.1	76 35	27.3	16 34.7	76 35	27.0	17 42.4	75 34	25.9	18 04.9	75 33	25.6	6
7	15 27.9	77 35	27.2	15 50.8	77 35	26.9	16 13.7	77 34	26.6	17 22.2	76 33	25.5	17 45.0	76 33	25.2	7
8	15 06.7	78 35	26.8	15 29.8	78 34	26.4	15 53.0	78 34	26.1	17 02.3	77 33	25.1	17 25.3	76 32	24.8	8
9	14 45.9	78 34	26.3	15 09.3	78 34	26.0	15 32.7	78 33	25.7	16 42.6	78 32	24.7	17 05.9	77 32	24.4	9
160	6 59.5	96 14	11.0	7 28.4	96 14	10.8	7 57.3	96 14	10.7	9 23.9	96 13	10.3	9 52.8	96 13	10.2	160
1	6 50.8	97 14	10.4	7 19.8	97 13	10.3	7 48.8	97 13	10.2	9 15.8	96 13	09.8	9 44.8	96 13	09.7	1
2	6 42.6	97 13	09.9	7 11.7	97 13	09.8	7 40.8	97 13	09.7	9 08.1	97 12	09.3	9 37.2	97 12	09.2	2
3	6 34.8	97 12	09.3	7 04.0	97 12	09.2	7 33.2	97 12	09.1	9 00.8	97 11	08.8	9 30.0	97 11	08.7	3
4	6 27.4	98 11	08.8	6 56.7	98 11	08.7	7 26.0	98 11	08.6	8 53.9	98 11	08.2	9 23.2	98 11	08.1	4
165	6 20.5	98 11	08.3	6 49.9	98 11	08.2	7 19.3	98 10	08.1	8 47.3	98 10	07.8	9 16.8	98 10	07.6	165
6	6 14.0	98 10	07.7	6 43.5	98 10	07.6	7 13.0	98 10	07.5	8 41.3	98 09	07.2	9 10.8	98 09	07.1	6
7	6 08.0	98 09	07.2	6 37.5	98 09	07.1	7 07.1	98 09	07.0	8 35.6	98 09	06.7	9 05.2	98 09	06.6	7
8	6 02.4	99 09	06.6	6 32.0	99 08	06.5	7 01.6	99 08	06.5	8 30.4	98 08	06.2	9 00.0	98 08	06.1	8
9	5 57.2	99 08	06.1	6 26.9	99 08	06.0	6 56.6	99 08	05.9	8 25.6	99 07	05.7	8 55.2	99 07	05.6	9
170	5 52.5	99 07	05.5	6 22.2	99 07	05.5	6 52.0	99 07	05.4	8 21.1	99 07	05.2	8 50.8	99 06	05.1	170
1	5 48.2	99 06	05.0	6 18.0	99 06	04.9	6 47.8	99 06	04.9	8 17.1	99 06	04.7	8 46.9	99 06	04.6	1
2	5 44.4	99 06	04.4	6 14.2	99 05	04.4	6 44.1	99 05	04.3	8 13.5	99 05	04.1	8 43.3	99 05	04.1	2
3	5 41.0	99 05	03.9	6 10.9	99 05	03.8	6 40.8	99 05	03.8	8 10.3	99 04	03.6	8 40.2	99 04	03.6	3
4	5 38.1	99 04	03.3	6 08.0	99 04	03.3	6 37.9	99 04	03.2	8 07.6	99 04	03.1	8 37.5	99 04	03.1	4
175	5 35.6	1.0 03	02.8	6 05.5	1.0 03	02.7	6 35.5	1.0 03	02.7	8 05.2	1.0 03	02.6	8 35.2	1.0 03	02.6	175
6	5 33.6	1.0 03	02.2	6 03.5	1.0 02	02.2	6 33.5	1.0 02	02.2	8 03.4	1.0 02	02.1	8 33.3	1.0 02	02.1	6
7	5 32.0	1.0 02	01.7	6 02.0	1.0 02	01.6	6 32.0	1.0 02	01.6	8 01.9	1.0 02	01.6	8 31.9	1.0 02	01.5	7
8	5 30.9	1.0 01	01.1	6 00.9	1.0 01	01.1	6 30.9	1.0 01	01.1	8 00.8	1.0 01	01.0	8 30.8	1.0 01	01.0	8
9	5 30.2	1.0 00	00.6	6 00.2	1.0 00	00.5	6 30.2	1.0 00	00.5	8 00.2	1.0 00	00.5	8 30.2	1.0 00	00.5	9
180	5 30.0	1.0 00	00.0	6 00.0	1.0 00	00.0	6 30.0	1.0 00	00.0	8 00.0	1.0 00	00.0	8 30.0	1.0 00	00.0	180

4—7666

FIG. 221.—Page from "H.O. 214."

With the value $K \sim d$, turn to Table B (Fig. 217*b*) and with the value to the nearest minute, 55°43′, find the number 24927, and add together *A* and *B* to get 26437. Now, entering Table B (Fig. 217*b*) at the bottom and looking in the column for this same number, the altitude is found to be

32°58′ as the calculated altitude. The difference between this altitude and the true altitude found with the sextant gives the altitude intercept 14 miles. Since the true altitude is greater than the calculated, the direction is toward the observed body.

Azimuth.—The azimuth is taken from the diagram shown in Fig. 217*c*. On the left-hand side of the page enter with 18°, cross on this horizontal line until it intersects the curved line of declination 21°, pass up this line in a vertical direction until the altitude curve 33° is intersected, then pass horizontally to the right side of diagram, and thus read the azimuth 20°. Since the sun bore southeast, the azimuth is S. 20°E.

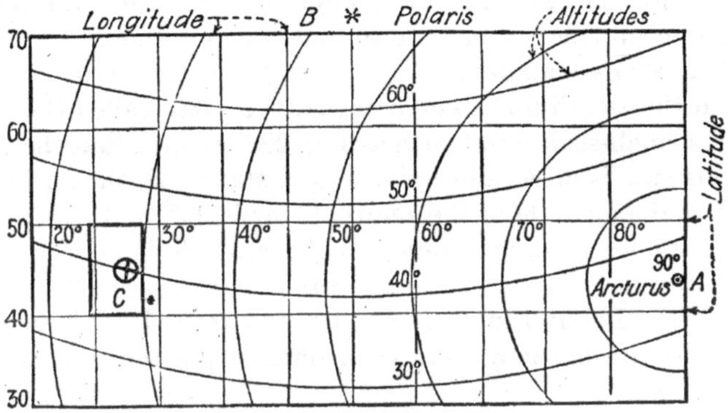

FIG. 222.—Substellar points.

For Meridian Altitude.—If the LHA is 0, the sun is on the meridian, and $A = 0$, and

$$K = \text{latitude}$$

Since $0 + B = B$, it is only necessary to enter Table B with $K \sim d$ and pick out from B the H_c direct. It is not necessary in this case to write down A, B, or log H_c. Also since the body is on the meridian, the azimuth is not required, and the a is applied to the assumed latitude to get the latitude direct.

Form Used.—The column form of work sheet as shown will be found convenient in the air, since several sights can be worked on one page and since the form need not be written each time. Also since similar terms for all sights appear on the same line, a ready check on the work is afforded. It is customary to make the small correction for refraction mentally.

"Star Altitude Curves."—From any given position on the earth, at any given instant of sidereal time, there is only one possible altitude for each fixed star. The "Star Altitude Curves" take advantage of this fact to do in advance a great part of the work otherwise required of the navigator.

The simultaneous altitudes of two stars, together with the Greenwich sidereal time of observation, definitely determine a point on the earth's surface. This may be put in graphic form by plotting the altitudes against the latitude and local sidereal time. The simultaneous altitudes thus determine by the curves a latitude and a corresponding local sidereal time. The local sidereal time found from the curves combined with the Greenwich sidereal time gives the longitude of the observer. Thus both latitude and longitude are determined without reference to the dead-reckoning position, right ascension, declination, hour angle, or azimuth. No plotting whatever is required to obtain a fix. The entire computation for the latitude and longitude of a definite position is reduced to one subtraction of time to find the longitude.

Graphical Representation.—At any given instant a star is directly over (*i.e.*, in the zenith for) some point on the earth called the substellar point. On a small-scale chart such as Fig. 222, assume that the substellar point for one star is at *A*, and for a second star at *B*, shown just off the chart. At *A* the altitude of the zenithal star is 90°, and at a distance of 600 miles (60 nautical miles equal 1°) the altitude is 80°, and at 1,200 miles the altitude is 70°, etc. In the same way, curves of altitudes for the second star (Polaris) may be constructed from *B*. These circles of equal altitudes are nothing more than lines of position laid down for a given instant of time.

Referring again to Fig. 222, suppose the altitude of star *A* is observed to be 28°, and of star *B*, 40°; then the intersection of these two altitude circles at *C* is the observer's position. There are two positions possible which are on the same two circles of altitude, but only the one for which the curves are constructed will give the proper value of the local sidereal time.

If the star *A* of 0° declination is on the prime vertical with the observer on the equator, its altitude at point *C* will increase at the rate of 1° for every 4 min. of elapsed sidereal time. The star *A* changes altitude from 0° to 90° in 6 hr., therefore, 1 hr. equals 15°, or 4 min. equals 1°. Four minutes after the altitude of star *A*, as observed from *C*, is 28°, it will have increased to 29°, and 4 min. later, to 30°, etc. This may be graphed by having the time scale increase toward the right as shown in Fig. 223. Instead of considering that the altitude increases for the passing of time, picture the time increasing, as represented by the local-sidereal-time scale, for greater altitudes. For other latitudes and declinations, the change of altitude would be less than 1° for 4 min. of time, but the figures given illustrate the principle. The "Δt" in "H.O. 214," shows the change of altitude for change of time.

Since the azimuth is at right angles to the line of position and since the altitude increases when the body is approached, the "Star Altitude

Curves" give the approximate azimuth at a glance. In Fig. 223 the azimuth of Vega will be seen at once to be rising because the altitude increases with time, and to be nearly east because the altitude curves for Vega run nearly north and south. Given the approximate local sidereal time and latitude, the curves give the name, azimuth, and approximate altitude of the star to be observed. The curves may be used conveniently for star finding.

Provision is also made for the accurate simple use of any edition of the curves for a date earlier or later than the date of publication. This is accomplished by applying to the *sextant* altitude a correction for the desired date. The figure below each star's name in Fig. 223 is the correction to be applied for the annual change in altitude, the sign showing how it is applied for a date *later* than the epoch for which the curves are computed and positioned. Figure 223 shows a sample page of the new curves reduced one-half.

Sidereal Time and Longitude.—*Local sidereal time* (LST) is found from the "Star Altitude Curves" by projecting the altitude intersection to the top or bottom scale. Longitude is the difference between *Greenwich sidereal time* (GST) and LST. GST may be determined by any of several different methods:

1. By GST watch showing GST in time units.
2. By GST watch showing GST in arc units.
3. By converting Greenwich civil time (GCT) to GST in arc by means of the *Air Almanac*, or by means of a mechanical time converter.

When using GST in time units, LST is taken from the top scale of the "Star Altitude Curves," and the difference is longitude in time units which should be converted to arc units. When GST in arc units is used, LST is taken from the bottom scale of the "Star Altitude Curves."

The *Air Almanac* gives GST in arc (GHA of ♈) for 10-min. intervals with a convenient interpolation table for minutes and seconds from 0 to 10 min. This is perhaps the most satisfactory way of finding longitude when the *Air Almanac* is available. Remember that the hour angle of Aries, ♈, is sidereal time.

Example.—At any time, any place, observed with an adjusted bubble sextant the altitude of Vega to be 39°35′ and the Greenwich sidereal time of observation to be $19^h15^m29^s$. Immediately thereafter observed the altitude of Polaris to be 37°58′. The star Vega is observed to be in the east and rising. Required, a fix.

Solution (Using GST watch).—(1) The altitude of curve of Polaris indicates the band of latitude (30° to 40°N.) in which the observer is located. (2) Follow through the curves until the altitude of the star Vega is approximately 40° and rising, or take the difference between the approximate longitude in time and the watch (GST) to get the approximate LST and turn to that page of the curves (Fig. 223). (3) Find the exact intersection of the curves for the two altitudes observed. This point projected *vertically* to the time scale at the top or bottom gives the local sidereal time ($14^h09^m38^s$)

of the place. The difference between the local sidereal time from the scale and the observed Greenwich sidereal time gives a longitude of $5^h05^m51^s$, this being in units of time, and when converted into arc gives a longitude of 76°28′W. (4) The point of intersection projected *horizontally* to either of the latitude scales gives a latitude of 38°57′.5N. Note that the Polaris altitude curves are not horizontal and should not be followed to pick latitude from the scale.

Figure 224 shows the solution of four examples, using the Air Almanac to find GHA ♈ (GST).

Practical Use of Adjusted Altitude Line.—Regardless of the method used, the difficulty of taking simultaneous altitudes of two or of three stars complicates celestial air navigation. This difficulty may be reduced by using the "Star Curves" in a manner similar to "Precomputed Altitudes" described later.

Suppose a plane making 300 m.p.h. on course 240° true is at *A*, Lat. 70°N., Long. 56°W. at 1800 GST, or 1416 LST (1800 less 3^h44^m for 56°W. longitude). In 10 min. the plane would travel 50 miles on 240° and, with a page of the curves used as a Mercator chart, as shown in Fig. 225, would arrive at point *B*, in Lat. 68°36′N. Because of the distance traveled in 10 min., the altitude of Vega would change from 44°36′ at *A* to 43°55′ at *B*. In the elapsed 10 min. Vega's altitude increases at point *B* from 43°55′ to 44°48′ at *C*. The combined effect of change of position and 10 min. of elapsed time would change Vega's altitude from 44°36′ to 44°48′. If Vega's altitude at 1800 GST is 44°36′ and at 1810 GST is 44°48′, then for 1805 GST Vega's altitude is shown at once to be 44°42′. In other words, the line *AE* is the locus of simultaneous altitudes of Vega, Capella, and Polaris, provided the plane remains on its schedule. If at 1830 GST Vega's altitude is 45°00′, the plane is at some point *D* about 12 miles off course to right. If an observation of Polaris gives an altitude of 67°40′ at 1830, the intersection with Vega gives *D* as the definite fix, and the plane is shown to be 13 miles 247° true from the scheduled 1830 position.

The adjusted altitude line may be laid down for any course and speed. By its use the problem of advancing lines may be greatly simplified.

Other Methods of Using Curves.—A transparent template may be used over the curves in the book to find a position without any calculation. The latitude and longitude are etched on the transparent template, making of it a Mercator chart to the same scale as that of the curves. Positions may be plotted on the template; also, courses and bearings. With this arrangement, the latitude and longitude may be determined without writing a single figure—simply by orienting the LST scale with a point on the transparent cover, and then by marking the intersection of the altitude curves. The intersection of the altitude curves shows the latitude and longitude on the transparent cover.

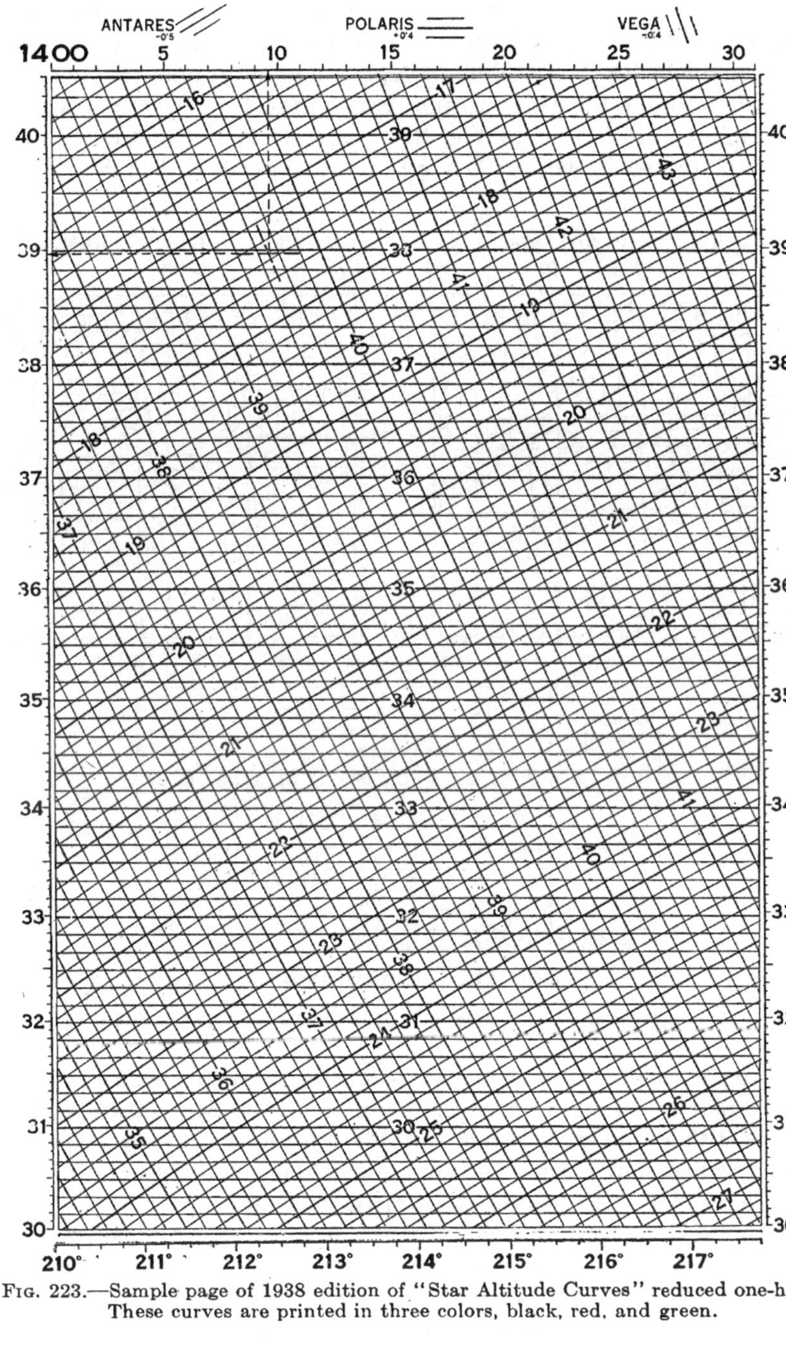

FIG. 223.—Sample page of 1938 edition of "Star Altitude Curves" reduced one-half. These curves are printed in three colors, black, red, and green.

Figure 226 shows a transparent template. The mid-longitude is marked 70°, 80°, etc., as desired. The mid-longitude of the template is then aligned with the last digit of degrees and exact minutes of GHA ♈, in which position the star altitude curve may be correctly traced on the template. Referring to Fig. 224, example 1, the mid-longitude of the template would be marked 80°, and aligned with (28) 3°38′, in

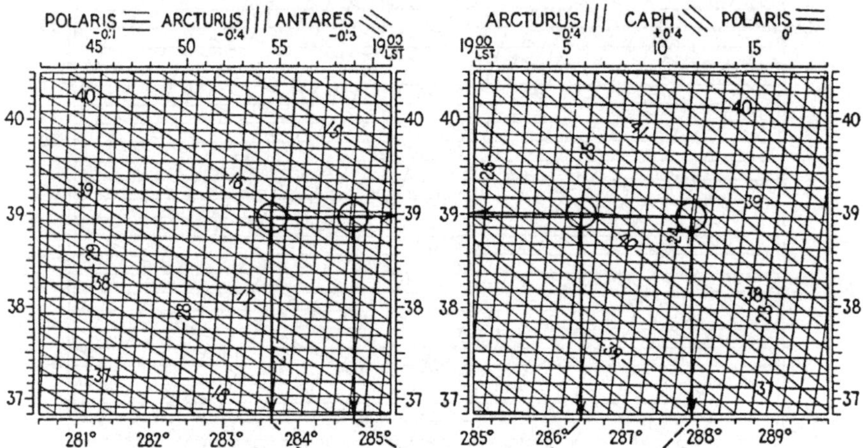

Hs POLARIS	38° 45′	38 47	38 52	38 51
Hs ARCTURUS	27 07	26 16	24 57	23 50
CORR. DATE	+ 02			
Hs ARCTURUS	27 09	26 18	24 59	23 52
G.C.T.	2 55 30	3 00 08	3 07 00	3 12 37
G.S.T.	358 44	361 14	361 14	363 44
CORR.	1 23	2	1 45	39
G.S.T.	360 07	361 16	362 59	364 23
L.S.T.	283 38	284 45	286 27	287 55
LONG.	76 29W	76 31W	76 32W	76 28W
LAT.	38 58N	38 58N	39 00N	38 58N

OBSERVERS POSITION-LAT. 38°58′N., LONG. 76°30′W

SAC TEST 8 AUG. 1941

FIG. 224.—Solution of four examples using the Air Almanac to find GHA ♈ (GST).

which position the Arcturus curve would be correctly positioned under the template. In addition to the book form, the Star Altitude Curves may be lithographed on strips suitable for use in a roller map holder with a transparent celluloid cover on which are etched the latitude and longitude to the same scale as that of the curves.

The slight advantage gained in the use of the template by saving the subtraction of time for longitude when working direct from the curves is offset by a slight loss in accuracy, and the necessity of carefully placing the longitude scale at a definite point on the LST scale.

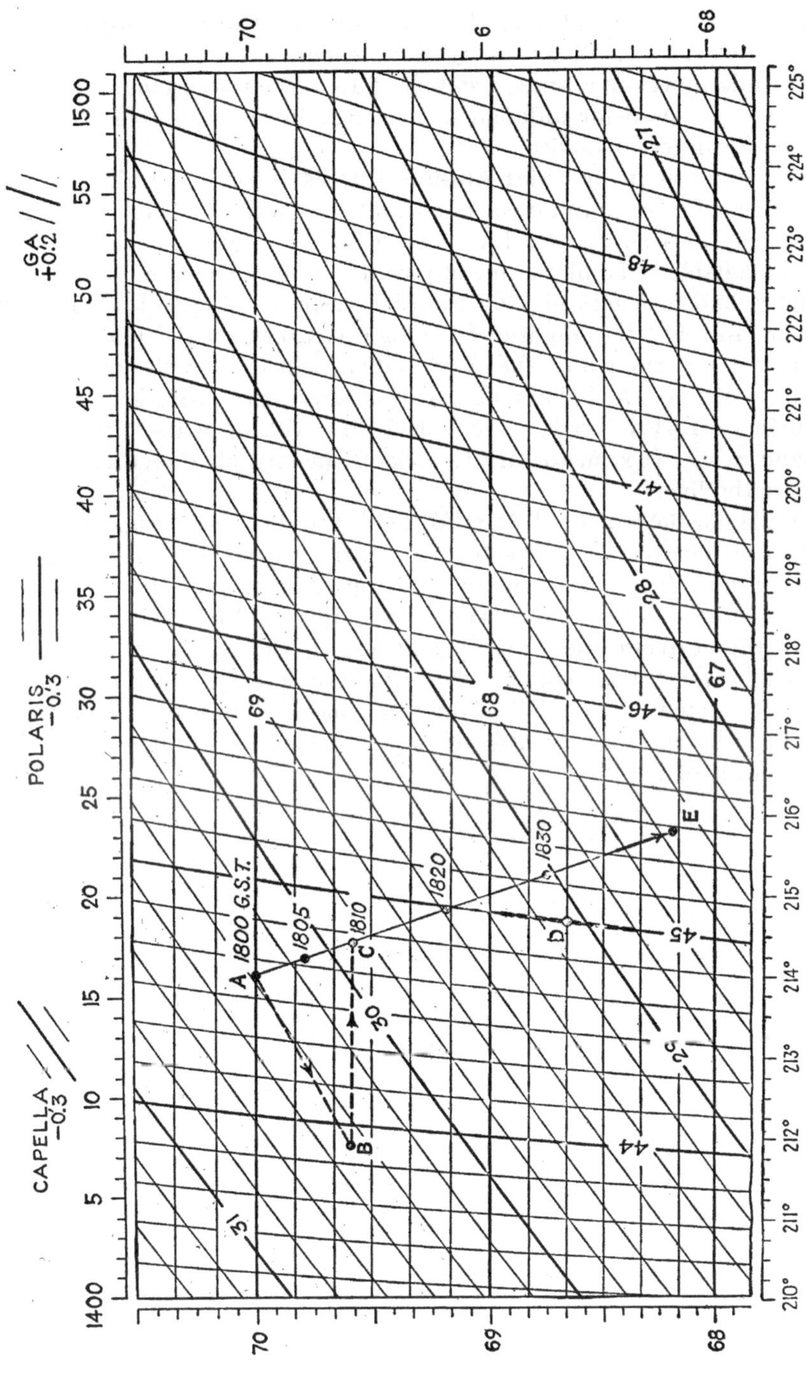

FIG. 225.—Practical use of adjusted altitude line. The curves in this figure are constructed for epoch 1940.

The same general idea of the Star Altitude Curves is used in the construction of the Baker navigation machine. With this machine, the altitude curves on a transparent sheet are passed over a map. Different sheets are used for different declinations in order to give a general solution for different bodies.

Line of Position by "Star Altitude Curves."—Although designed to determine a fix by simultaneous observations of two or more stars, the curves may also be used to lay down a line of position when only one star is available. To plot a line of position from the curves, assume two latitudes, and for each, pick off the LST corresponding to the observed altitude, and find the longitudes for each latitude. Then plot the two positions so determined on the chart and connect them with the required line of position. A Polaris observation gives the latitude when the approximate LST is used. The Polaris line will, of course, run nearly east and west. It is customary to pick off the latitude direct and not to consider the line.

Problems Solved by "Star Altitude Curves."—In Fig. 224, the solution of four problems is shown together with portions of two pages of the curves. The correction for date appearing on the third line of the solution is ignored for Polaris, and is four times 0.4′ for Arcturus. The sign of the correction is reversed and added because the observations were made about four years in advance of the date for which the curves were constructed. The GST in the sixth line of the solution is given in arc and is taken from the *Air Almanac.* The LST is taken from the bottom scale of Fig. 224.

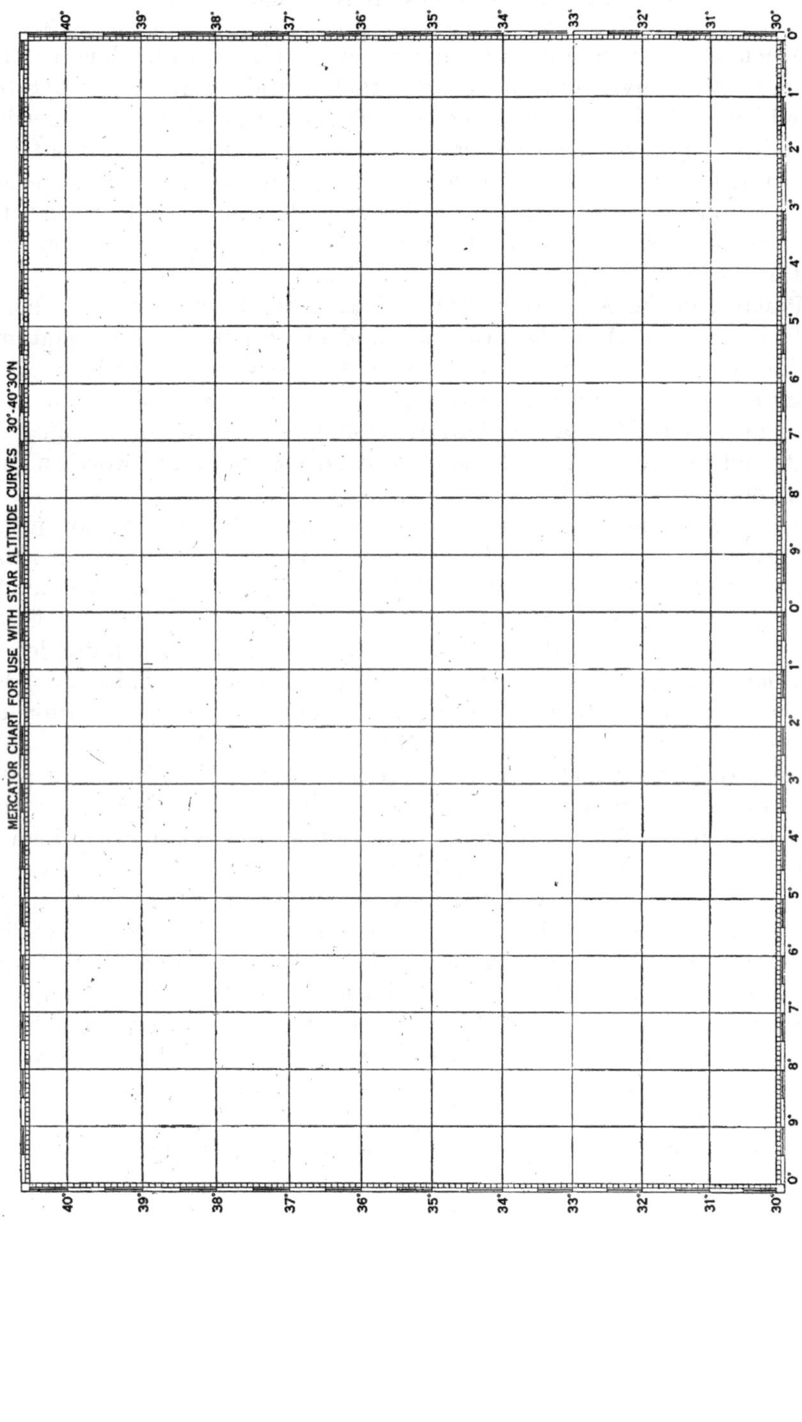
MERCATOR CHART FOR USE WITH STAR ALTITUDE CURVES 30°-40°30'N
40° 39° 38° 37° 36° 35° 34° 33° 32° 31° 30°
0° 1° 2° 3° 4° 5° 6° 7° 8° 9° 0° 1° 2° 3° 4° 5° 6° 7° 8° 9° 0°

FIG. 226.

CHAPTER XIII

CELESTIAL NAVIGATION—PRACTICE

Much of the computation for navigation in flight can be done before the take-off. Obviously, the charts, tables, and equipment must be assembled and the courses selected. The courses, distances, speeds, and parts of the computations can be prepared in advance. Since time consumed before the take-off is not pressing and since practice is beneficial, as much of the work as possible should be done and checked in advance and condensed in a neat, orderly manner for the most convenient used in the air.

Practice in the Air.—In a text such as this, the most we can hope to do is to make clear the principles and procedure of air navigation. Obviously, the practical air navigator must follow this up with practice in the air. A few experiences trying to tune in a radio station, a few hours in a dense fog, a few hundred celestial observations applied in flight, and practice in air pilotage and dead reckoning are worth many weeks of classroom study.

Value of a Single Line of Position.—A single line of position gives the navigator some valuable information. If such a line intersects the course at right angles or nearly so, it shows the navigator the speed made good since the last fix; if it runs parallel to the course, it shows a definite set across the course; if it runs north and south, it gives a definite determination of longitude and similarly; if east and west, a definite latitude. A single line of position represents the locus of the ship's position at the time of sight, and the *point* on this line that is the most probable position (called estimated position, abbreviation E.P.) will vary with conditions. At times its intersection with the D.R. course gives the most probable position; at others, it gives the intersection with the line that represents the D.R. course plus the drift. In the final analysis, the point accepted as the estimated position must take into consideration the probable effect of such drift as has been experienced since the last fix.

Intersecting Lines of Position; a Fix.—Two or more sights taken simultaneously or, nearly so, give two or more lines of position, the intersection of which is called a fix. Simultaneous sights are seldom taken, as usually the navigator likes to take his own sights. Therefore the sights are necessarily taken at different times, the time between the first and last sights varying by several minutes. As each sight must be worked out for the exact time of sight, the resulting lines of position must be advanced or retarded for the distance between the sights that the ship has run on the course, in order to get a fix for any particular time.

Advancing or Retarding Lines of Position.—To advance or retard a line of position, it is necessary to carry such a line forward or backward parallel to itself along the course followed by the plane for the distance run during the time interval.

When a line of position is run up, say for 100 miles or more, it partakes appreciably of the inaccuracies of dead reckoning during that time and consequently may be subject to slight error. Under such conditions, a fix obtained in this manner is called a "running fix," to distinguish it from the accurate fix obtained by nearly simultaneous observations.

Combining Celestial and Terrestrial Lines of Position.—In the foregoing discussion of plotting a fix we have discussed only lines of position as determined from celestial observations. Corrected compass bearings of charted objects and radio bearings are also lines of position and at the time of taking them represent the locus of the plane's position. These bearing lines may be handled in the same manner as the lines of position heretofore described in this chapter. A radio bearing or compass bearing may be advanced or retarded similarly to intersect with a line of position determined from the observation of a celestial body to obtain a fix.

Precomputed Altitude Curves.—Normally, the altitude and azimuth of an observed heavenly body are computed for the time of observation, and the position line plotted on a chart. The run of 30 to 60 miles while the plane's position is being fixed detracts from the value of the fix. It is most desirable to know *where we are*, not *where we were*.

By judicious application of the known methods of navigation we may fix the plane's position almost instantaneously after taking observations of two bodies. We may do this by:

1. Calculating the altitude and azimuth for a predetermined time and dead-reckoning position.

2. Use of methods such as the "Star Altitude Curves."

3. Plotting curves of altitude and azimuth in advance; usually only the altitude curve will be required, which we shall now discuss.

The rapid changes in the right ascension and declination of the sun, moon, and planets render it impracticable to precompute their altitudes and plot them as permanent altitude curves as we have done in the case of the stars.

On the other hand, it is entirely practical to compute in advance the altitude of any heavenly body for any given time and place. Now by estimating the course and speed to be made good, it is possible to read from the chart for any instant of the flight the estimated position of the plane. By selecting a series of such positions at reasonable intervals of time, these positions may be used to compute the altitudes of the sun or other bodies. These computed altitudes may be plotted on cross-section paper as ordinates, with the times as abscissas. A smooth curve may be drawn through them, thus giving a curve of precomputed altitudes for any instant during the flight.

When an altitude is observed during the flight, it is compared with the precomputed altitude read from the curve for the moment of observation. If the observed altitude is the same as the computed altitude, the plane is at the estimated position or on the line of position passing through it. If the observed altitude differs from the computed altitude, we know that the plane is on a line of position parallel to the line of position through the estimated position but displaced by the difference between the observed and the computed altitudes. If the observed altitude is 20′ of arc greater than the computed altitude, we know from our study of the line of position that the plane is not at the scheduled position for that instant of time but is at a point 20 miles toward the observed body. Similarly, if the observed altitude is 20′ of arc less than the computed altitude, the plane is not at the scheduled position but is at a point 20 miles farther away from the observed body.

Thus, once the curve of precomputed altitudes has been made for a particular flight, the navigator may make time and altitude observations as he wishes and, by a glance at the precomputed altitude curves, get an altitude difference without writing a single figure. This tells him at once whether he is at his estimated position, or, if not, how far he is toward or away from the observed heavenly body.

To save the trouble of correcting each sextant altitude for refraction (and parallax in the case of the moon), the necessary correction may be applied to the computed altitude with the reversed sign.

Figure 227 shows a precomputed altitude curve of the sun as described in the "Line of Position Book Supplement." Computations are made both from the D.R. position and from an assumed position. In working from the D.R. position more computation and less plotting are required. In working from an assumed position, an additional correction is made to allow for the distance of the D.R. position to the line of position through the assumed position.

Application of the Curves.—This application of the principle of the line of position makes it possible for a ground computing staff to precompute the desired altitude curves and hand them to the navigator before he takes off. The advantage of transferring the bulk of the computing work from the navigator in the plane to professional computers in an office, with all the facilities of extended tables, calculating machines, and freedom from hustle, is self-evident.

24 April 1939 — PRE-COMPUTED ALTITUDE OF SUN — FOR LAT. 38°58'N, LONG. 76°30'W

	1	2	3	4	5	6	
	USING D. R. POSITION			USING ASSUMED POSITION			
GCT	15h 00m 00s	15 08 00	15 16 00	15 00 00	15 08 00	15 16 00	GCT
GHA	30° 26'.1	30° 26'.1	30° 26'.1	30° 26'.1	30° 26'.1	30° 26'.1	GHA
CORR.	15 00.0	17 00	19 00	15 00	17 00	19 00	CORR.
GHA	45 26.1	47 26.1	49 26.1	45 26.1	47 26.1	49 26.1	GHA
LONG.	76 30	76 30	76 30	76 26.1	76 26.1	76 26.1	LONG.
LHA(t)	31 03.9 E	29 03.9 E	27 03.9 E	31° E	29 E	27 E	LHA (A)
LAT.	38 58 N	38 58 N	38 58 N	39 N	39 N	39 N	LAT. (A)
t csc	28734	31354	34198	43 22.3 N	42 47.7 N	42 16.0 N	K
L sec+	10929	10929	10929	12 41.9 N	12 42.0 N	12 42.1 N	d
N csc	39663	42283	45127	30 40.4	30 05.7	29 33.9	K~d
N	23° 39'.1	22° 11'.5	20° 43'.1				
A	3811	3343	2904	3791	3325	2887	A
L csc−	20144	20144	20144	6546	6289	6058	B
K csc	16333	16801	17240	10337	9614	8945	l Hc
K	43° 21'.5 N	42° 46'.8 N	42° 14'.9	52° 01'.0	53° 16'.0	54° 28'.4	Hc
d	12 41.9 N	12 42.0 N	12 42.1	(+) .7	(+) .6	(+) .6	REFR. REVERSED
K~d	30 39.6	30 04.8	29 32.8	52 01.7	53 16.6	54 29.0	Hs (ASS'D. POS.)
				(−) 1.5	(−) 1.2	(−) 1.0	POS. CORR.
A	3811	3343	2904	52 00.2	53 15.4	54 28.0	Hs
B+	6540	6282	6051				
l Hc	10351	9625	8955	S 54°.5 E	S 52°.0 E	S 49°.5 E	Z
Hc	51° 59'.6	53° 14'.8	54° 27'.3				
REFR. REVERSED	(+) .7	(+) .6	(+) .6				
Hs	52 00.3	53 15.4	54 27.9				

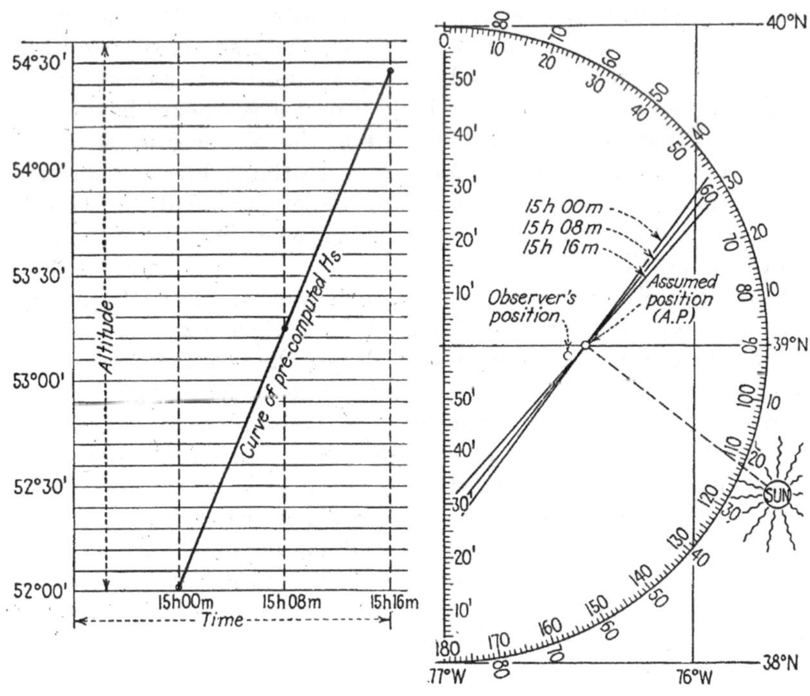

FIG. 227.—Precomputed altitude of the sun.

The student at once asks the question, "What happens if I do not follow the course and timetable for which the altitudes were computed?" The answer is that the precomputed altitude curves are designed to show not only if the plane is on schedule, but also if it is off, and, if so, by how much. If the prearranged route and timetable were always followed, neither the precomputed altitudes nor any other method would be required to supplement the dead reckoning.

With only one observed body, the best that can be done is to determine a line of position for a given instant of time, and for an *assumed position*. The distance of the observer from this assumed position is indicated by the altitude difference and may, and does, differ by amounts up to 60′ or more, depending on how close the assumed position is to the actual position of the observer. The assumed position may differ from the observer's position by as much as 100′ or more, and the observed data still be of value.

In the same way, the observer using precomputed altitudes may be off schedule 100 miles or more and find the curves of practical use. The worst that could happen would be for the flight to be delayed, say overnight, due to weather or motor trouble, in which case the altitude curves would be recomputed for the new schedule.

The azimuth may, of course, be precomputed and used if desired, but this is not usually necessary. If the observer is 10 miles "away" with the body bearing ahead of the plane, the plane is simply 10 miles behind schedule, and is speeded up if it is desired to keep to schedule. In the same way, if the observed body is on the starboard beam and the observed altitude is 20′ above the curve, the observer is 20 miles too close to the body, or 20 miles off course to the right. To correct the error the plane would change course to the left, gradually working back to the scheduled course and determining when it was on course by further observations.

When the observed body is on the bow or quarter, the navigator judges his position as based on his knowledge of the *line of position*.

Altitude Curves for Two Bodies.—When the altitude curves for two bodies are plotted for the same time, the plane's position may be checked continuously. If the altitude differences for both bodies as found from the altitude curves are kept at zero, the plane is on the estimated course and making the scheduled speed. This, of course, is the same condition as having two lines of position, the intersection being the fix.

Using Precomputed Curves for Making Landfall.—An illustration of one of the uses of a precomputed altitude curve is given below. This ingenious application to a particular problem, that of determining position relative to a landfall, has been worked out by Lieutenant W. C. Bentley, Jr., Air Corps, U. S. Army.

The conditions of this problem assume that only one body is available for observations and that the navigator after a long overwater flight with no recent fixes is in doubt as to his position and, therefore, not sure on which side of his present course his objective lies. He therefore precomputes a curve of altitudes of the body for the position of the objective and lays off lines of position for half-hourly intervals of time. He takes observations of the body and plots the altitudes against his precomputed curve (Fig. 228) until he determines that he has arrived on a line of position through his objective. The objective not then being in sight, he turns and flies along the LP, heading along the azimuth

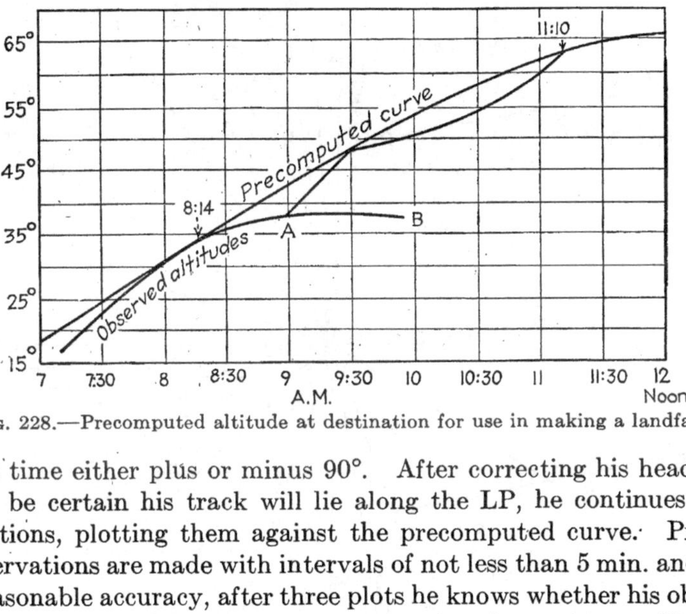

FIG. 228.—Precomputed altitude at destination for use in making a landfall.

for that time either plus or minus 90°. After correcting his heading for drift to be certain his track will lie along the LP, he continues taking observations, plotting them against the precomputed curve. Provided the observations are made with intervals of not less than 5 min. and taken with reasonable accuracy, after three plots he knows whether his objective lies ahead or behind and also, approximately, the position. He determines the direction of the objective by the following rules:

1. When heading azimuth plus 90°:
   a. If altitudes plotted are less than curve, destination is ahead.
   b. If altitudes plotted are greater than curve, destination is behind.
2. When heading azimuth minus 90°:
   a. If altitudes plotted are less than curve, destination is behind.
   b. If altitudes plotted are greater than curve, destination is ahead.

In Fig. 229 is shown how a navigator would determine the position of a landfall (labeled "Destination" in the figure) using the foregoing method. His course would place him to the right of his destination, therefore, on reaching a line of position through his objective at 0814 he turned left on the LP heading azimuth minus 90°. On getting successive plots less than the precomputed curve, he applied the appropriate rule for heading azimuth minus 90° and determined that his objective was behind him. He then turned right at 0900 onto the azimuth for 0930 as a heading. At 0929 he again arrived on a LP of the destination. He then turned right, knowing from his previous reckoning the direction of his destination. The run between 0900 and 0929 was moved along until it just equaled the distance between the position lines for 0900 and 0929. This gave the approximate distance to destination and E.T.A. He corrected

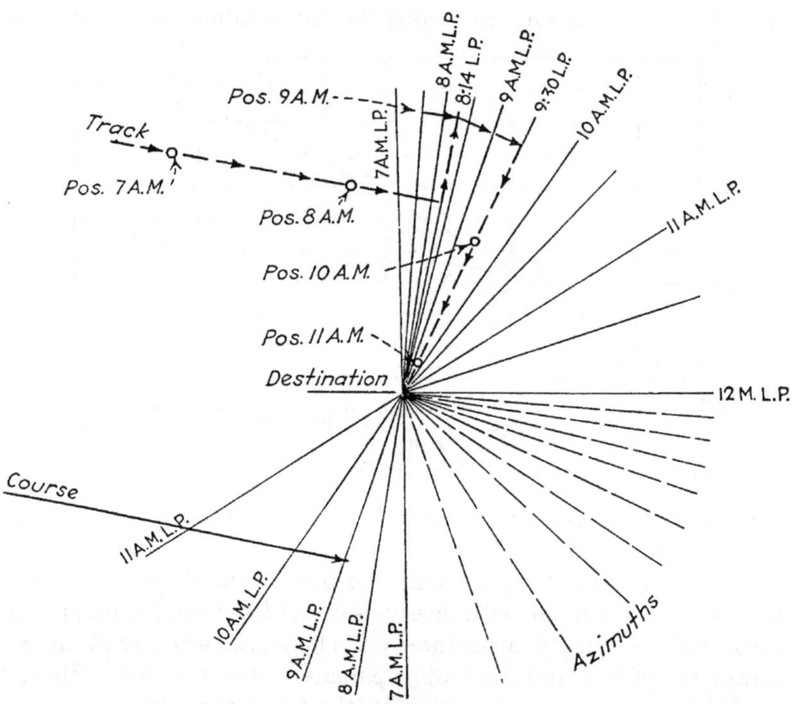

FIG. 229.—Process of finding destination by means of precomputed altitude curves.

heading for drift on this new course so that his track would lie along the LP. He continued taking observations and plotting them as shown in Fig. 228. By drawing a smooth curve through the first five plots made and continuing the curve until it met the precomputed curve, he was able to estimate the time of his arrival at his destination independently of previous estimate found by plotting the run between 0900 and 0929.

This method, of course, has its limitations, yet should be valuable when understood and used properly.

Finding a Destination.—During the daylight hours it is often impossible to get more than one LP, *viz.*, that given by the sun. If pilotage cannot be used (as when flying over water) or radio bearings are not available, this single position line may be utilized for finding a destination.

The air navigator, having found a position line as he approaches his destination, continues flying on his course until the position line carried forward by D.R. passes through the destination. He then turns right or left and follows the LP. If, after a reasonable time, the destination is not sighted, he infers that he has turned the wrong way, and so reverses his track. Some navigators approaching a landfall purposely keep to one side so that they know the proper direction to turn.

Special Cases.—There are two special cases for finding latitude, each of which may be handled by a special short solution peculiar to the case. These are

1. Meridian altitudes.
2. Altitudes of Polaris.

1. When the observed body is on the meridian, at which time its LHA is zero, the latitude may be found by the meridian altitude method. The general formula when the body is at upper transit is

$$L = z + d$$

where L is the latitude.
d is the declination.
z is the zenith distance, or $90° - H_0$.
Name z south when the body bears south, and north when the body bears north, and then combine z and d algebraically; *i.e.*, use their sum when the signs are unlike and the difference when alike. At lower transit,

$$L = H_0 + p$$

where p is the polar distance = 90° − declination.

Figure 230 will help the student to understand the different cases.

2. The altitude of the north pole is equal to the latitude in the northern hemisphere. Since Polaris is only about 63′ from the pole and revolves about it, a small correction depending on the hour angle of Polaris may be applied to its H_0 to find the latitude directly. A table giving these corrections is printed on the back of the star chart in the *Air Almanac*.

Circles of Equal Altitude Plotted on Chart.—The only cases where the LP must be plotted as the arc of a circle are those where the body observed is close to the zenith of the observer. For these high altitudes, between 85° and 90°, the assumption that the LP is a straight line does not hold, for the circles of equal altitude are so small that errors would be introduced if they were plotted as a straight line. In these cases the procedure for plotting the LP is as follows:

The subsolar, substellar, or sublunar geographical position of the body observed is plotted by data obtained from the "Nautical Almanac" for the GCT of sight. The GHA of the body observed is plotted as the longitude and the declination as the latitude. From this point as a center, an arc is drawn with a radius in nautical miles equal to the body's zenith distance in *minutes* of arc. It is not necessary to draw the entire circle of altitude but only that section (arc) which plots in the vicinity of the D.R. position. The arc thus drawn may be advanced or retarded by advancing or retarding the subsolar, substellar, or sublunar point parallel to the ship's course and again drawing the arc.

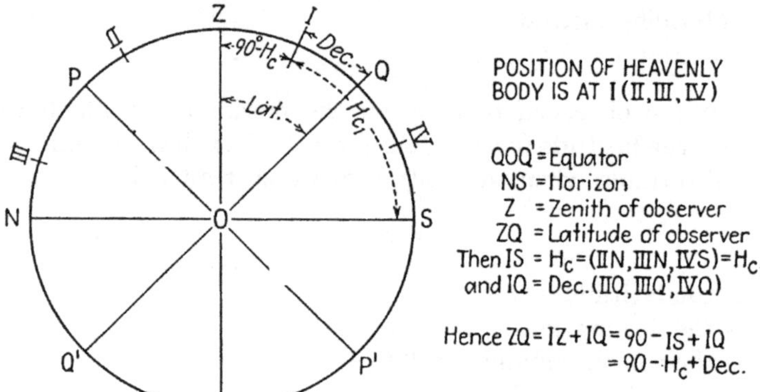

FIG. 230.—Meridian altitude figure.

Such high altitudes as those necessitating plotting are in general avoidable, except in the tropics when the latitude of the ship's position approximates the declination of the sun. In any case, they are not desirable, inasmuch as it is difficult to measure such high altitudes with accuracy.

Polar Celestial Navigation.—For an observer at the pole, the Greenwich hour angle in effect becomes the azimuth, and the declination of the sun becomes its true altitude. Theoretically there is only one direction, south, for an observer at the north pole. However, by measuring azimuth from a standard meridian, say Greenwich, we may indicate direction and this is then identical with GHA. This allows the pole to be used as the assumed position when within about 15° of the pole and permits the use of a convenient graphical method of laying down lines of position. Plot the sun's GHA as its azimuth on a suitable polar chart and consider the sun's declination as the computed altitude. The line of position is then plotted in the usual way, using the difference (laid off from the pole) between the observed and the computed altitudes. If the observer is more than 100 miles from the point where the line of position crosses the sun's meridian, a correction should be made for curvature of the line. The accompanying table (which was made specially for Ellsworth) gives, for distances from the intersection of the line of position and the sun's meridian, the corrections to be applied to the altitude and azimuth to permit a portion of the *circle of position* to be plotted.

CORRECTIONS FOR RECTIFYING LINES OF POSITION

True alt.	100′		200′		300′		400′		500′	
	ΔH	ΔZ	ΔH	ΔZ	ΔH	ΔZ	ΔH	ΔZ	ΔH	ΔZ
°	′	°	′	°	′	°	′	°	′	°
10	0.3	0.3	0.7	0.6	2.1	0.9	3.8	1.2	6.2	1.5
15	0.4	0.5	1.5	0.9	3.5	1.4	6.2	1.8	9.6	2.2
20	0.5	0.6	2.1	1.2	4.8	1.8	8.6	2.3	13.3	3.0
25	0.7	0.7	2.7	1.6	6.2	2.3	11.0	3.2	16.9	3.9
30	0.8	1.0	3.5	1.9	7.8	2.9	13.5	3.8	21.0	4.8
35	0.9	1.2	4.0	2.3	8.8	3.5	16.1	4.7	25.3	5.8

True alt.	500′		550′		600′		650′		700′	
	ΔH	ΔZ	ΔH	ΔZ	ΔH	ΔZ	ΔH	ΔZ	ΔH	ΔZ
°	′	°	′	°	′	°	′	°	′	°
10	6.2	1.5	7.5	1.6	9.1	1.8	10.6	1.9	12.7	2.1
15	9.6	2.2	11.7	2.5	13.2	2.7	16.2	2.9	18.9	3.1
20	13.3	3.0	15.8	3.3	18.9	3.6	22.2	3.9	26.2	4.2
25	16.9	3.9	20.5	4.3	24.2	4.6	28.5	5.0	32.7	5.4
30	21.0	4.8	25.3	5.3	30.1	5.7	35.9	6.2	41.4	6.7
35	25.3	5.8	30.4	6.4	36.1	6.9	42.3	7.5	49.0	8.1

The table of corrections may be replaced by templates giving the proper curve for various altitudes. The student should give a free rein to the imagination in order to visualize that this is merely a special method of laying down the line of position.

In order to show how the curvature of the line of position may be rectified by this table, Fig. 231 is drawn to represent to scale the condition where an observer at Lat. 80°N., Long. 135°W., observes the sun's altitude to be 25° when its declination is 22°N., and its GHA is 100°. The pole is used as the assumed position. It is also the observer's zenith for this assumed position, and the azimuth is equal to the GHA of the sun. Since the declination should equal the altitude at the pole, and the measured altitude is 3° greater, the observer is 180 miles toward the sun. Since the observer's position is about 400 miles from the intersection of the line of position with the sun's meridian, the corrections as given in the table should be applied to the plotted straight line to get a segment of the 25° circle of altitude.

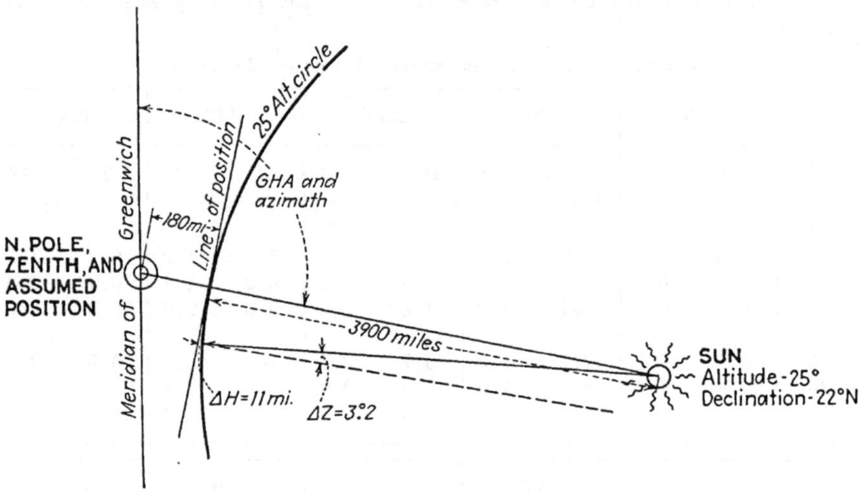

FIG. 231.—Polar celestial navigation.

The sun gives only one line. During the summer months, the moon will often be in a position to give a second line. Also, Venus should often be visible. During the six months of night, stars will be available and should be used with the "Star Altitude Curves" to fix a position.

PRACTICAL HINTS TO THE NAVIGATOR

1. Take every opportunity to practice navigation when the position of the plane is known as this will give increased confidence in the results obtained.
2. Avoid the tedious computations for dead reckoning required in older texts; instead, measure courses and distances directly and accurately on charts.
3. For accomplishing dead reckoning such as speed-time-distance and wind-drift problems, use the Dalton Mark VII computer or other suitable mechanical device to save mental effort and speed up the work.
4. Use a good magnetic compass, preferably the aperiodic type.
5. Have a working knowledge of meteorology, radio, and instrument flying, and make full use of this knowledge.

6. Make every effort not to get "lost." Once the sequence of navigation is broken, it is often difficult to determine the plane's position, not to mention the increased danger, work, and worry.
7. Collect the necessary navigation equipment and keep it intact and in good condition, *i.e.*, the sextant in adjustment, the watches carefully rated. Keep a current Almanac, know the compass deviation, etc.
8. Install and use a good drift indicator. The Gatty Periscopic Drift Indicator is a good type.
9. Keep the necessary charts, including large-scale charts such as the U.S. Aeronautical Sectional Charts for contact flying, and smaller scale charts such as the U. S. Aeronautical Regional and WSN (1:5,000,000) skeleton charts for long-range navigation.
10. Make full use of radio and celestial navigation, and for the latter, take full advantage of the "Star Altitude Curves." A prominent authority on aviation has stated that "The plane of the future will fly looking up"—meaning that stratosphere flying will require more celestial navigation, and that contact flying will not ordinarily be possible, due to the high altitudes at which the planes will operate.
11. Practical adjustment of the bubble sextant:
    a. With the sextant held at a convenient point, sight at an object such as a telephone pole about 100 yd. away, noting where the center of the direct image bubble appears to be. Ignore the reflected image of the bubble.
    b. Go to the point sighted and shoot back at the point where the sextant was first held.
    c. If in adjustment, the center of the bubble should coincide with the first position of the sextant. If it doesn't, remove *half* the error by means of the bubble adjustment and test again. It should be in adjustment.
    d. With the bubble in adjustment, bring the direct image of a distant body into coincidence with its reflected image. If the index then reads zero, there is no index error. If the index does not indicate zero, set it at zero and then tighten the lock screw.
    Also see detailed instructions furnished with the instrument, and Chap. XII.
12. Converting standard or zone time to Greenwich civil time:

Civil time over about 15° of longitude is reckoned from the same meridian. For example, Eastern standard time (EST) is reckoned from the 75th meridian; Central standard time from the 90th meridian, etc. Therefore, Eastern standard time is 5 hr. (75°) behind GCT. To find

GCT when EST is given, simply add 5 hr. to EST. In the same way add 6 hr. to CST; 7 hr. to Mountain time; and 8 hr. to Pacific time. The data in the "Almanac" are given for GCT 0 to 24 hr. Add 12 hr. to watch reading in the afternoon.

13. Setting and rating watches:

For celestial navigation, it is suggested that a reliable watch, preferably with the second-setting feature, be carefully regulated and set to Greenwich civil time (GCT). The exact EST in New York may be obtained by dialing "Mermaid." Other cities have similar time services. The Naval Observatory, or Longines broadcast over WOR, Newark, gives the correct time. With the watch set to the exact GCT, wind it regularly and avoid rough treatment. Daily, or as often as desired, note the error of the watch and from these errors establish the watch rate. See also Chap. XII.

The following table shows a convenient method of rating a watch:

RECORD OF LONGINES SECOND-SETTING GCT WATCH NO. 2

Date, 1942	Watch error, sec.	Daily rate, sec.	Dial setting	Remarks
1 May	......	.....	10	Set by radio tick
2 May	+5	+5	5	Set by radio tick
4 May	+8	+4	57	Checked by Longines broadcast
8 May	+18	+4.5	39	Checked by local time service
9 May	......	.....	....	Forgot to wind. Ran down
10 May	......	.....	31.5	Checked by Longines broadcast
20 May	+38.5	+3.9	53	Checked by radio

Rate established as gaining 4 sec. per day.

NAVIGATION IN FLIGHT

Figure 233 shows how celestial navigation data may be arranged on the pages of the navigator's notebook. This particular flight was made to test the Link sextant and the Link timer for indicating the average time. More details about the dead reckoning would have been entered had time permitted. The air was quite bumpy during the latter part of the flight, but the errors are about what might be expected under normal conditions. Particular note should be made of the fact that the *average error* was 4.8 miles and that the actual navigation error, allowing for the plus and minus errors, was only 0.8 mile. "H.O. 214" was used for solving the lines of position. The plot of this flight is shown in Fig. 232.

Figure 235 shows the solution of 12 lines of position by the "Line of Position Book." The first nine of the problems represent the first attempt at celestial navigation by the author. The original observations were made from a Navy mailplane on a run from San Diego to Los

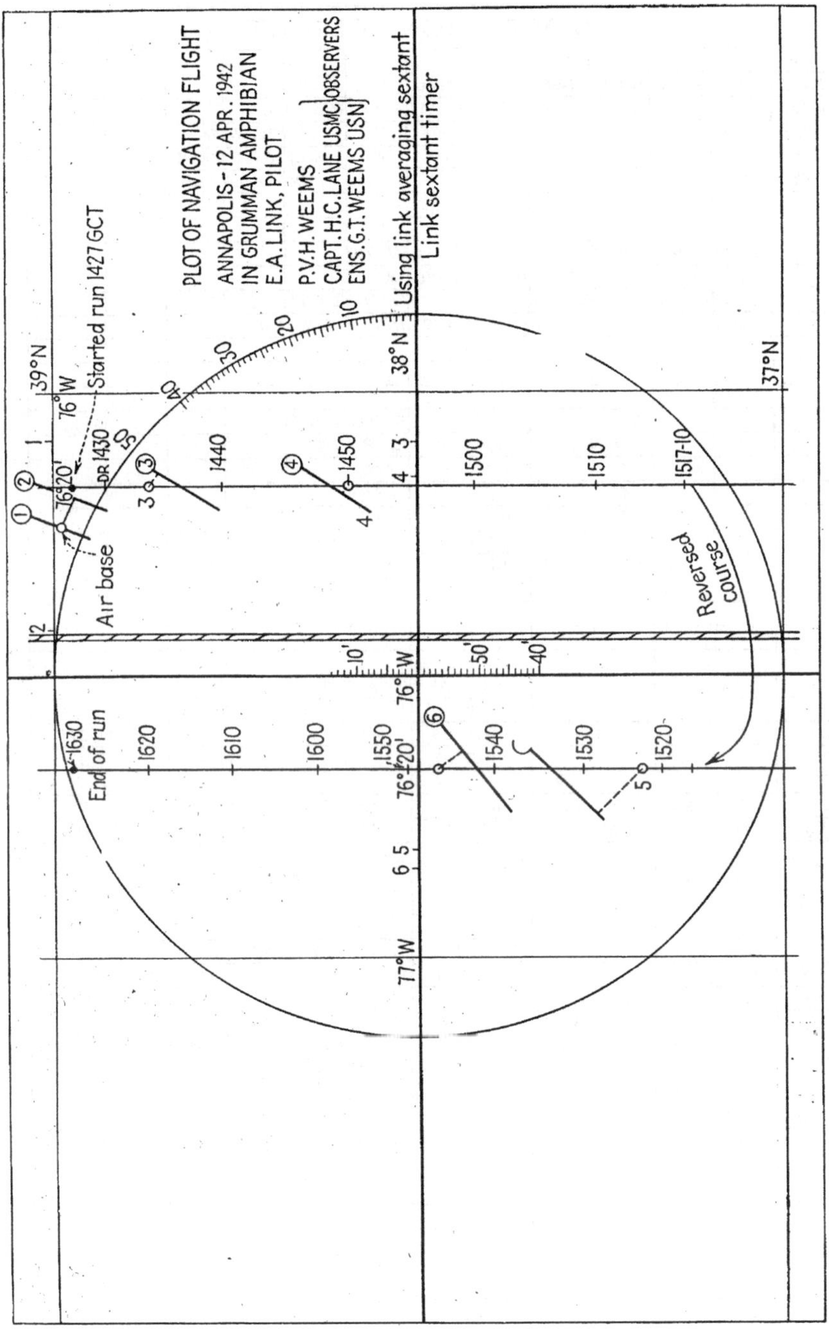

FIG. 232.—Navigation test flight as plotted.

NAVIGATION FLIGHT

April 12, 1942

TIME G.C.T.	POSITION LAT. N	POSITION LONG. W	COMPASS HEADING	TRUE HEADING	DRIFT OR LEEWAY	TRACK	G/S	I. A. S. OR D. R. SP.			
1353+	38°59′	76°29′	TOOK TWO SINGLE SHOTS AT SEAPLANE BASE								
1426^{55}	38°57′	76°20′				180	120	STARTED RUN			
1517^{10}	37°16′	76°20′				0	84	REVERSED COURSE			
1630	38°59′	76°20′						COMPLETED RUN			

NOTES

LINK AND LANE OBSERVED CHECK POSITIONS
SEXTANT OBSERVATIONS BY P. V. H. WEEMS
20 OBSERVATIONS FOR EACH GROUP TAKEN IN FLIGHT
USED LINK AVERAGING SEXTANT AND LINK SEXTANT TIMER

RESULTS

NO. OF POSITION	ERROR	WITH CORIOLIS CORR. APPLIED
3	+2.5	+1.0
4	−1.5	−3.0
5	−9.5	−8.0
6	+5.5	+7.0
AVERAGE ERROR	4.8	4.8
MEAN ERROR	−0.8	−0.8

SUN LINES

	1	2	3	4	5	6
G.C.T.	13-53-29	13-56-16	14-33-35	14-50-05	15-23-20	15-47-34
G.H.A.	27 17	27 17	37 17	42 17	49 47	54 47
CORR.	52	1 34	54	1	0 50	1 54
G.H.A.	28 09 W	28 51	38 11	42 18	50 37	56 41
λ Ⓐ	76 09 W	76 51	76 11	76 18	76 37	76 41
L.H.A.	48E	48	38	34	26	20
DEC.	8 35 N	8 35 N	8 36	8 36	8 36	8 37
LAT Ⓐ	39 N	39	38	38		38
H_T	37 23.8		44 50.5	47 29.2	52 19.4	55 25.3
Δd	68 3.4		73 +4.4	75 +4.5	82 +4.9	88 +6.1
H_C	37 27.2	37 27.2	44 55	47 33.7	52 24.3	55 31.4
H_O	37 14.0	37 49.0	44 29	47 25.0	52 49.0	55 49.0
A	13.2 A	21.8 T	26 A	8.7 A	24.7 T	17.6 T
Z	N 112 E	112	121	125	135	143
ERROR	+0′.5	−5.5	+1.0	−3.0	−8.0	+7.0
	← STATIONARY →		← USING TIMER →			

FIG. 233.—Navigation flight to test the Link sextant and Link timer for indicating average time.

Angeles on Nov. 28, 1927. The work has been brought up-to-date by making the necessary alterations. Three additional problems are included in order to show the solution with 0° and westerly local hour angle.

Given: The data shown in Fig. 235, for GCT, H_0, and the assumed position.

Required: Work and plot the lines of position using the "Air Almanac."

Solutions: See Fig. 235 for computations and Fig. 234 for the plot.

Figure 236 shows the solution of two problems by means of the "Star Altitude Curves." Note that in the first problem the observer's approximate position is not known. In this case, the information that *Arcturus is rising* is sufficient information to determine a fix. In the second problem the approximate longitude is given but the fact that *Procyon is setting* is not stated. The two problems are stated and solved as follows:

On Jan. 1, 1940, at $5^h20^m55^s$ GCT simultaneous altitudes taken with adjusted bubble sextants were: Arcturus, 47°13′, rising; Polaris, 50°38′.

Required: A fix.

Solution:

Arcturus...........	47°10′	(47°13′ less correction for date, 5 × 0.6)
Polaris............	50°40′	(50°38′ plus correction for date, 5 × 0.3)
GHA ♈...........	179°58′	From "Air Almanac" for $5^h20^m55^s$
LHA ♈...........	176°53′	From Fig. 236
Long..............	3°05′W.	
Lat...............	51°30′N.	Fix.

On Feb. 5, 1940 (approximate Long. 24°E.), the navigator of a plane on track 60°, speed 240 knots, observes with an adjusted bubble sextant as follows:

$0^h55^m22^s$ GT, Procyon, 24°23′
$1^h00^m23^s$ GT, Polaris, 50°04′

Required: A fix at $1^h00^m23^s$.

Correction for date for each star is +2′ (5 × 0.3), giving corrected altitudes: Procyon 24°25′; Polaris 50°06′.

Referring to Fig. 236, the plotted altitude of Procyon for $0^h55^m22^s$ is shown at *A*, which point is moved horizontally to the right 5^m for time between observations to point *B*. Point *B* is then moved on track 60°, 20 miles, to allow for run between observations, to point *C*. The Procyon curve through *C* crossed with Polaris curve determines point *D*, which projected vertically to bottom scale gives LHA ♈ = 173°07′; and, horizontally to the right scale, gives latitude of 50°55′N.

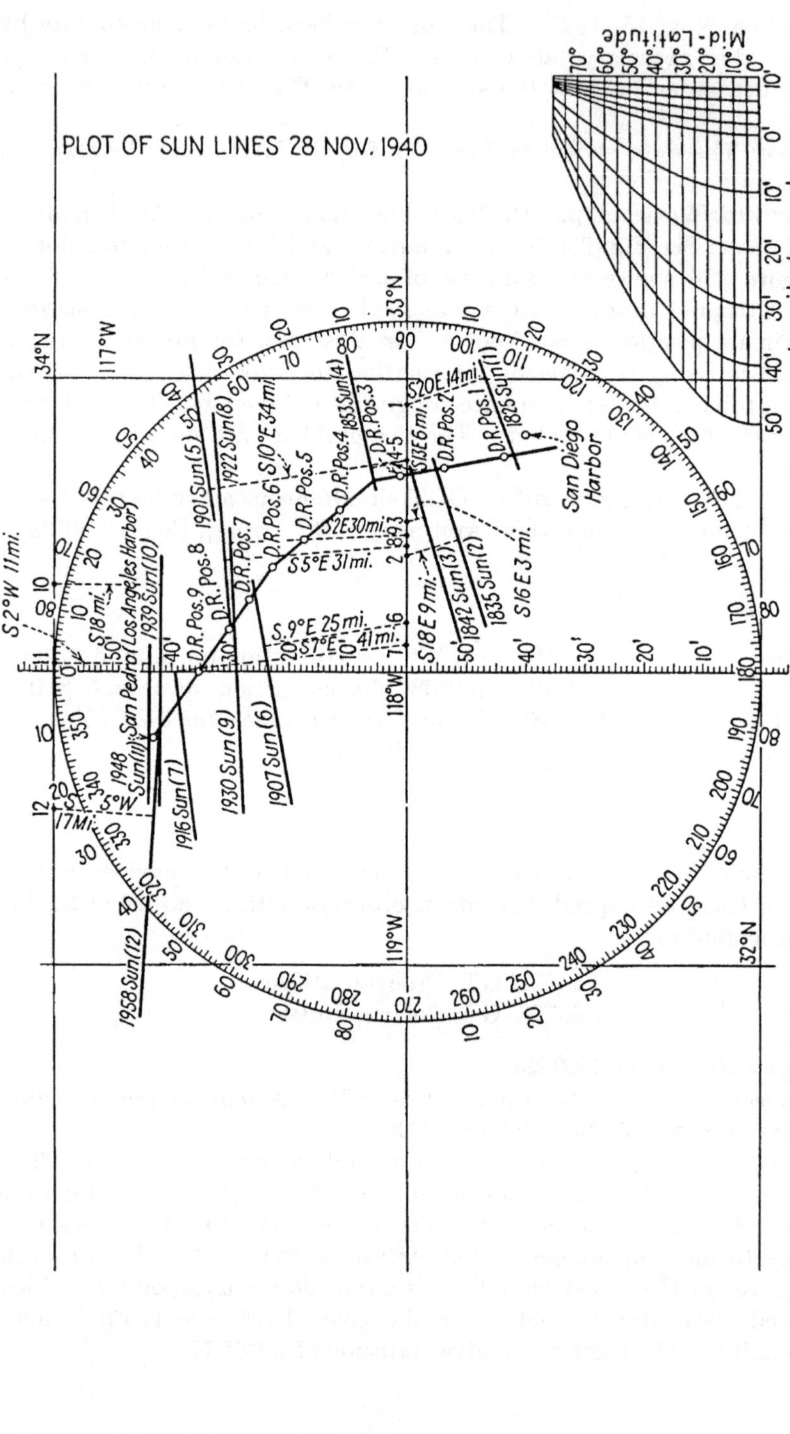

FIG. 234. Plot of 12 lines of position in flight.

SUN LINES 28 NOV. 1940						
	1	2	3	4	5	6
GCT	18h 24m 32s	18 34 30	18 42 02	18 53 14	19 01 14	19 07 26
GHA	97° 58'	100 28	102 58	105 28	107 58	107 58
CORR	1 08	1 08	0 31	0 49	0 19	1 52
GHA	99° 06'	101 36	103 29	106 17	108 17	109 50
LONG(A)	117 06	117 36	117 29	117 17	117 17	117 50
LHA	18 E	16 E	14 E	11 E	9 E	8 E
LAT(A)	33 N	33 N	33 N	33 N	33 N	33 N
DEC	21 23S	21 23S	21 23S	21 23S	21 23S	21 23S
K	34 20N	34 03N	33 48N	33 29N	33 20N	33 15N
K~DEC	55 43	55 26	55 11	54 52	54 43	54 38
A	1510	1193	913	563	377	298
+B	24927	24614	24340	23997	23836	23747
LOG H_c	26437	25807	25253	24560	24213	24045
H_c	32° 58'	33 30	34 00	34 37	34 56	35 05
H_o	33 12	33 39	34 03	34 31	34 22	34 40
a	14T	9T	3T	6A	34A	25A
Z	S 20° E	S 18 E	S 16 E	S 13 E	S 10 E	S 9 E
	7	8	9	10	11	12
GCT	19h 15m 42s	19 22 22	19 30 14	19 39 00	19 48 00	19 58 00
GHA	110° 28'	112 58	115 28	115 28	117 58	120 28
CORR	1 26	0 36	0 04	2 15	2 00	2 00
GHA	111° 54'	113 34	115 32	117 43	119 58	122 28
LONG(A)	117 54	117 34	117 32	117 43	117 58	118 28
LHA	6E	4E	2E	0	2W	4W
LAT(A)	33N	33N	33N	34N	34N	34N
DEC	21 23 S	21 23 S	21 23 S	21 23 S	21 23 S	21 23 S
K	33 09 N	33 04 N	33 01 N	34 00 N	34 01 N	34 04 N
K~DEC	54 32	54 27	54 24	55 23	55 24	55 27
A	168	74	19		18	73
+B	23640	23552	23499		24577	24632
LOG H_c	23808	23626	23518		24595	24705
H_c	35 19	35 29	35 35	34 37	34 35	34 29
H_o	34 38	34 58	35 05	34 55	35 46	34 46
a	41A	31A	30A	18T	11T	17T
Z	S 7° E	S 5 E	S 2 E	S	S 2 W	S 5 W

FIG. 235.—Solution of 12 lines of position in flight by "Line of Position Book."

GHA ♈ for $1^h00^m23^s$	149°09′
LHA ♈ from point *D*, Fig. 236	173°07′
Long	23°58′E.

Fix: Lat., 50°55′N.
Long., 23°58′W.

We shall now give the solution of two lines of position to determine the position of an observer in flight. This necessitates "handling lines"

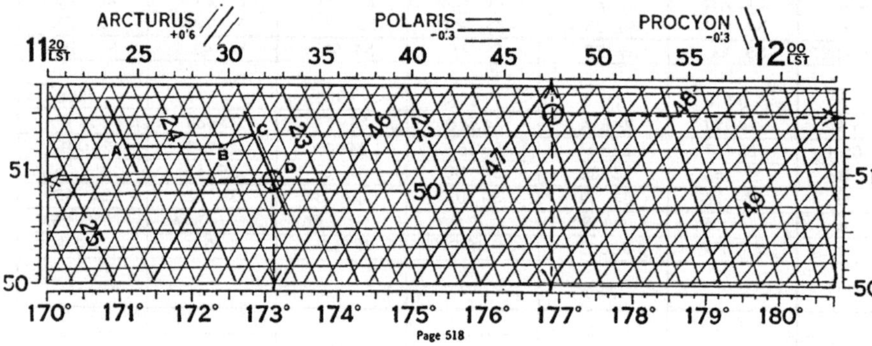

FIG. 236.—Solution of two problems by means of "Star Altitude Curves." (Extract from Band 50°–60° N.)

by advancing or retarding to allow for the run between observations as previously described. In the two following problems, the run between observations is so short that definite fixes as opposed to "running fixes" are obtained. Also, the altitude corrections are shown in detail where in practice this is ordinarily done mentally.

Problem: At about GCT 2204, Mar. 3, 1941, the navigator of a plane whose D.R. position was Lat. 39°45′N., Long. 30°00′W., took sights as follows:

Star	GCT	Hs
Polaris	$22^h03^m52^s$	40°02′
Capella	$22^h04^m37^s$	71°28′

The plane was making good a track of 000° true and a GS (ground speed) of 240 knots. Find the plane's position (fix) at the time of the Capella sight.

For plotting this problem, construct a chart on a universal plotting chart for Lat. 39°N. to 41°N. on the left of the center meridian. Number the center meridian 30°W. and the left-hand meridian 31°W. See Fig. 237.

Solution:

GCT 2200 GHA ♈	131°17′	H_s	40°02′
Int. for 3^m52^s	58	Corr. (−)	1
GCT $22^h03^m52^s$ GHA ♈	132°15′	H_o	40°01′
D.R. Long	30 00W.	LHA corr. (−)	14
LHA ♈	102°15′	Lat.	39°47′N.
Capella, Mar. 3, 1941:			
GCT 2200 GHA ♈	131°17′		
Int. for 4^m37^s	1 09		
GCT $22^h04^m37^s$ GHA ♈	132°26′		
SHA Capella	281 55		
GHA Capella	414°21′		
Long. (A)	30 21W.		
LHA Capella	384° or		
t	24°W.		
L (A)	40°N.		
K	42°34.0′N.	H_s	71°28′
d	45 56.0N.	Corr.	0 (AAA tables)
$K \sim d$	3°22′.0	H_o	71°28′
A	2218		
B	75		
log H_c	2293		
H_c	71°32′.8		
H_o	71 28		
a	4′.8A.		
Z	N. 63°W.		
Z_n	297°		

Plotting: First plot the Polaris latitude line as a dashed line. Then advance it along the track 000° (T) for the distance run between the times of the Polaris and Capella sights: 45^s = 3 miles. Plot the Capella sight from the *AP* along the azimuth 48′.*A* and draw the *LP* as a solid line. Now enter tables either in pamphlet or on page 6, *LPB*, and find the Coriolis correction. The table is entered opposite Lat. 40° and since the GS for 240 knots is not given, interpolate between columns, tabulating 200 and 250 knots GS to find the correction 4.0 nautical miles for 240 knots. The intersection of the advanced Polaris line and the Capella *LP* gives the *uncorrected* fix. To obtain the corrected fix, we apply the Coriolis correction, 4 miles, to the right, perpendicular to the track. The corrected fix is Lat. 39°50′N., Long. 30°15′W. See the accompanying diagram for plotting.

Problem: At about GCT 1726, Mar. 21, 1941, the navigator of a plane whose D.R. position was Lat. 40°35′N., Long. 24°30′E., took sights as follows:

Star	GCT	Hs
Polaris	$17^h25^m00^s$	40°44′
Regulus	$17^h26^m42^s$	40°46′

The plane was making good a track of 045° true and a GS of 180 knots. Find the plane's position (fix) at the time of the Regulus sight.

For plotting this problem, use the right-hand part of the chart constructed for the preceding problem for Lat. 39°N. to 41°N. For this problem, call the center meridian 24°E. and the right-hand meridian 25°E. See Fig. 237.

Solution:

Polaris, Mar. 21, 1941:

GCT 1720 GHA ♈	78°50′	H_s	40°44′
Int. for 5^m	1 15	Corr. (−)	1
GCT 1725 GHA ♈	80°05′	H_o	40°43′
D.R. Long	24 30E.	LHA ♈ corr. (−)	12
LHA ♈	104°35′	Lat	40°31′N.

Regulus, Mar. 21, 1941:

GCT 1720 GHA ♈	78°50′	H_s	40°46′
Int. for 6^m42^s	1 41	Corr. (−)	1
GCT $17^h26^m42^s$ GHA ♈	80°31′	H_o	40°45′
SHA Regulus	208 41		
LHA Regulus	289°12′		
Long. (A)	24 48E.		
LHA*	314 00		
t	46°E.		
L (A)	41°N.		
K	51°22′.3N.		
d	12 15 .0N.		
$K \sim d$	39°07′.3		
A	7583		
B	11025		
log H_c	18608		
H_c	40°39′.3		
H_o	40°45′.0		
a	5′.7T.		
Z	S 67°.5 E.		
Z_n	112°.5		

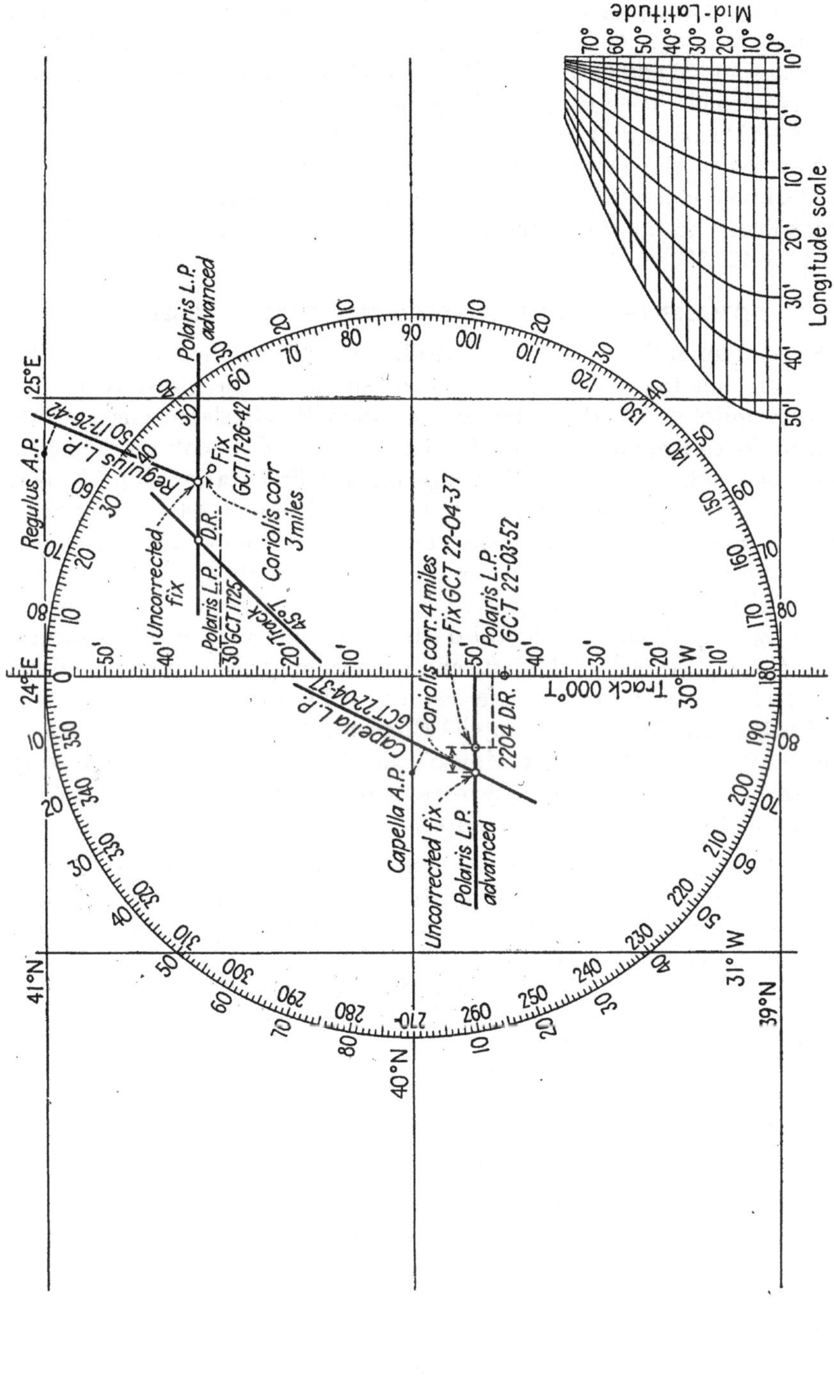

FIG. 237.—Handling lines of position in the air.

Plotting: First plot the Polaris LP as a dashed line. Now advance this line along the course 45° (*T*) for the distance run between the time of Polaris and Regulus sights: 1^m42^s at 3 miles a minute = 5.1 miles and plot as a solid line. Plot the Regulus LP. Its intersection with the advanced Polaris LP gives the *uncorrected* fix.

Now enter the table to find the Coriolis correction. This will be found to be 3 miles. This is laid off from the *uncorrected* fix to the right, perpendicular to the track, and gives the corrected fix as Lat. 40°32′.5N., Long. 24°45′.5E.

COMMENTS ON OUTSTANDING NAVIGATION FLIGHTS

First Atlantic Crossing, *N.C.4*, 1919.—A total distance of 4,513 miles was flown, but the navigation was made simple by having station ships every 50 miles across the Atlantic. Captain Read related one unusual incident in connection with navigation. The *N.C.4* took off from the Azores in rough water. It was noted that the first station ship 50 miles to the east was left well to the north; the second ship at 100 miles was not sighted. Shortly after this it was noted that on taking off the compass had bounced out of its gimbals, and the forward lubber line had moved about 30° to the left. The plane was headed for Africa! The compass was reset, and at the time the third ship (150 miles east) was due abeam, the *N.C.4* turned 90° to the left, and soon sighted the station ship dead ahead; when over it, she reset her course for Lisbon.

Hawker and Grieve's Attempted Non-stop Atlantic Crossing, 1919.—Shortly after the *N.C.4* flight, Hawker and Grieve attempted a non-stop Atlantic crossing in a Sopwith two-seater plane. Speaking of navigation on this plucky attempt, Grieve states,

> I preferred to navigate chiefly by celestial observations, and my position by the stars when picked up was practically correct. I used a cloud horizon instead of the sea horizon, as the sea was hardly visible at any part of the time we were in the air. For the first four hours after leaving we were passing over fog banks, and the clouds below were like the sea, giving a perfect horizon. I had only to judge our height above them and take the sun, as on a sea horizon, and about seven o'clock Greenwich mean time I saw the sea for a few seconds through a hole in the fog and cloud bank. I obtained the drift of the machine by noting the breaking waves through the drift indicator, and we were then at 4,000 feet and climbing. The drift was 10° to the right of our course, which I had already allowed for on starting, owing to the northeast wind then blowing from St. John's.

Alcock-Brown First Non-stop Atlantic Crossing, 1919.—Alcock and Brown in a Vickers-Vimy made the first non-stop Atlantic crossing in 1919. They flew non-stop 1,890 miles from Newfoundland to Ireland in 15 hr. 57 min., at 117 m.p.h. They followed the approximate great-circle course.

Coutinho South Atlantic Crossing, 1922.—Admiral Coutinho, of the Portuguese Navy, navigated a plane on the first non-stop South Atlantic crossing in 1922. He had good weather, and used the sea horizon for sextant observations and specially prepared tables for reducing the observations to position. He claims to have known his exact position within 4 or 5 miles continuously on the flight, and picked up St. Paul's Rock, a mere speck in the sea. This is an excellent example of efficient navigation.

Dole Race, 1927.—Several planes in 1927 competed in a race from San Francisco to Honolulu. Lieutenant Davis, navigator of the winning plane, *Woolaroc*, used all four methods of navigation effectively, especially radio and celestial navigation, and flew direct to his destination. Perhaps more lessons may be gained from a study of the methods used by Schluter, the navigator of the plane winning second place. He was a merchant-marine navigator with little air experience before the flight. He used a marine sextant and the old time-sight method of fixing his longitude. After flying through the night he took a careful observation with the sun bearing 90°, and from it computed his longitude. As the time-sight method could not be used to determine latitude, he had to depend on getting his latitude by a meridian altitude of the sun. Unfortunately, the plane reached the longitude of Honolulu before the sun did. Schluter told his pilot to "circle till noon," which he did for more than half an hour. He then took a careful observation, told the pilot to head south, and reached Honolulu with 20 min. of gasoline left! We must give Schluter credit for sticking to his guns. His feat is a classic example of saving the plane by accurate, if antiquated, methods of navigation. Actually, a plotted line of position about 11 A.M. would have led him direct to Honolulu.

Lindbergh New York to Paris Flight, 1927.—On this flight Lindbergh used only pilotage and dead reckoning, but he did it skillfully. He had previously spent a week carefully computing his great-circle course and other data, and during the flight followed his course carefully. At about mid-distance he ran into a storm, and ice started to form on his plane. He turned back, cleared the storm, flew east past it, then north to get on his course again. Finally he sighted Ireland within a few miles of the intended point. After this flight Lindbergh did not make any long overseas flights without means (radio and celestial navigation) of definitely fixing his position. He tells of flying for hours above fog over the Newfoundland Banks, and how "the fact that I might have been blown in any direction without knowing it preyed on my mind." He came down through a hole in the clouds till he could see the surface of the water, and got some estimate of the wind. In doing this he left a good tail wind and lost about an hour.

Post-Gatty Round-the-world Flight, 1931.—On this flight Gatty used pilotage, dead reckoning, and celestial navigation. He carried a radio set but made no practical use of it. Except for the fact that full use was not made of radio, the navigation on this flight might be considered the best to that date. He measured drift and ground speed on his own instrument designed for the flight. He constructed his own navigation charts. He used celestial navigation effectively and claims to have been within 25 miles of his scheduled course at all times. Of particular interest is his feat of making a landfall at Blagoveschensk, Russia, after dark by means of the "Star Altitude Curves." His best second-setting watch was out only 4 sec. in 10 days.

Lindbergh 1933 South Atlantic Flight.—In 1933, with Mrs. Lindbergh as copilot and navigator, Lindbergh flew 3,000 miles across the south Atlantic Ocean. This flight is the best example to date of efficient navigation. There was perfect teamwork between Colonel and Mrs. Lindbergh. Normally he piloted while she operated the radio, observed the ground speed and drift on the Gatty instrument, and kept the log. Lindbergh accomplished the celestial navigation while Mrs. Lindbergh piloted the plane. This is clearly the best combination, since each person may at times have short periods of relaxation. Mrs. Lindbergh was in communication with South America within an hour after the take-off from Dakar, near Bathurst, Africa. She obtained three wireless bearings on the *S.S. Westfalen*, and directed the course to her. Later Lindbergh set a new course to the island of Fernando Noronha, and from there to Natal, Brazil. In other words, they showed the ability to make any desired contacts at sea or on islands by celestial navigation, radio, dead reckoning, and air pilotage—thus making full use of all methods for the safe navigation of aircraft.

Navigation on Scheduled Transoceanic Planes.—The navigation of scheduled transoceanic aircraft is accomplished with the assurance and accuracy approaching that of ocean liners. The large clipper planes have roomy and fully equipped navigation compartments and full use is made of all available methods for safe navigation. Many of the skilled navigators were formerly licensed marine navigators and usually are permitted to use their own preferred methods and equipment.

With the twofold purpose of giving a valuable technical description of the navigation as accomplished on a clipper plane, and also as a tribute to the navigator of the Earhart plane, the following confidential letter to the author is published with the knowledge of Pan American Airways:

I hope you will pardon this long delay in acknowledging your congratulatory and greatly appreciated letter of April 1st. Preparatory work prior to the flight to Hawaii, and subsequent arrangement and study of data gathered during the flight, has so occupied my time that I am afraid all my friends consider me an extremely poor correspondent.

For reasons which I am certain you will understand, we are not permitted to discuss the particulars of the flight for dissemination among the general public. However, there can be no objection to an informal discussion that will not pass beyond the second party. Having long considered you the foremost authority on the subject of aerial navigation, and appreciating the interest you naturally would have in the Hawaiian flight of the "Pan American Clipper," I am exercising that privilege.

Due to the spacious chart room and large chart table aboard the Clipper, the navigation equipment need not be so severely limited as in smaller planes, hence the choice of equipment may be governed entirely by the individual's personal preference or the Company's desires in the matter. To date the Company has not decided upon any standard equipment, and therefore I chose the equipment used on the subject flight. My choice was not necessarily based upon a conviction that the particular type of any instrument chosen was superior to any other type. As a matter of fact, several factors influenced the selection. Preeminent among them was the fact that most of the instruments had been used extensively by the writer and had proven satisfactory; in some instances a choice was governed entirely by the nature of the work involved—as for instance, parallel rulers versus protractors—and I suspect that plain prejudice, which actuates so many of us, carried some weight.

A set of marine charts, general, coastwise, and harbor, was carried; also aviation strip charts of the California coast. The actual chart work was carried out on VP-3 and 4 Aircraft Plotting Sheets. By working along the track from Alameda to the left-hand border of the chart, then transferring that termination of the track back to the right-hand border in the same latitude, and continuing in this manner, two sheets sufficed for the entire crossing.

Timepieces carried were a Longines Civil Time chronometer, and a Longines second-setting watch. The latter was set to correct G.C.T. at all times by checking with the chronometer. This watch was of the arm type, but the strap was removed, and the watch clips on the octant were adjusted to accommodate the beckets on each side of the watch. I prefer such arrangement to carrying the watch on the arm.

Two sextants were carried—a Pioneer bubble octant, and a mariner's sextant. The former was used for all sights; the latter carried as a "preventer."

Originally a Pioneer Universal protractor was installed on the chart table but experience convinced me that parallel rulers, where room permits their use, are more satisfactory for rapid plotting of long-range D.F. bearings because of their greater scope. Protractors, such as yours, are more convenient for plotting short bearings and lines of position. However, a minimum of instruments lessens confusion, so I decided upon the rulers only. Those carried were of the Captain Fields Improved type—graduated in degrees—and consequently the greatest objection to their use in aircraft; namely, creeping, when referring to compass roses, was removed.

I also carried a Dalton Mark VII Aircraft Navigational Computer, which I find a great convenience.

The actual navigation was comparable with such as would be practiced afloat—fixes were determined entirely by stellar observations at night. These fixes were more reliable than would be possible by crossing a line of position with a D.F. bearing, due to the amount of error which would be introduced by even a small angular error in the long-range D.F. bearings. By day, having only the single heavenly body for determination of lines of position, we did cross the bearings. However, during daylight hours we were nearer the radio stations and consequently the error introduced was generally considerably reduced.

The accuracy of fixes was very gratifying. By that, an accuracy of approximately ten miles is implied. My experience is that such a degree of accuracy is about the average one may expect in aerial navigation. A comparison of our expected time of arrival over Kaneohe Bay, Oahu, where our D.F. station is located, and the time of our actual landfall affords a good indication of the reliability of our sights. At 0457 L.C.T., while still above a solid cumulus bank, our fix by ⋆ Polaris and ⋆ Deneb was latitude 24°04′N., longitude 153°14′W. That was our last observation, and on the strength of it we advised our station we would pass over at 0700. We were then cruising at reduced speed so as to arrive at Pearl Harbor not earlier than 0800. Going below the clouds shortly after establishing the fix, we encountered light mist and scattered showers. The visibility varied between two to twelve miles, which prevented us sighting Molakai, as we would have done with normal visibility. At 0653 we sighted Makapuu Point slightly on our port bow, with Kaneohe Bay directly ahead. At 0700 we were directly off the radio station. This accuracy was due to smooth flying conditions at the time of sight, and of course it could not be cited as an example of accuracy consistently possible.

The greatest difficulty is, of course, the determination of drift angle. We carried smoke bombs and water flares for this purpose. The latter are of an improved pattern and are unusually effective when the surface of the water is visible. However, during both flights—westbound and eastbound—we were above solid cumulus banks approximately 90 per cent of the time. The smoke bombs are not entirely satisfactory. Although a special pattern has been developed, we find that the smoke blends too closely with the water color to afford a good reference mark.

Consequently, the difference between "no wind" positions and fixes established by observations were utilized entirely for determination of drift angle, and, of course, wind direction and velocity for laying new courses. This method proved to have been quite accurate, as indicated by the very nearly direct track we made for the entire westbound flight. However, it would not be so desirable in a region where sudden wind shifts could be expected. Then reliance would necessarily have to be placed on D.F. bearings despite their lack of extreme accuracy.

In addition to the actual navigation, I maintained a very detailed log during both flights. In addition to recording courses, variation, deviation, track made good, indicated and true air speeds, ground speed, etc., a complete meteorological record was kept. As you may imagine, each hour represented sixty very busy minutes.

I consider the development of the Greenwich hour-angle idea the greatest contribution to the science of navigation since Sumner, and have used it exclusively since first published in the Air Almanac. The second-setting watch runs it a close second as a time saver and an aid tending to minimize errors. Navigators owe you a debt of gratitude for those contributions to the science.

I suppose you wonder what method I use for computation of observations. I use Dreisonstok exclusively. Probably another prejudice, but I have used it since it first became available in 1927 or 1928, and still prefer it.

I would appreciate further communication with you upon any navigational matters which might be of mutual interest.

Thanking you very much for your letter, which really was greatly appreciated despite my tardiness in replying, I am

Very truly yours,
FREDERICK J. NOONAN

Special Training Devices.—Obviously, navigation in flight is the best training for the beginner. However, this training is expensive and slow, owing to delays caused by weather, material, etc. A great deal may be accomplished by "synthetic training" where special training devices such as the "Navitrainer," the Link Celestial Navigation Trainer, and improvised devices that simulate navigation as accomplished in flight. The worth of an instructor hinges largely on his ability to give efficient training in the shortest time and with a minimum time in training flights.

Experiences in the Second World War.—Personal reports from former pupils of the author giving their experiences and describing the methods being used have substantiated the general procedure followed in this book. Owing to the rapid expansion of long-range transocean flying over both the Atlantic and the Pacific, navigation becomes increasingly important. It now appears that beginners are first classified to select the best navigators before the majority start pilot training. In any case, the most critical need in aviation today is for efficient long-distance navigators.

All four general methods of navigation are in use, although radio is restricted in war zones. On long overwater flights the principal dependence is placed on celestial navigation, and numerous reports indicate that excellent results are being obtained.

Having progressed through the various steps of fixing position by celestial navigation, we refer the student to the folder in the book pocket which gives the actual work accomplished by Captain Lewis A. Yancey, navigator of the Archbold Expedition plane *Guba* on its flight across the Pacific Ocean. The author believes this is the most complete available example of skilled navigation. It will be noted that the *Guba* made all three landfalls with the destination "sighted dead ahead"; that she passed around a storm; and, having arrived off Hollandia ahead of schedule, she "lay off and on until daylight." It is suggested that the student reconstruct the *Guba* navigation in detail, using Yancey's observed data.

APPENDIX A

AVIATOR'S DEAD-RECKONING TABLES

I. Ground-speed and drift table (sample only).
II. Speed-time-distance table.
III. Course-correction table.
IV. Ratio of ground speed to air speed.
V. Conversion angle for converting great-circle courses to rhumb-line courses and vice versa (see Chap. VI).
VI. Course errors for distances off the required course.
VII. Formulas for radius-of-action problems (Chap. VIII).
VIII. Meridional parts and length of 1° of longitude and latitude.
IX. Dead-reckoning formulas for various sailings.

When the first edition of this book was published in 1931, there were no suitable aircraft computers. To simplify dead reckoning, the author consulted aviators on the *U.S.S. Lexington*, *U.S.S. Saratoga*, etc., and prepared 23 pages of aviators' tables. With the development of the Dalton Mark VII and other dead-reckoning computers, the tables so laboriously prepared have decreased in value. Since these tables are in type, some of them are included here to provide a check on the accuracy of the various computers and to help the student get a thorough understanding of the problems for which the tables were designed.

Table I.—This table of ground speed and drift for various winds covered speeds from 80 to 200 m.p.h. A sample only is given here since any of several computers give the same results with less effort.

Table II.—This was one of the first tables designed especially for the aviator and originally appeared in the "Line of Position Book," 1927.

Tables III, IV, and VI.—These are condensed tables designed to solve special problems as listed.

Table V.—This table is in general use at present, but should be replaced by the graphical method described in Chap. V.

Table VII.—These formulas are in general use. See Chap. VIII.

Tables VIII and IX.—These tables are not to be used by practical navigators, but are included because of their condensed form and for the rare occasions when they might be of interest.

TABLE I.—GROUND-SPEED AND DRIFT TABLE

Wind angle	10-mile wind		15-mile wind		20-mile wind		25-mile wind		30-mile wind		40-mile wind		Wind angle
	Wind corr.	Track speed, m.p.h.	Wind corr.	Track speed, m.p.h.	Wind corr.	Track speed, m.p.h.	Wind corr.	Track speed, m.p.h.	Wind corr.	Track speed, m.p.h.	Wind corr.	Track speed, m.p.h.	
Air Speed 180 Miles per Hour													
0°	0°	170	0°	165	0°	160	0°	155	0°	150	0°	140	0°
10°	1°	170	1°	165	1°	160	1°	155	2°	150	2°	140	10°
20°	1°	170	2°	166	2°	161	3°	156	3°	151	4°	142	20°
30°	2°	171	2°	167	3°	163	4°	158	5°	153	6°	144	30°
40°	2°	172	3°	168	4°	164	5°	160	6°	156	8°	147	40°
50°	2°	173	4°	170	5°	167	6°	163	7°	159	10°	152	50°
60°	3°	175	4°	172	6°	169	7°	166	8°	163	11°	156	60°
70°	3°	176	5°	174	6°	172	8°	170	9°	167	12°	162	70°
80°	3°	178	5°	177	6°	176	8°	174	9°	172	13°	168	80°
90°	3°	180	5°	180	6°	179	8°	178	9°	177	13°	175	90°
100°	3°	182	5°	182	6°	183	8°	183	9°	182	13°	182	100°
110°	3°	184	5°	185	6°	186	8°	185	9°	188	12°	190	110°
120°	3°	185	4°	187	6°	189	7°	191	8°	193	11°	197	120°
130°	2°	186	4°	189	5°	192	6°	195	7°	198	10°	203	130°
140°	2°	188	3°	191	4°	195	5°	198	6°	202	8°	209	140°
150°	2°	189	2°	193	3°	197	4°	201	5°	205	6°	214	150°
160°	1°	190	2°	194	2°	198	3°	203	3°	208	4°	217	160°
170°	1°	190	1°	195	1°	200	1°	205	2°	209	2°	219	170°
180°	0°	190	0°	195	0°	200	0°	205	0°	210	0°	220	180°
Air Speed 190 Miles per Hour													
0°	0°	180	0°	175	0°	170	0°	165	0°	160	0°	150	0°
10°	1°	180	1°	175	1°	170	1°	165	2°	160	2°	150	10°
20°	1°	180	2°	176	2°	171	3°	166	3°	162	4°	152	20°
30°	2°	181	2°	177	3°	173	4°	168	4°	164	6°	155	30°
40°	2°	182	3°	178	4°	175	5°	170	6°	166	8°	158	40°
50°	2°	183	3°	180	5°	177	6°	173	7°	170	9°	162	50°
60°	3°	185	4°	182	5°	180	7°	176	8°	173	11°	167	60°
70°	3°	186	4°	184	6°	183	7°	180	9°	178	11°	173	70°
80°	3°	188	4°	187	6°	186	7°	184	9°	182	12°	179	80°
90°	3°	190	5°	190	6°	190	8°	188	9°	188	12°	186	90°
100°	3°	192	4°	192	6°	192	7°	193	9°	193	12°	193	100°
110°	3°	194	4°	194	6°	196	7°	197	8°	198	11°	200	110°
120°	3°	196	4°	197	5°	200	7°	202	8°	204	11°	207	120°
130°	2°	197	3°	199	5°	202	6°	206	7°	208	9°	213	130°
140°	2°	198	3°	201	4°	204	5°	209	6°	212	8°	219	140°
150°	2°	199	2°	203	3°	207	4°	212	4°	216	6°	225	150°
160°	1°	200	2°	204	2°	209	3°	214	3°	218	4°	228	160°
170°	1°	200	1°	205	1°	210	1°	215	2°	220	2°	229	170°
180°	0°	200	0°	205	0°	210	0°	215	0°	220	0°	230	180°

TABLE II.—AVIATOR'S SPEED-TIME-DISTANCE TABLE

Time in minutes	Speed in miles per hour															
	50	60	70	80	90	100	110	120	130	140	150	160	170	180	190	200
	Distance															
1	0.8	1	1.2	1.3	1.5	1.7	1.8	2	2.2	2.3	2.5	2.7	2.8	3	3.2	3.3
2	1.7	2	2.3	2.7	3.0	3.3	3.7	4	4.3	4.7	5.0	5.3	5.7	6	6.3	6.7
3	2.5	3	3.5	4.0	4.5	5.0	5.5	6	6.5	7.0	7.5	8.0	8.5	9	9.5	10.0
4	3.3	4	4.7	5.3	6.0	6.7	7.3	8	8.7	9.3	10.0	10.7	11.3	12	12.7	13.3
5	4.2	5	5.8	6.7	7.5	8.3	9.2	10	10.8	11.7	12.5	13.3	14.2	15	15.8	16.7
6	5.0	6	7.0	8.0	9.0	10.0	11.0	12	13.0	14.0	15.0	16.0	17.0	18	19.0	20.0
7	5.8	7	8.2	9.3	10.5	11.7	12.8	14	15.2	16.3	17.5	18.7	19.8	21	22.2	23.3
8	6.7	8	9.3	10.7	12.0	13.3	14.7	16	17.3	18.7	20.0	21.3	22.7	24	25.3	26.7
9	7.5	9	10.5	12.0	13.5	15.0	16.5	18	19.5	21.0	22.5	24.0	25.5	27	28.5	30.0
10	8.3	10	11.7	13.3	15.0	16.7	18.3	20	21.7	23.3	25.0	26.7	28.3	30	31.7	33.3
11	9	11	13	15	17	18	20	22	24	26	28	29	31	33	35	37
12	10	12	14	16	18	20	22	24	26	28	30	32	34	36	38	40
13	11	13	15	17	20	22	24	26	28	30	33	35	37	39	41	43
14	12	14	16	19	21	23	26	28	30	33	35	37	40	42	44	47
15	13	15	18	20	23	25	27	30	33	35	38	40	43	45	48	50
16	13	16	19	21	24	27	29	32	35	37	40	43	45	48	51	53
17	14	17	20	23	26	28	31	34	37	40	43	45	48	51	54	57
18	15	18	21	24	27	30	33	36	39	42	45	48	51	54	57	60
19	16	19	22	25	29	32	35	38	41	44	48	51	54	57	60	63
20	17	20	23	27	30	33	37	40	43	47	50	53	57	60	63	67
21	18	21	25	28	32	35	39	42	46	49	53	56	60	63	66	70
22	18	22	26	29	33	37	40	44	48	51	55	59	62	66	70	73
23	19	23	27	31	35	38	42	46	50	54	58	61	65	69	73	77
24	20	24	28	32	36	40	44	48	52	56	60	64	68	72	76	80
25	21	25	29	33	38	42	46	50	54	58	63	67	71	75	79	83
26	22	26	30	35	39	43	48	52	56	61	65	69	74	78	82	87
27	23	27	32	36	41	45	50	54	59	63	68	72	77	81	85	90
28	23	28	33	37	42	47	51	56	61	65	70	75	80	84	89	93
29	24	29	34	39	44	48	53	58	63	68	73	77	82	87	92	97
30	25	30	35	40	45	50	55	60	65	70	75	80	85	90	95	100
31	26	31	36	41	47	52	57	62	67	72	78	83	88	93	98	103
32	27	32	37	43	48	53	59	64	69	75	80	85	91	96	101	107
33	28	33	39	44	50	55	61	66	72	77	83	88	94	99	105	110
34	28	34	40	45	51	57	62	68	74	79	85	91	96	102	108	113
35	29	35	41	47	53	58	64	70	76	82	88	93	99	105	111	117
36	30	36	42	48	54	60	66	72	78	84	90	96	102	108	114	120
37	31	37	43	49	56	62	68	74	80	86	93	99	105	111	117	123
38	32	38	44	51	57	63	70	76	82	89	95	101	108	114	120	127
39	33	39	46	52	59	65	72	78	85	91	98	104	110	117	123	130
40	33	40	47	53	60	67	73	80	87	93	100	107	113	120	127	133
41	34	41	48	55	62	68	75	82	89	96	103	109	116	123	130	137
42	35	42	49	56	63	70	77	84	91	98	105	112	119	126	133	140
43	36	43	50	57	65	72	79	86	93	100	108	115	122	129	136	143
44	37	44	51	59	66	73	81	88	95	103	110	117	125	132	139	147
45	38	45	53	60	68	75	83	90	98	105	113	120	128	135	143	150
46	38	46	54	61	69	77	84	92	100	107	115	123	130	138	146	153
47	39	47	55	63	71	78	86	94	102	110	118	125	133	141	149	157
48	40	48	56	64	72	80	88	96	104	112	120	128	136	144	152	160
49	41	49	57	65	74	82	90	98	106	114	123	131	139	147	155	163
50	42	50	58	67	75	83	92	100	108	117	125	133	142	150	158	167
51	43	51	60	68	77	85	94	102	111	119	128	136	145	153	162	170
52	43	52	61	69	78	87	95	104	113	121	130	139	147	156	165	173
53	44	53	62	71	80	88	97	106	115	124	133	141	150	159	168	177
54	45	54	63	72	81	90	99	108	117	126	135	144	153	162	171	180
55	46	55	64	73	83	92	101	110	119	128	138	147	156	165	174	183
56	47	56	65	75	84	93	103	112	121	131	140	149	159	168	177	187
57	48	57	67	76	86	95	105	114	123	133	143	152	162	171	181	190
58	48	58	68	77	87	97	106	116	126	135	145	155	164	174	184	193
59	49	59	69	79	89	98	108	118	128	138	148	157	167	177	187	197
60	50	60	70	80	90	100	110	120	130	140	150	160	170	180	190	200
Time in hours																
2	100	120	140	160	180	200	220	240	260	280	300	320	340	360	380	400
3	150	180	210	240	270	300	330	360	390	420	450	480	510	540	570	600
4	200	240	280	320	360	400	440	480	520	560	600	640	680	720	760	800
5	250	300	350	400	450	500	550	600	650	700	750	800	850	900	950	1,000
6	300	360	420	480	540	600	660	720	780	840	900	960	1,020	1,080	1,140	1,200
7	350	420	490	560	630	700	770	840	910	980	1,050	1,120	1,190	1,260	1,330	1,400
8	400	480	560	640	720	800	880	960	1,040	1,120	1,200	1,280	1,360	1,440	1,520	1,600
9	450	540	630	720	810	900	990	1,080	1,170	1,260	1,350	1,440	1,530	1,620	1,710	1,800

Course-correction Table.—This table tabulates the drift angle for various wind forces and directions. Given the direction and force of the wind, the table gives the amount to be allowed to make good a certain course. In other words, if the force and direction of the wind are known or if they can be fairly closely estimated, this little table will give the same data as are obtained by a drift observation.

TABLE III.—COURSE-CORRECTION TABLE

Ratio of $\frac{WF}{AS}$	Inclination of wind to course to be made good								
	10° or 170°	20° or 160°	30° or 150°	40° or 140°	50° or 130°	60° or 120°	70° or 110°	80° or 100°	90°
	°	°	°	°	°	°	°	°	°
0.05	0.5	1.0	1.4	1.9	2.2	2.5	2.7	2.8	2.9
0.10	1.0	2.0	2.9	3.7	4.4	5.0	5.4	5.7	5.8
0.15	1.5	3.0	4.3	5.5	6.6	7.5	8.1	8.5	8.6
0.20	2.0	3.9	5.8	7.4	8.8	10.0	10.8	11.4	11.6
0.25	2.5	4.9	7.2	9.3	11.1	12.5	13.6	14.3	14.5
0.30	3.0	5.9	8.6	11.1	13.3	15.1	16.4	17.2	17.5
0.35	3.5	6.9	10.1	13.0	15.5	17.6	19.2	20.2	20.5
0.40	4.0	7.9	11.5	14.9	17.8	20.2	22.1	23.2	23.6
0.45	4.5	8.9	13.0	16.8	20.2	22.9	25.0	26.3	26.8
0.50	5.0	9.9	14.5	18.8	22.5	25.7	28.2	29.5	30.0
0.55	5.5	10.9	16.0	21.2	24.9	28.5	31.1	32.8	33.4
0.60	6.0	11.9	17.5	22.7	27.4	31.3	34.3	36.3	36.9
0.65	6.5	12.9	19.0	24.7	29.9	34.3	37.6	39.8	40.6
0.70	7.0	13.9	20.5	26.8	32.4	37.3	41.8	43.9	44.6

The ratio WF is equal to $\frac{\text{force of wind}}{\text{speed of plane}}$ and is shown in left margin. Enter with ratio in left margin and inclination of wind to course to be made good at the top and pick out the angle to be allowed for the wind to make good the course.

TABLE IV.—GROUND SPEED IN PER CENT OF AIR SPEED

Ratio $\frac{WF}{AS}$	Inclination of wind to course to be made good																		
	0°	10°	20°	30°	40°	50°	60°	70°	80°	90°	100°	110°	120°	130°	140°	150°	160°	170°	180°
	Percentage																		
0.05	95	95	95	96	96	97	97	98	99	100	101	102	102	103	104	104	105	105	105
0.10	90	90	92	91	92	93	95	96	98	100	101	103	105	106	107	109	109	110	110
0.15	85	85	86	87	88	90	92	94	96	99	102	105	107	109	111	113	114	115	115
0.20	80	81	82	83	84	86	89	92	95	98	102	105	109	112	115	117	119	120	120
0.25	75	75	76	78	80	82	85	89	93	97	101	106	110	114	118	121	123	124	125
0.30	70	70	72	73	75	78	82	86	90	95	101	106	112	116	121	125	128	129	130
0.35	65	65	66	68	71	74	78	82	88	94	100	106	113	119	124	129	132	134	135
0.40	60	60	62	63	66	70	74	79	85	92	99	106	114	121	127	133	137	139	140
0.45	55	55	56	58	61	65	69	75	82	89	97	106	114	123	130	136	141	144	145
0.50	50	50	52	53	56	60	65	71	78	86	95	105	115	124	133	140	146	149	150
0.55	45	45	46	48	51	55	60	67	75	83	94	104	116	126	136	144	150	154	155
0.60	40	40	42	43	46	50	55	62	70	80	91	103	115	127	138	147	154	159	160
0.65	35	35	36	38	41	45	50	57	66	76	88	101	115	128	141	151	159	164	165
0.70	30	30	32	33	36	40	45	52	60	72	84	99	114	130	143	155	163	168	170

Enter this table as in the table preceding and pick out the corresponding number. This number is the percentage of the air speed that equals the ground speed. Thus if ratio of wind force to air speed is 0.25 and the inclination of wind to track is 120°, then the ground speed is 110 per cent of the air speed.

TABLE V.—CONVERSION ANGLE FOR CONVERTING GREAT-CIRCLE COURSES TO RHUMB-LINE COURSES AND VICE VERSA

Conversion angle = ½ DLo. sin L_m

Mid-lat.	Difference of longitude													
	1°	2°	3°	4°	5°	6°	7°	8°	9°	10°	11°	12°	13°	14°
°	°	°	°	°	°	°	°	°	°	°	°	°	°	°
5	0.0	0.1	0.1	0.2	0.2	0.3	0.3	0.3	0.4	0.4	0.5	0.5	0.6	0.6
10	0.1	0.2	0.3	0.3	0.4	0.5	0.6	0.7	0.8	0.9	1.0	1.0	1.1	1.2
15	0.1	0.3	0.4	0.5	0.6	0.8	0.9	1.0	1.2	1.3	1.4	1.6	1.7	1.8
20	0.2	0.3	0.5	0.7	0.9	1.0	1.2	1.4	1.5	1.7	1.9	2.1	2.2	2.4
25	0.2	0.4	0.6	0.8	1.1	1.3	1.5	1.7	1.9	2.1	2.3	2.5	2.7	3.0
30	0.2	0.5	0.8	1.0	1.2	1.5	1.8	2.0	2.2	2.5	2.8	3.0	3.2	3.5
35	0.3	0.6	0.9	1.1	1.4	1.7	2.0	2.3	2.6	2.9	3.2	3.4	3.7	4.0
40	0.3	0.6	1.0	1.3	1.6	1.9	2.2	2.6	2.9	3.2	3.5	3.9	4.2	4.5
45	0.4	0.7	1.1	1.4	1.8	2.1	2.5	2.8	3.2	3.5	3.9	4.2	4.6	4.9
50	0.4	0.8	1.1	1.5	1.9	2.3	2.7	3.1	3.4	3.8	4.2	4.6	5.0	5.4
55	0.4	0.8	1.2	1.6	2.0	2.5	2.9	3.3	3.7	4.1	4.5	4.9	5.3	5.7
60	0.4	0.9	1.3	1.7	2.2	2.6	3.0	3.5	3.9	4.3	4.8	5.2	5.6	6.1
65	0.5	0.9	1.4	1.8	2.3	2.7	3.2	3.6	4.1	4.5	5.0	5.4	5.9	6.3

Mid-lat.	Difference of longitude											
	14°	15°	16°	17°	18°	19°	20°	21°	22°	23°	24°	25°
°	°	°	°	°	°	°	°	°	°	°	°	°
5	0.6	0.7	0.7	0.7	0.8	0.8	0.9	0.9	1.0	1.0	1.0	1.1
10	1.2	1.3	1.4	1.5	1.6	1.6	1.7	1.8	1.9	2.0	2.1	2.2
15	1.8	1.9	2.1	2.2	2.3	2.5	2.6	2.7	2.8	3.0	3.1	3.2
20	2.4	2.6	2.7	2.9	3.1	3.2	3.4	3.6	3.8	3.9	4.1	4.3
25	3.0	3.2	3.4	3.6	3.8	4.0	4.2	4.4	4.6	4.9	5.1	5.3
30	3.5	3.8	4.0	4.2	4.5	4.8	5.0	5.2	5.5	5.8	6.0	6.2
35	4.0	4.3	4.6	4.9	5.2	5.4	5.7	6.0	6.3	6.6	6.9	7.2
40	4.5	4.8	5.1	5.5	5.8	6.1	6.4	6.7	7.1	7.4	7.7	8.0
45	4.9	5.3	5.7	6.0	6.4	6.7	7.1	7.4	7.8	8.1	8.5	8.8
50	5.4	5.7	6.1	6.5	6.9	7.3	7.7	8.0	8.4	8.8	9.2	9.6
55	5.7	6.1	6.6	7.0	7.4	7.8	8.2	8.6	9.0	9.4	9.8	10.2
60	6.1	6.5	6.9	7.4	7.8	8.2	8.7	9.1	9.5	10.0	10.4	10.8
65	6.3	6.8	7.3	7.7	8.2	8.6	9.1	9.5	10.0	10.4	10.9	11.3

If the difference of longitude is greater than 25°, it may be divided by 2, and the tabular entry doubled.

TABLE VI.—COURSE ERRORS FOR DISTANCE OFF COURSE

Miles flown	Miles off course															
	1	2	3	4	5	6	7	8	9	10	15	20	25	30	40	50
	Compass correction to parallel track course															
	°	°	°	°	°	°	°	°	°	°	°	°	°	°	°	°
10	6	12	17	24	30	37	44	53	64	90						
20	3	6	9	12	14	17	20	24	27	30	49	90				
30	2	4	6	8	10	12	14	15	17	19	30	42	56	90		
40	1	3	4	6	7	9	10	12	13	14	22	30	39	49	90	
50	1	2	3	5	6	7	8	9	10	12	17	24	30	37	53	90
60	1	2	3	4	5	6	7	8	9	10	14	19	25	30	42	56
70	1	2	3	3	4	5	6	7	7	8	12	17	21	25	35	46
80	1	1	2	3	4	4	5	6	6	7	11	14	18	22	30	39
90	1	1	2	3	3	4	4	5	6	6	10	13	16	19	26	34
100	1	1	2	2	3	3	4	5	5	6	9	12	14	17	24	30
110	1	1	2	2	3	3	4	4	5	5	8	10	13	16	21	27
120	0	1	1	2	2	3	3	4	4	5	7	10	12	14	19	25
130	0	1	1	2	2	3	3	4	4	4	7	9	11	13	18	23
140	0	1	1	2	2	2	3	3	4	4	6	8	10	12	17	21
150	0	1	1	2	2	2	3	3	3	4	6	8	10	12	15	19
160	0	1	1	1	2	2	3	3	3	4	5	7	9	11	14	18
170	0	1	1	1	2	2	2	3	3	3	5	7	8	10	14	17
180	0	1	1	1	2	2	2	3	3	3	5	6	8	10	13	16
190	0	1	1	1	2	2	2	2	3	3	5	6	8	9	12	15
200	0	1	1	1	1	2	2	2	3	3	4	6	7	9	12	14

Example.—After flying 11½ miles the perpendicular distance to the required track is 1½ miles. What is the error of the track made good? Entering at 115 miles and 15 miles, we find the required value, namely, 7½°.

TABLE VII.—FORMULAS FOR RADIUS-OF-ACTION PROBLEMS

The problem here considered is that of determining how far a plane may fly from its base under different wind conditions (see also pages 157, 166).

R = distance from base at time of turning back, *i.e.*, radius of action.
t_1 = time of outward flight.
t_2 = time of return flight.
T = total time of flight $= t_1 + t_2$.
S_1 = rate of departure (ground speed) from base.
S_2 = rate of returning (ground speed) to base.

Since

$$t_1 = \frac{R}{S_1} \qquad \text{and} \qquad t_2 = \frac{R}{S_2}$$

we have

$$T = \frac{R}{S_1} + \frac{R}{S_2} = \frac{R(S_1 + S_2)}{S_1 S_2}$$

from which

$$R = \frac{TS_1S_2}{S_1 + S_2}$$

But $R = t_1S_1$, so

$$t_1 = \frac{TS_2}{S_1 + S_2}$$

$$t_2 = \frac{TS_1}{S_1 + S_2} = T - t_1$$

TABLE VIII.—MERIDIONAL PARTS AND LENGTH OF 1° OF LONGITUDE AND LATITUDE

Lat. L	Meridional parts M		Length of 1° of		Lat. L	Meridional parts M		Length of 1° of	
			Long.	Lat.				Long.	Lat.
°					°				
0	0	60	60.1	59.7	40	2608	78	46.1	59.9
1	60	59	60.1	59.7	41	2686	80	45.4	59.9
2	119	60	60.0	59.7	42	2766	81	44.7	59.9
3	179	60	60.0	59.7	43	2847	83	44.0	59.9
4	239	59	59.9	59.7	44	2930	83	43.3	60.0
5	298	60	59.8	59.7	45	3013	86	42.5	60.0
6	358	60	59.7	59.7	46	3099	87	41.8	60.0
7	418	60	59.6	59.7	47	3186	88	41.0	60.0
8	478	61	59.5	59.7	48	3274	90	40.3	60.0
9	539	60	59.3	59.7	49	3364	93	39.5	60.0
10	599	61	59.2	59.7	50	3457	94	38.7	60.0
11	660	61	59.0	59.7	51	3551	96	37.9	60.0
12	721	61	58.8	59.7	52	3647	98	37.1	60.0
13	782	61	58.5	59.7	53	3745	101	36.2	60.1
14	843	61	58.3	59.7	54	3846	103	35.4	60.1
15	904	62	58.0	59.7	55	3949	106	34.5	60.1
16	966	63	57.8	59.7	56	4055	108	33.7	60.1
17	1029	62	57.5	59.7	57	4163	111	32.8	60.1
18	1091	63	57.1	59.7	58	4274	115	31.9	60.1
19	1154	63	56.8	59.7	59	4389	118	31.0	60.1
20	1217	64	56.5	59.7	60	4507	122	30.1	60.1
21	1281	64	56.1	59.7	61	4629	125	29.2	60.1
22	1345	65	55.7	59.7	62	4754	130	28.3	60.1
23	1410	65	55.3	59.8	63	4884	134	27.3	60.1
24	1475	65	54.9	59.8	64	5018	140	26.4	60.2
25	1540	66	54.5	59.8	65	5158	144	25.5	60.2
26	1606	67	54.0	59.8	66	5302	150	24.5	60.2
27	1673	67	53.6	59.8	67	5452	157	23.5	60.2
28	1740	68	53.1	59.8	68	5609	164	22.6	60.2
29	1808	69	52.6	59.8	69	5773	171	21.6	60.2
30	1877	69	52.1	59.8	70	5944	180	20.6	60.2
31	1946	70	51.5	59.8	71	6124	189	19.6	60.2
32	2016	71	51.0	59.8	72	6313	199	18.6	60.2
33	2087	71	50.4	59.8	73	6512	211	17.6	60.2
34	2158	73	49.9	59.9	74	6723	225	16.6	60.2
35	2231	73	49.3	59.9	75	6948	239	15.6	60.2
36	2304	75	48.7	59.9	76	7187	257	14.6	60.2
37	2379	75	48.0	59.9	77	7444	278	13.6	60.2
38	2454	76	47.4	59.9	78	7722	301	12.5	60.2
39	2530	78	46.7	59.9	79	8023	329	11.5	60.3
40	2608		46.1	59.9	80	8352		10.5	60.3

TABLE IX.—DEAD-RECKONING FORMULAS FOR VARIOUS SAILINGS

Given	To find	Mid-latitude Sailing	Mercator sailing
Both latitudes and both longitudes	Departure Course Distance	$p = \text{DLo.} \cos L_m$ $\tan C = \frac{p}{l} = \frac{\text{DLo.} \cos L_m}{l}$ $D^* = l \sec C = p \operatorname{cosec} C$	$p = \frac{l\ \text{DLo.}}{m}$ $\tan C = \frac{\text{DLo.}}{m}$ $D = l \sec C = l \tan C \operatorname{cosec} C$†
Both latitudes and departure	Course Distance Diff. of Long.	$\tan C = \frac{p}{l}$ $D^* = l \sec C = p \operatorname{cosec} C$ $\text{DLo.} = p \sec L_m$	$\tan C = \frac{p}{l}$ $D^* = l \sec C = p \operatorname{cosec} C$ $\text{DLo.} = m \tan C = \frac{pm}{l}$
One latitude, course, and distance	Diff. of Lat. Departure Diff. of Long.	$l = D \cos C$ $p = D \sin C$ $\text{DLo.} = p \sec L_m = D \sin C \sec L_m$	$l = D \cos C$ $p = D \sin C$ $\text{DLo.} = m \tan C = \frac{pm}{l}$
Both latitudes and course	Departure Distance Diff. of Long.	$p = l \tan C$ $D^* = l \sec C = p \operatorname{cosec} C$ $\text{DLo.} = p \sec L_m = l \tan C \sec L_m$	$p = l \tan C$ $D = l \sec C$ $\text{DLo.} = m \tan C$
Both latitudes and distance	Course Departure Diff. of Long.	$\cos C = \frac{l}{D}$ $p = D \sin C$ $\text{DLo.} = p \sec L_m = D \sin C \sec L_m$	$\cos C = \frac{l}{D}$ $p = D \sin C$ $\text{DLo.} = m \tan C$
One latitude, course, and departure	Diff. of lat. Distance Diff. of long.	$l = p \cos C$ $D = p \operatorname{cosec} C$ $\text{DLo.} = p \sec. L_m$	$l = p \cot C$ $D = p \operatorname{cosec} C$ $\text{DLo.} = m \tan C = \frac{pm}{l}$
One latitude and departure	Course Diff. of lat. Diff. of long.	$\sin C = \frac{p}{D}$ $l = D \cos C$ $\text{DLo.} = p \sec L_m$	$\sin C = \frac{p}{D}$ $l = D \cos C$ $\text{DLo.} = m \tan C = \frac{pm}{l}$

* If both l and p are known, use the greater.

† Use this form if $\tan C$ is greater than 1.

APPENDIX B

NAVIGATION AUXILIARY TABLES

Bubble Correction for Sun, Stars, and Planets.—This table is similar to those in various navigation tables, except that corrections for low altitudes are included. Mariners prefer not to use altitudes less than about 10°, owing to the variations in altitude with changes in temperature and pressure. When the bubble sextant is used, extreme accuracy is not expected, and, as will be seen by a study of the four additional corrections, observations may be made down to the horizon without serious loss of accuracy. See page 372 for these table.

Correction for Temperatures.—Refraction tables are made for a standard temperature of 50°F. For variations from this temperature, an additional correction should be made to the refraction table given above as shown in the table Correction for Temperatures.

Correction for Pressure.—Variations from the standard pressure of 30″ cause the refraction to change. This should be corrected as shown in the table Correction for Pressure.

Correction for Height.—The refraction tables are computed for sea level. Observations taken at, say, 20,000 ft., are subject to only a portion of the total refraction at sea level. The table Refraction Subcorrection gives the subcorrections that must be applied for various heights.

Correction for Earth's Rotation.—For high latitudes and high speeds the earth's rotation affects the bubble as shown in the table Correction for Earth's Rotation.

It will be noted that, except for altitudes less than about 4° and for extremes in temperature and pressure, the tables correcting for variations will not be required. Also, the correction for height is only 2′ for 10° observed altitude up to a height of 20,000 ft.

Common sense should indicate when to use the various auxiliary tables. The sheet is printed on heavy paper to permit its use as a bookmark in the "Navigation Note Book."

Warner Course-distance-conversion Table.—This table, prepared by L. A. Warner, is included (page 373) to permit rapid conversion of azimuths from "H.O. 214," etc., to Z_n, azimuth measured from north to the right, from 0° to 360°. This table also indicates the cardinal and intercardinal points, and converts integral degrees of arc to nautical miles.

Extracts from Air Almanac

Interpolation for GHA.—This table (page 374) and the table of Position of the Stars (page 384) (both taken from the Air Almanac) are included as a convenience to the navigator.

Correction for Earth's Rotation
Ground Speed (M.P.H.)

Lat.	100	200	300	400
0	0	0	0	0
20	0.9	1.8	2.7	3.6
40	1.7	3.4	5.1	6.8
60	2.3	4.6	6.8	9.1
75	2.5	5.1	7.6	10.17

Translate all lines $\frac{R}{L}$ in $\frac{N}{S}$ hemisphere, perpendicular to track.

Bubble Correction for Sun, Stars and Planets

Hs	REF (−)
0° 0′	34′
20	30
40	27
1 00	24
30	21
2 00	18
30	16
3 00	14
4 00	12
5 00	10
6 00	8
7 00	7
9 00	6
11 00	5
13 00	4
18 00	3
25 00	2
40 00	1
90 00	0

Correction for Temperatures
(Fahrenheit)

Hs	10°	30°	50°F	70°	90°
0	−3	−2	0	+2	+4
1	−3	−1	0	+1	+2
2	−2	−1	0	+1	+2
4	−1	−1	0	0	+1
6	−1	0	0	0	+1

Refraction Sub-correction
Subtract from Bubble Correction

Height in feet	Observed altitude			
	10°	20°	45°	75°
0	0	0	0	0
10,000	1	1	0	0
20,000	2	2	0	0
30,000	3	2	1	0
40,000	4	2	1	0

Correction for Pressure—Inches

Hs	28.5″	29.0″	29.5″	30.0″	30.5″
0	+2	+1	+1	0	−1
3	+1	0	0	0	0
6	0	0	0	0	0

WARNER COURSE-DISTANCE-CONVERSION TABLE

AZ		*ZN*				*AZ*		*ZN*			
Deg.	*Dist.*	*N-W*	*S-E*	*S-W*		*Deg.*	*Dist.*	*N-W*	*S-E*	*S-W*	
1	60	359	179	181		91	5460	269	89	271	
2	120	358	178	182		92	5520	268	88	272	
3	180	357	177	183		93	5580	267	87	273	
4	240	356	176	184		94	5640	266	86	274	
5	300	355	175	185		95	5700	265	85	275	
6	360	354	174	186		96	5760	264	84	276	
7	420	353	173	187		97	5820	263	83	277	
8	480	352	172	188		98	5880	262	82	278	
9	540	351	171	189		99	5940	261	81	279	
10	600	350	170	190		100	6000	260	80	280	
11	660	349	169	191		101	6060	259	79	281	
12	720	348	168	192		102	6120	258	78	282	
13	780	347	167	193		103	6180	257	77	283	
14	840	346	166	194		104	6240	256	76	284	
15	900	345	165	195		105	6300	255	75	285	
16	960	344	164	196		106	6360	254	74	286	
17	1020	343	163	197		107	6420	253	73	287	
18	1080	342	162	198		108	6480	252	72	288	
19	1140	341	161	199		109	6540	251	71	289	
20	1200	340	160	200		110	6600	250	70	290	
21	1260	339	159	201		111	6660	249	69	291	
22	1320	338	158	202		112	6720	248	68	292	
NNE 23	1380	337	157	203	SSW	ESE 113	6780	247	67	293	WNW
24	1440	336	156	204		114	6840	246	66	294	
25	1500	335	155	205		115	6900	245	65	295	
26	1560	334	154	206		116	6960	244	64	296	
27	1620	333	153	207		117	7020	243	63	297	
28	1680	332	152	208		118	7080	242	62	298	
29	1740	331	151	209		119	7140	241	61	299	
30	1800	330	150	210		120	7200	240	60	300	
31	1860	329	149	211		121	7260	239	59	301	
32	1920	328	148	212		122	7320	238	58	302	
33	1980	327	147	213		123	7380	237	57	303	
34	2040	326	146	214		124	7440	236	56	304	
35	2100	325	145	215		125	7500	235	55	305	
36	2160	324	144	216		126	7560	234	54	306	
37	2220	323	143	217		127	7620	233	53	307	
38	2280	322	142	218		128	7680	232	52	308	
39	2340	321	141	219		129	7740	231	51	309	
40	2400	320	140	220		130	7800	230	50	310	
41	2460	319	139	221		131	7860	229	49	311	
42	2520	318	138	222		132	7920	228	48	312	
43	2580	317	137	223		133	7980	227	47	313	
44	2640	316	136	224		134	8040	226	46	314	
NE 45	2700	315	135	225	SW	SE 135	8100	225	45	315	NW
46	2760	314	134	226		136	8160	224	44	316	
47	2820	313	133	227		137	8220	223	43	317	
48	2880	312	132	228		138	8280	222	42	318	
49	2940	311	131	229		139	8340	221	41	319	
50	3000	310	130	230		140	8400	220	40	320	
51	3060	309	129	231		141	8460	219	39	321	
52	3120	308	128	232		142	8520	218	38	322	
53	3180	307	127	233		143	8580	217	37	323	
54	3240	306	126	234		144	8640	216	36	324	
55	3300	305	125	235		145	8700	215	35	325	
56	3360	304	124	236		146	8760	214	34	326	
57	3420	303	123	237		147	8820	213	33	327	
58	3480	302	122	238		148	8880	212	32	328	
59	3540	301	121	239		149	8940	211	31	329	
60	3600	300	120	240		150	9000	210	30	330	
61	3660	299	119	241		151	9060	209	29	331	
62	3720	298	118	242		152	9120	208	28	332	
63	3780	297	117	243		153	9180	207	27	333	
64	3840	296	116	244		154	9240	206	26	334	
65	3900	295	115	245		155	9300	205	25	335	
66	3960	294	114	246		156	9360	204	24	336	
67	4020	293	113	247		157	9420	203	23	337	
ENE 68	4080	292	112	248	WSW	SSE 158	9480	202	22	338	NNW
69	4140	291	111	249		159	9540	201	21	339	
70	4200	290	110	250		160	9600	200	20	340	
71	4260	289	109	251		161	9660	199	19	341	
72	4320	288	108	252		162	9720	198	18	342	
73	4380	287	107	253		163	9780	197	17	343	
74	4440	286	106	254		164	9840	196	16	344	
75	4500	285	105	255		165	9900	195	15	345	
76	4560	284	104	256		166	9960	194	14	346	
77	4620	283	103	257		167	10020	193	13	347	
78	4680	282	102	258		168	10080	192	12	348	
79	4740	281	101	259		169	10140	191	11	349	
80	4800	280	100	260		170	10200	190	10	350	
81	4860	279	99	261		171	10260	189	9	351	
82	4920	278	98	262		172	10320	188	8	352	
83	4980	277	97	263		173	10380	187	7	353	
84	5040	276	96	264		174	10440	186	6	354	
85	5100	275	95	265		175	10500	185	5	355	
86	5160	274	94	266		176	10560	184	4	356	
87	5220	273	93	267		177	10620	183	3	357	
88	5280	272	92	268		178	10680	182	2	358	
89	5340	271	91	269		179	10740	181	1	359	
E 90	5400	270	90	270	W	S 180	10800	180	0	360	N

INTERPOLATION OF GHA

SUN, PLANETS, ♈

Int. m s	Corr. ° ′	Int. m s	Corr. ° ′	Int. m s	Corr. ° ′
00 00	0 00	03 17	0 50	06 37	1 40
01	0 01	21	0 51	41	1 41
05	0 02	25	0 52	45	1 42
09	0 03	29	0 53	49	1 43
13	0 04	33	0 54	53	1 44
17	0 05	37	0 55	57	1 45
21	0 06	41	0 56	07 01	1 46
25	0 07	45	0 57	05	1 47
29	0 08	49	0 58	09	1 48
33	0 09	53	0 59	13	1 49
37	0 10	57	1 00	17	1 50
41	0 11	04 01	1 01	21	1 51
45	0 12	05	1 02	25	1 52
49	0 13	09	1 03	29	1 53
53	0 14	13	1 04	33	1 54
57	0 15	17	1 05	37	1 55
01 01	0 16	21	1 06	41	1 56
05	0 17	25	1 07	45	1 57
09	0 18	29	1 08	49	1 58
13	0 19	33	1 09	53	1 59
17	0 20	37	1 10	57	2 00
21	0 21	41	1 11	08 01	2 01
25	0 22	45	1 12	05	2 02
29	0 23	49	1 13	09	2 03
33	0 24	53	1 14	13	2 04
37	0 25	57	1 15	17	2 05
41	0 26	05 01	1 16	21	2 06
45	0 27	05	1 17	25	2 07
49	0 28	09	1 18	29	2 08
53	0 29	13	1 19	33	2 09
57	0 30	17	1 20	37	2 10
02 01	0 31	21	1 21	41	2 11
05	0 32	25	1 22	45	2 12
09	0 33	29	1 23	49	2 13
13	0 34	33	1 24	53	2 14
17	0 35	37	1 25	57	2 15
21	0 36	41	1 26	09 01	2 16
25	0 37	45	1 27	05	2 17
29	0 38	49	1 28	09	2 18
33	0 39	53	1 29	13	2 19
37	0 40	57	1 30	17	2 20
41	0 41	06 01	1 31	21	2 21
45	0 42	05	1 32	25	2 22
49	0 43	09	1 33	29	2 23
53	0 44	13	1 34	33	2 24
57	0 45	17	1 35	37	2 25
03 01	0 46	21	1 36	41	2 26
05	0 47	25	1 37	45	2 27
09	0 48	29	1 38	49	2 28
13	0 49	33	1 39	53	2 29
17	0 50	37	1 40	57	2 30
21		41		10 00	

MOON

Int. m s	Corr. ° ′	Int. m s	Corr. ° ′	Int. m s	Corr. ° ′
00 00	0 00	03 20	0 49	06 39	1 37
02	0 01	24	0 50	43	1 38
06	0 02	29	0 51	47	1 39
10	0 03	33	0 52	52	1 40
14	0 04	37	0 53	56	1 41
18	0 05	41	0 54	07 00	1 42
22	0 06	45	0 55	04	1 43
26	0 07	49	0 56	08	1 44
31	0 08	53	0 57	12	1 45
35	0 09	58	0 58	16	1 46
39	0 10	04 02	0 59	20	1 47
43	0 11	06	1 00	25	1 48
47	0 12	10	1 01	29	1 49
51	0 13	14	1 02	33	1 50
55	0 14	18	1 03	37	1 51
01 00	0 15	22	1 04	41	1 52
04	0 16	27	1 05	45	1 53
08	0 17	31	1 06	49	1 54
12	0 18	35	1 07	54	1 55
16	0 19	39	1 08	58	1 56
20	0 20	43	1 09	08 02	1 57
24	0 21	47	1 10	06	1 58
29	0 22	51	1 11	10	1 59
33	0 23	56	1 12	14	2 00
37	0 24	05 00	1 13	18	2 01
41	0 25	04	1 14	23	2 02
45	0 26	08	1 15	27	2 03
49	0 27	12	1 16	31	2 04
53	0 28	16	1 17	35	2 05
58	0 29	20	1 18	39	2 06
02 02	0 30	25	1 19	43	2 07
06	0 31	29	1 20	47	2 08
10	0 32	33	1 21	52	2 09
14	0 33	37	1 22	56	2 10
18	0 34	41	1 23	09 00	2 11
22	0 35	45	1 24	04	2 12
26	0 36	49	1 25	08	2 13
31	0 37	54	1 26	12	2 14
35	0 38	58	1 27	16	2 15
39	0 39	06 02	1 28	21	2 16
43	0 40	06	1 29	25	2 17
47	0 41	10	1 30	29	2 18
51	0 42	14	1 31	33	2 19
55	0 43	18	1 32	37	2 20
03 00	0 44	23	1 33	41	2 21
04	0 45	27	1 34	45	2 22
08	0 46	31	1 35	50	2 23
12	0 47	35	1 36	54	2 24
16	0 48	39	1 37	58	2 25
20	0 49	43		10 00	
24					

Correction to be added to GHA for interval of GCT

EXPLANATION AND EXAMPLES

(See also condensed explanation on back cover)

1. The object of this volume is to provide in convenient form the astronomical data required for aerial navigation. The two sides of a single leaf give complete data for a single day. Auxiliary tables are given inside the front and back covers, on the flap near the back, and on the outside back cover; these tables give values to the nearest minute without interpolation.

2. Columns 2–7 of the daily sheets give the Greenwich Hour Angles at ten-minute intervals for the Sun, Vernal Equinox, the three planets most suitable for observation at that time, and the Moon, and declinations at ten-minute intervals for the Moon and at hourly intervals for the Sun and planets. The magnitudes and symbols of the planets are given in the headings with their names.

Example: For Jan. 1, 1942, at 17ʰ47ᵐ16ˢ, Greenwich Civil Time find the GHA and Dec. of Sun, Moon and Venus:

	Sun	Moon	Venus
(From the daily sheet for Jan. 1) GHA at 17ʰ40ᵐ	84°06′	275°13′	44°43′
(From table on flap) GHA Int. of 7ᵐ16ˢ	1 49	1 45	1 49
(From table on A. M. side of daily sheet) 7ᵐ Corr. HA☾ .		0	
The required GHA	85°55′	276°58′	46°32′
(By inspection from daily sheet) The required Dec. . . .	S 23°01′	N 18°35′	S 14°39′

3. The GHA of a star is found by adding the Greenwich Hour Angle of the Vernal Equinox to the star's Sidereal Hour Angle, i. e.

$$\text{GHA}^* = \text{GHA}♈ + \text{SHA}^*$$

On the inside of the back cover are given the Name, Mag., SHA, Dec., and RA of each of the 55 principal navigational stars. Two separate lists are given: one in alphabetical order, and the other in order of SHA. The numbers in the first column are the index numbers in *Astronomical Navigation Tables* (H. O. 218) and in the British publication of the same name (Air Publication 1618).

Example: For Jan. 1, 1942, at 17ʰ47ᵐ16ˢ find the GHA and Dec. of Aldebaran:

(From the daily sheet for Jan. 1) GHA♈ at 17ʰ40ᵐ	5°44′
(From table on flap) GHA♈ Int. of 7ᵐ16ˢ	1 49
(From inside back cover) SHA*	291 51
The required GHA*	299°24′
(From inside back cover) The required Dec.	N 16°23′

4. The semidiameters of the Sun and Moon and the correction for Moon's parallax are given on the A. M. side of the daily sheets; the values given are for the middle of the day and the first correction value in the parallax table is the Moon's Horizontal Parallax. Two correction tables are given on the outside of the back cover, one for refraction and one for dip. The correction for refraction, which must be applied to all observed altitudes, depends on the height of the observer in feet and on the observed altitude. The correction for dip, which must be applied to altitudes measured from the sea horizon, depends on the height of the observer.

Example: Correct the following observed altitudes taken at a height of 5,000 feet with a bubble sextant.

	Sun	Moon	Polaris
Observed Alt.	40°25′	25°11′	15°49′
Refraction	−1	−1	−3
Parallax		+49	
Corrected Alt.	40°24′	25°59′	15°46′

GREENWICH A. M. 1942 JANUARY 1 (THURSDAY) V

GCT h m	☉ SUN GHA ° ′	SUN Dec. ° ′	♈ GHA ° ′	♀ VENUS −4.4 GHA ° ′	VENUS Dec. ° ′	♂ MARS 0.0 GHA ° ′	MARS Dec. ° ′	♃ JUPITER −2.2 GHA ° ′	JUPITER Dec. ° ′	☽ MOON GHA ° ′	MOON Dec. ° ′
0 00	179 12	S23 04	100 01	139 15	S14 52	77 18	N10 21	27 55	N21 50	18 48	N18 11
10	181 41		102 31	141 45		79 48		30 25		21 13	11
20	184 11		105 02	144 16		82 18		32 55		23 38	12
30	186 41	• •	107 32	146 46	• •	84 48	• •	35 26	• •	26 03	• 12
40	189 11		110 02	149 16		87 18		37 56		28 28	12
50	191 41		112 33	151 46		89 49		40 27		30 54	13
1 00	194 11	S23 04	115 03	154 17	S14 52	92 19	N10 21	42 57	N21 50	33 19	N18 13
10	196 41		117 34	156 47		94 49		45 28		35 44	13
20	199 11		120 04	159 17		97 19		47 58		38 09	14
30	201 41	• •	122 34	161 47	• •	99 50	• •	50 29	• •	40 34	• 14
40	204 11		125 05	164 18		102 20		52 59		42 59	14
50	206 41		127 35	166 48		104 50		55 30		45 25	15
2 00	209 11	S23 04	130 06	169 18	S14 51	107 20	N10 21	58 00	N21 50	47 50	N18 15
10	211 41		132 36	171 48		109 51		60 30		50 15	15
20	214 11		135 07	174 19		112 21		63 01		52 40	16
30	216 41	• •	137 37	176 49	• •	114 51	• •	65 31	• •	55 05	• 16
40	219 11		140 07	179 19		117 21		68 02		57 30	16
50	221 41		142 38	181 49		119 52		70 32		59 56	17
3 00	224 11	S23 04	145 08	184 20	S14 50	122 22	N10 22	73 03	N21 50	62 21	N18 17
10	226 41		147 39	186 50		124 52		75 33		64 46	17
20	229 11		150 09	189 20		127 22		78 04		67 11	18
30	231 40	• •	152 39	191 50	• •	129 52	• •	80 34	• •	69 36	• 18
40	234 10		155 10	194 21		132 23		83 05		72 01	18
50	236 40		157 40	196 51		134 53		85 35		74 27	18
4 00	239 10	S23 04	160 11	199 21	S14 49	137 23	N10 22	88 06	N21 50	76 52	N18 19
10	241 40		162 41	201 52		139 53		90 36		79 17	19
20	244 10		165 11	204 22		142 24		93 06		81 42	19
30	246 40	• •	167 42	206 52	• •	144 54	• •	95 37	• •	84 07	• 20
40	249 10		170 12	209 22		147 24		98 07		86 32	20
50	251 40		172 43	211 53		149 54		100 38		88 58	20
5 00	254 10	S23 03	175 13	214 23	S14 48	152 25	N10 23	103 08	N21 50	91 23	N18 20
10	256 40		177 44	216 53		154 55		105 39		93 48	21
20	259 10		180 14	219 23		157 25		108 09		96 13	21
30	261 40	• •	182 44	221 54	• •	159 55	• •	110 40	• •	98 38	• 21
40	264 10		185 15	224 24		162 26		113 10		101 03	21
50	266 40		187 45	226 54		164 56		115 41		103 29	22
6 00	269 10	S23 03	190 16	229 24	S14 48	167 26	N10 23	118 11	N21 50	105 54	N18 22
10	271 40		192 46	231 55		169 56		120 41		108 19	22
20	274 10		195 16	234 25		172 26		123 12		110 44	23
30	276 40	• •	197 47	236 55	• •	174 57	• •	125 42	• •	113 09	• 23
40	279 10		200 17	239 25		177 27		128 13		115 34	23
50	281 40		202 48	241 56		179 57		130 43		118 00	23
7 00	284 09	S23 03	205 18	244 26	S14 47	182 27	N10 24	133 14	N21 50	120 25	N18 24
10	286 39		207 48	246 56		184 58		135 44		122 50	24
20	289 09		210 19	249 27		187 28		138 15		125 15	24
30	291 39	• •	212 49	251 57	• •	189 58	• •	140 45	• •	127 40	• 24
40	294 09		215 20	254 27		192 28		143 16		130 05	25
50	296 39		217 50	256 57		194 59		145 46		132 30	25
8 00	299 09	S23 03	220 20	259 28	S14 46	197 29	N10 24	148 16	N21 50	134 56	N18 25
10	301 39		222 51	261 58		199 59		150 47		137 21	25
20	304 09		225 21	264 28		202 29		153 17		139 46	25
30	306 39	• •	227 52	266 58	• •	205 00	• •	155 48	• •	142 11	• 26
40	309 09		230 22	269 29		207 30		158 18		144 36	26
50	311 39		232 53	271 59		210 00		160 49		147 01	26
9 00	314 09	S23 03	235 23	274 29	S14 45	212 30	N10 25	163 19	N21 50	149 27	N18 26
10	316 39		237 53	276 59		215 00		165 50		151 52	27
20	319 09		240 24	279 30		217 31		168 20		154 17	27
30	321 39	• •	242 54	282 00	• •	220 01	• •	170 51	• •	156 42	• 27
40	324 09		245 25	284 30		222 31		173 21		159 07	27
50	326 39		247 55	287 00		225 01		175 52		161 32	27
10 00	329 09	S23 02	250 25	289 31	S14 45	227 32	N10 25	178 22	N21 50	163 57	N18 28
10	331 39		252 56	292 01		230 02		180 52		166 23	28
20	334 08		255 26	294 31		232 32		183 23		168 48	28
30	336 38	• •	257 57	297 02	• •	235 02	• •	185 53	• •	171 13	• 28
40	339 08		260 27	299 32		237 33		188 24		173 38	28
50	341 38		262 57	302 02		240 03		190 54		176 03	29
11 00	344 08	S23 02	265 28	304 32	S14 44	242 33	N10 26	193 25	N21 50	178 28	N18 29
10	346 38		267 58	307 03		245 03		195 55		180 53	29
20	349 08		270 29	309 33		247 34		198 26		183 19	29
30	351 38	• •	272 59	312 03	• •	250 04	• •	200 56	• •	185 44	• 29
40	354 08		275 30	314 33		252 34		203 27		188 09	30
50	356 38		278 00	317 04		255 04		205 57		190 34	30
12 00	359 08	S23 02	280 30	319 34	S14 43	257 34	N10 26	208 27	N21 50	192 59	N18 30

☾ Par.

Alt. °	Corr. + ′
0	
10	54
15	53
18	52
21	51
24	50
26	49
29	48
31	47
33	46
35	45
36	44
38	43
40	42
41	41
43	40
44	39
46	38
47	37
49	36
50	35
51	34
53	33
54	32
55	31
57	30
58	29
59	28
60	27
62	26
63	25
64	24
65	23
66	22
67	21
68	20
70	19
71	18
72	17
73	16
74	15
75	14
76	13
77	12
78	11
79	10

SD, ☉ ′ 16

SD, ☾ ′ 15

Corr. HA ☾

Int. m	Corr. ′
0	
10	0

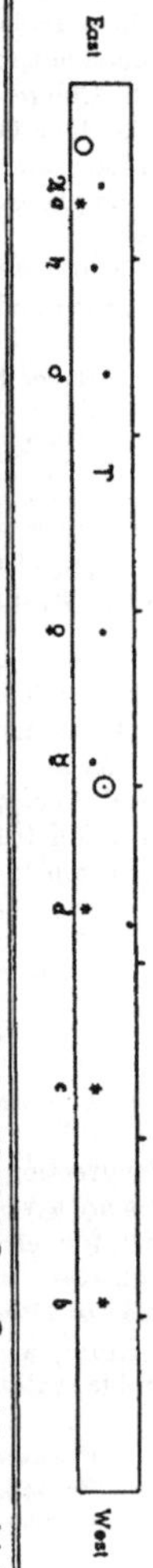

VI

GREENWICH P. M. 1942 JANUARY 1 (THURSDAY)

GCT h	GCT m	☉ SUN GHA	☉ SUN Dec.	♈ GHA	♀ VENUS –4.4 GHA	♀ VENUS Dec.	♂ MARS 0.0 GHA	♂ MARS Dec.	♃ JUPITER –2.2 GHA	♃ JUPITER Dec.	☽ MOON GHA	☽ MOON Dec.
h	m	° ′	° ′	° ′	° ′	° ′	° ′	° ′	° ′	° ′	° ′	° ′
12	00	359 08	S23 02	280 30	319 34	S14 43	257 34	N10 26	208 27	N21 50	192 59	N18 30
	10	1 38		283 01	322 04		260 05		210 58		195 24	30
	20	4 08		285 31	324 34		262 35		213 28		197 49	30
	30	6 38	• •	288 02	327 05	• •	265 05	• •	215 59	• •	200 15	• 30
	40	9 08		290 32	329 35		267 35		218 29		202 40	31
	50	11 38		293 02	332 05		270 06		221 00		205 05	31
13	00	14 08	S23 02	295 33	334 36	S14 42	272 36	N10 27	223 30	N21 50	207 30	N18 31
	10	16 38		298 03	337 06		275 06		226 01		209 55	31
	20	19 08		300 34	339 36		277 36		228 31		212 20	31
	30	21 38	• •	303 04	342 06	• •	280 07	• •	231 02	• •	214 45	• 31
	40	24 07		305 34	344 37		282 37		233 32		217 10	32
	50	26 37		308 05	347 07		285 07		236 03		219 36	32
14	00	29 07	S23 02	310 35	349 37	S14 42	287 37	N10 27	238 33	N21 50	222 01	N18 32
	10	31 37		313 06	352 07		290 08		241 03		224 26	32
	20	34 07		315 36	354 38		292 38		243 34		226 51	32
	30	36 37	• •	318 07	357 08	• •	295 08	• •	246 04	• •	229 16	• 32
	40	39 07		320 37	359 38		297 38		248 35		231 41	32
	50	41 37		323 07	2 08		300 08		251 05		234 06	33
15	00	44 07	S23 01	325 38	4 39	S14 41	302 39	N10 27	253 36	N21 50	236 31	N18 33
	10	46 37		328 08	7 09		305 09		256 06		238 57	33
	20	49 07		330 39	9 39		307 39		258 37		241 22	33
	30	51 37	• •	333 09	12 10	• •	310 09	• •	261 07	• •	243 47	• 33
	40	54 07		335 39	14 40		312 40		263 38		246 12	33
	50	56 37		338 10	17 10		315 10		266 08		248 37	33
16	00	59 07	S23 01	340 40	19 40	S14 40	317 40	N10 28	268 38	N21 50	251 02	N18 34
	10	61 37		343 11	22 11		320 10		271 09		253 27	34
	20	64 07		345 41	24 41		322 41		273 39		255 52	34
	30	66 37	• •	348 11	27 11	• •	325 11	• •	276 10	• •	258 18	• 34
	40	69 07		350 42	29 41		327 41		278 40		260 43	34
	50	71 37		353 12	32 12		330 11		281 11		263 08	34
17	00	74 06	S23 01	355 43	34 42	S14 39	332 42	N10 28	283 41	N21 50	265 33	N18 34
	10	76 36		358 13	37 12		335 12		286 12		267 58	34
	20	79 06		0 43	39 43		337 42		288 42		270 23	34
	30	81 36	• •	3 14	42 13	• •	340 12	• •	291 13	• •	272 48	• 34
	40	84 06		5 44	44 43		342 42		293 43		275 13	35
	50	86 36		8 15	47 13		345 13		296 13		277 38	35
18	00	89 06	S23 01	10 45	49 44	S14 39	347 43	N10 29	298 44	N21 50	280 04	N18 35
	10	91 36		13 16	52 14		350 13		301 14		282 29	35
	20	94 06		15 46	54 44		352 43		303 45		284 54	35
	30	96 36	• •	18 16	57 14	• •	355 14	• •	306 15	• •	287 19	• 35
	40	99 06		20 47	59 45		357 44		308 46		289 44	35
	50	101 36		23 17	62 15		0 14		311 16		292 09	35
19	00	104 06	S23 01	25 48	64 45	S14 38	2 44	N10 29	313 47	N21 50	294 34	N18 35
	10	106 36		28 18	67 16		5 15		316 17		296 59	35
	20	109 06		30 48	69 46		7 45		318 48		299 25	35
	30	111 36	• •	33 19	72 16	• •	10 15	• •	321 18	• •	301 50	• 35
	40	114 06		35 49	74 46		12 45		323 49		304 15	35
	50	116 36		38 20	77 17		15 16		326 19		306 40	36
20	00	119 06	S23 00	40 50	79 47	S14 37	17 46	N10 30	328 49	N21 50	309 05	N18 36
	10	121 36		43 20	82 17		20 16		331 20		311 30	36
	20	124 06		45 51	84 47		22 46		333 50		313 55	36
	30	126 35	• •	48 21	87 18	• •	25 16	• •	336 21	• •	316 20	• 36
	40	129 05		50 52	89 48		27 47		338 51		318 45	36
	50	131 35		53 22	92 18		30 17		341 22		321 11	36
21	00	134 05	S23 00	55 53	94 49	S14 36	32 47	N10 30	343 52	N21 50	323 36	N18 36
	10	136 35		58 23	97 19		35 17		346 23		326 01	36
	20	139 05		60 53	99 49		37 48		348 53		328 26	36
	30	141 35	• •	63 24	102 19	• •	40 18	• •	351 24	• •	330 51	• 36
	40	144 05		65 54	104 50		42 48		353 54		333 16	36
	50	146 35		68 25	107 20		45 18		356 24		335 41	36
22	00	149 05	S23 00	70 55	109 50	S14 36	47 49	N10 31	358 55	N21 50	338 06	N18 36
	10	151 35		73 25	112 20		50 19		1 25		340 31	36
	20	154 05		75 56	114 51		52 49		3 56		342 57	36
	30	156 35	• •	78 26	117 21	• •	55 19	• •	6 26	• •	345 22	• 36
	40	159 05		80 57	119 51		57 50		8 57		347 47	36
	50	161 35		83 27	122 21		60 20		11 27		350 12	36
23	00	164 05	S23 00	85 57	124 52	S14 35	62 50	N10 31	13 58	N21 50	352 37	N18 36
	10	166 35		88 28	127 22		65 20		16 28		355 02	36
	20	169 05		90 58	129 52		67 50		18 59		357 27	36
	30	171 35	• •	93 29	132 23	• •	70 21	• •	21 29	• •	359 52	• 36
	40	174 05		95 59	134 53		72 51		24 00		2 17	36
	50	176 34		98 30	137 23		75 21		26 30		4 42	36
24	00	179 04	S23 00	101.00	139 53	S14 34	77 51	N10 32	29 00	N21 50	7 08	N18 36

Lat.	Sunrise h	Sunrise m	Twilight	Moonrise h	Moonrise m	Diff.
N °	h	m	m	h	m	m
60	9	03	57	14	52	50
58	8	46	51	15	05	50
56		32	47		16	50
54		19	43		26	49
52	8	08	40		34	50
50	7	59	38		42	49
45		38	34	15	58	49
40		22	31	16	11	50
35	7	08	28		23	49
30	6	56	26		33	49
20		35	24	16	50	49
10		17	23	17	05	49
0	6	00	22		19	49
10	5	43	23		33	48
20		24	25	17	48	48
30	5	03	27	18	05	48
35	4	50	29		15	48
40		35	32		27	47
45	4	18	36		40	48
50	3	55	43	18	57	47
52		45	48	19	05	46
54		33	53		13	47
56		19	61		23	46
58	3	03	72		34	46
60	2	43	95	19	47	46
S						

Lat.	Sunset h	Sunset m	Twilight	Moonset h	Moonset m	Diff.
N °	h	m	m	h	m	m
60	15	04	57	7	19	50
58		21	51	7	06	50
56		36	47	6	56	49
54		48	43		46	50
52	15	59	40		38	49
50	16	08	38		30	50
45		28	34		14	49
40		45	31	6	00	50
35	16	59	28	5	49	49
30	17	11	26		39	49
20		32	24		22	49
10	17	50	22	5	08	48
0	18	07	22	4	54	48
10		25	23		40	48
20	18	43	25		25	48
30	19	05	27	4	08	48
35		18	29	3	58	48
40		32	32		47	47
45	19	50	37		34	47
50	20	12	44		18	46
52		22	49		10	47
54		34	54	3	02	46
56	20	48	61	2	52	46
58	21	04	73		41	46
60	21	24	96	2	29	45
S						

VII

Example: Correct the following observed altitudes taken at a height of 2,000 feet with the sea horizon:

	Sun (lower limb)	Moon (upper limb)	Jupiter
Observed Alt.	48°32′	19°51′	8°39′
Refraction	−1	−3	−5
Semidiameter	+16	−15	
Parallax		+51	
Dip	−44	−44	−44
Corrected Alt.	48°03′	19°40′	7°50′

5. The narrow diagram on the A. M. side of the daily sheet shows the region along the ecliptic circle within which the Sun, Moon and Planets are always found. The four bright stars Aldebaran (*a*), Regulus (*b*), Spica (*c*) and Antares (*d*) are also near the Ecliptic and are shown in the diagram except when they are within 5° of the Sun. The Moon is shown in its proper phase. The five planets, Mercury ☿, Venus ♀, Mars ♂, Jupiter ♃, and Saturn ♄, are included except when they are within 5° of the Sun.

Example: At Sunset on Jan. 1, 1942, one finds from the diagram that the Full Moon is rising in the East. Jupiter, Aldebaran, Saturn and Mars are above the Moon. Venus is in the West about 40° from the Sun. At Sunrise, on the other hand, one finds Antares and Spica about 30° and 75° respectively from the Sun, while Regulus is about 130° west of the Sun. The position of the Vernal Equinox (♈) is also shown on the diagram to aid in estimating approximate SHA. This shows SHA ♀ =39°.

6. The Polaris table found on the back of the flap gives for various values of the Local Hour Angle (LHA) of the Vernal Equinox the correction which must be applied to an observed altitude of Polaris to determine the latitude.

Example: The corrected altitude of Polaris found in §4 was 15°46′. If this observation was made in longitude 99°27′ W at $12^h16^m27^s$ GCT, the observer's latitude is found as follows:

GHA ♈ at 12^h10^m	283°01′	Corrected alt	15°46′
GHA ♈ Int. of 6^m27^s	1 37	Corr. from table for LHA ♈	+57
GHA ♈	284 38	Latitude	16°43′ N
Longitude W	−99 27		
LHA ♈	185°11′		

7. Tables for finding the times of Sunrise, Sunset, beginning and ending of Civil Twilight, Moonrise and Moonset for latitudes between 60° S and 60° N are given on the P. M. side of the daily sheets. The columns under Sunrise and Sunset give the local civil times of these phenomena. The columns under Twilight (Twlt.) give the duration of Civil Twilight. It is assumed that morning Civil Twilight begins when the Sun is 6° below the horizon and ends at Sunrise and that evening Civil Twilight begins at Sunset and ends when the Sun is 6° below the horizon. The time of beginning of morning Civil Twilight is obtained by subtracting the duration of Twilight from the time of Sunrise; the ending of evening Twilight is obtained by adding the duration of Twilight to the time of Sunset.

Examples: Find the Local Civil Time (LCT) of Sunrise, Sunset, beginning of morning Civil Twilight and ending of evening Civil Twilight in latitude 39° N on Jan. 1. Interpolations between the values for latitudes 40° N and 35° N in the table give 7^h19^m for the time of Sunrise and 16^h48^m for the time of Sunset. Similar interpolation for the duration of Twilight gives 30^m for the duration of both morning and evening Twilight; combining the 30^m with the above times of Sunrise and Sunset it is found that for latitude 39° N the

beginning of morning Civil Twilight is $7^h19^m-30^m=6^h49^m$ LCT
and ending of evening Civil Twilight is $16^h48^m+30^m=17^h18^m$ LCT.

VIII

8. Except for high latitudes, values for the approximate durations of Nautical Twilight (beginning or ending when Sun is 12° below horizon) and Astronomical Twilight (beginning or ending when Sun is 18° below horizon) may be obtained by multiplying the duration of Civil Twilight by factors of two and three respectively.

9. The duration of Civil Twilight may also be used to compute the effect of height of the observer on the time of Sunrise or Sunset. Half the duration is equivalent to a height of 32,000 feet and one-fourth to 8,000 feet.

Example: The time of Sunrise observed from a height of 32,000 ft. on Jan. 1, latitude 40° N, will be 16^m earlier than on the ground.

10. The columns under Moonrise and Moonset give the Local Civil Time of these phenomena for the meridian of Greenwich. Since the times of Moonrise and Moonset are considerably later on succeeding days, it is necessary to interpolate for the longitude of the observer; the last column (Diff.) is provided for this purpose.

Example: Find the time of Moonrise in longitude 95° W and latitude 41° N on Jan. 1. By interpolation it is found that the time of Moonrise for the meridian of Greenwich and latitude 41° N is 16^h08^m. From the Diff. column it is found that the time of rising will be 50^m later the following day. Since 95° is about one-fourth of 360°, one adds 13^m:

$$16^h08^m+13^m=16^h21^m\text{LCT}.$$

11. The times of Moonrise and Moonset as given on the daily sheets are sometimes greater than 24^h. This means that the phenomenon really occurs on the following day but the time is given in this form to facilitate the interpolation for longitude. For any given meridian the time of Moonrise or Moonset is about an hour later (on the average) each succeeding day. If then, on a given day, one of these phenomena occurs near midnight, the next one will occur about 25 hours later which carries it over into the second day. For example, if the Moon should rise at Greenwich at 23^h10^m GCT on Jan. 10 it would rise at 24^h10^m on Jan. 11 which is 0^h10^m on Jan. 12. The daily sheet would therefore give 24^h10^m for Jan. 11 and 0^h10^m for Jan. 12. In this case, there would be no Moonrise at Greenwich on Jan. 11. However, for a meridian 6^h east of Greenwich the time would be 15^m earlier or 23^h55^m on Jan. 11.

12. The LCT found in the above example is the local time for the observer's local meridian. GCT is obtained from the LCT by applying the observer's longitude from Greenwich; the longitude is first converted from arc to time and then added to the LCT for an observer in west longitude or subtracted for an observer in east longitude.

Example: Change 16^h21^m LCT in longitude 95° W to GCT.

Longitude 95° W converted to time is longitude 6^h20^m W;

16^h21^m LCT$+6^h20^m=22^h41^m$ GCT.

13. The tables, Interpolation of GHA, Dip, Polaris, ☾'s Par. and Corr. HA☾, are the so-called "critical" or "turning point" type; i. e., the values of the argument given are those for which the function changes from one unit to the next. The value of the function is therefore found to the nearest unit without interpolation. If the required value of the argument is one of the printed values of the table, the upper of the two adjacent values of the function should be taken.

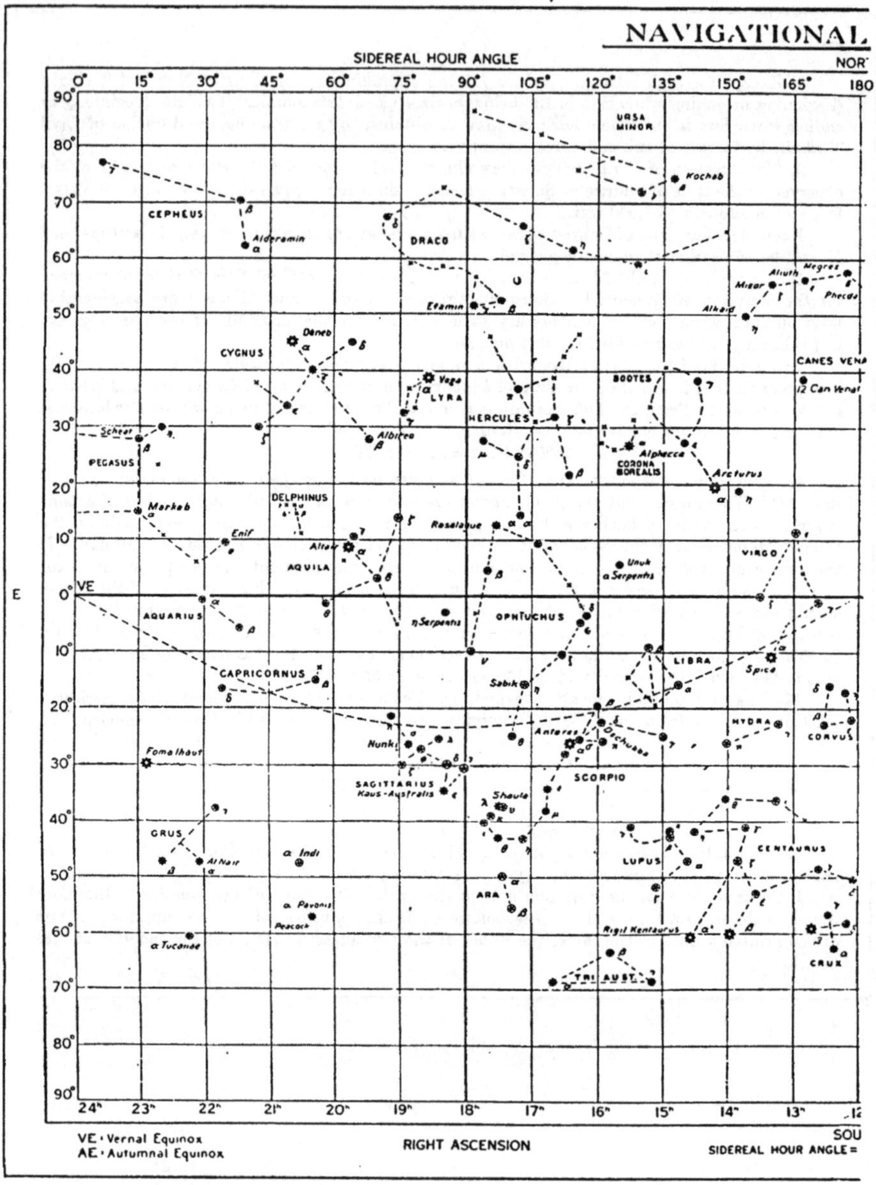
NAVIGATIONAL
SIDEREAL HOUR ANGLE
NORT
RIGHT ASCENSION
SOU
VE: Vernal Equinox
AE: Autumnal Equinox
SIDEREAL HOUR ANGLE =
URSA MINOR
Kochab
CEPHEUS
Alderamin
DRACO
Etamin
CYGNUS
Deneb
Vega
LYRA
HERCULES
BOOTES
CANES VENA
12 Can Venat
Mizar
Alioth
Megrez
Phecda
Alkaid
Scheat
PEGASUS
Albireo
CORONA BOREALIS
Alphecca
Arcturus
DELPHINUS
Markab
Enif
Altair
AQUILA
Rasalague
Unuk
a Serpentis
VIRGO
AQUARIUS
η Serpentis
OPHIUCHUS
CAPRICORNUS
Sabik
LIBRA
Spica
HYDRA
CORVUS
Fomalhaut
Nunki
Antares
SAGITTARIUS
Kaus-Australis
SCORPIO
Shaula
GRUS
Al Nair
α Indi
LUPUS
CENTAURUS
ARA
α Pavonis
Peacock
α Tucanae
Rigil Kentaurus
CRUX
TRI. AUST.

STAR CHART

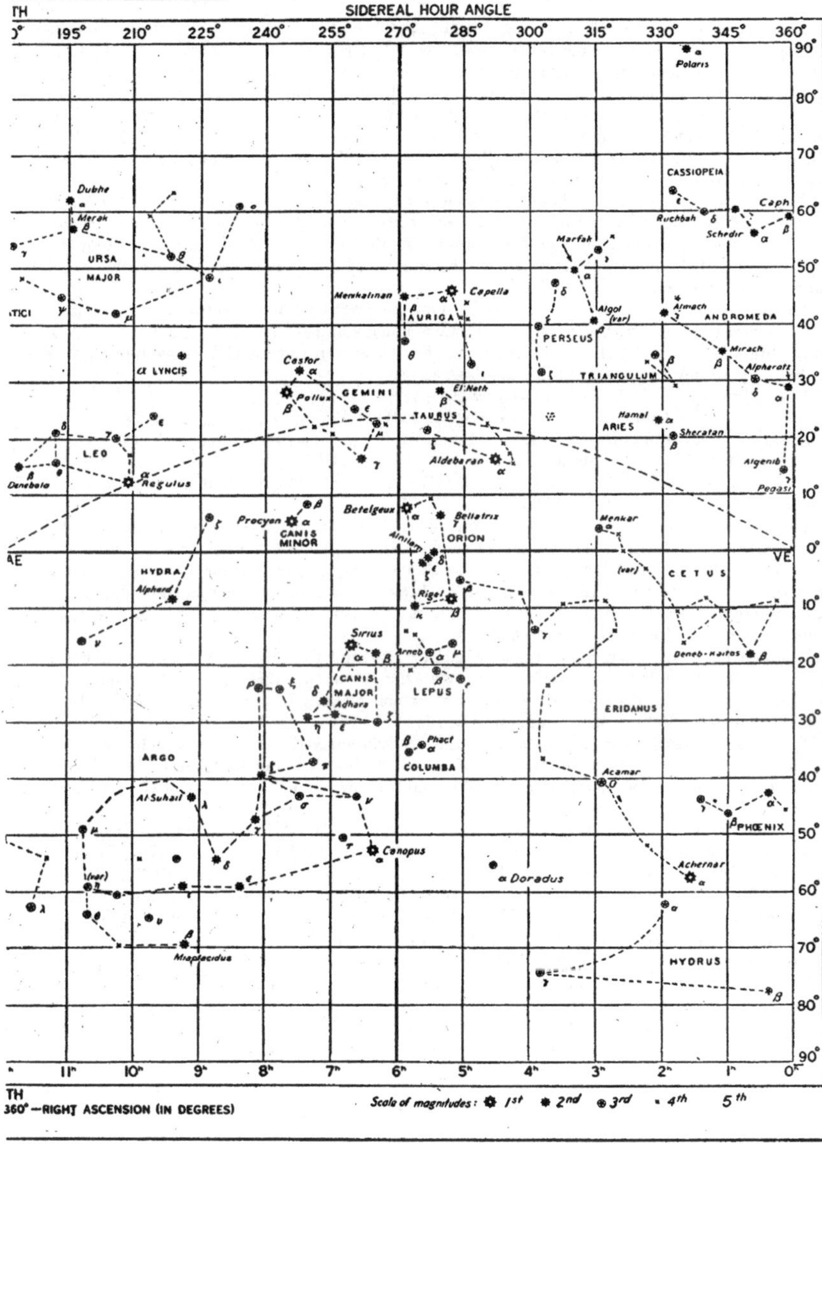
SIDEREAL HOUR ANGLE
195° 210° 225° 240° 255° 270° 285° 300° 315° 330° 345° 360°
90° 80° 70° 60° 50° 40° 30° 20° 10° 0° 10° 20° 30° 40° 50° 60° 70° 80° 90°
NORTH
DECLINATION
SOUTH
W
VE
AE
11h 10h 9h 8h 7h 6h 5h 4h 3h 2h 1h 0h
Polaris
CASSIOPEIA
Caph
Ruchbah
Schedir
Dubhe
Merak
URSA MAJOR
Marfak
Capella
Menkalinan
AURIGA
Algol (var)
PERSEUS
Almach
ANDROMEDA
Mirach
Alpheratz
TRIANGULUM
α LYNCIS
Castor
Pollux
GEMINI
El Nath
TAURUS
Hamal
ARIES
Sheratan
LEO
Denebola
Regulus
Aldebaran
Algenib
Pegasi
Procyon
CANIS MINOR
Betelgeux
Bellatrix
ORION
Menkar
HYDRA
Alphard
Rigel
CETUS
Sirius
CANIS MAJOR
Adhara
Arneb
LEPUS
Deneb-Kaitos
ERIDANUS
Phact
COLUMBA
ARGO
Acamar
Al Suhail
PHŒNIX
Canopus
Achernar
α Doradus
Miaplacidus
HYDRUS
Scale of magnitudes: 1st 2nd 3rd 4th 5th
360° — RIGHT ASCENSION (IN DEGREES)

NAVIGATIONAL STAR CHART

The purpose of the Star Chart facing this page is to assist the navigator in identifying stars for navigation. The stars of each constellation are connected by dotted lines; the bright stars are identified with their Greek letters and the principal stars with their common names. The Sidereal Hour Angle and declination of each star can be determined from the map by means of the network of vertical and horizontal lines drawn upon it for this purpose. The SHA's are measured (0° to 360°) from the vertical line passing through the Vernal Equinox at 0°. The declinations are measured North and South (N 90° to S 90°) from the celestial equator which is represented by a heavy horizontal line through the center.

The user of the star chart should be forewarned that the rectangular shape of the chart distorts the relative positions of the stars in the polar regions. A globe would give a better representation, and an observer from the inside would see the constellations as they appear in the sky. Then the SHA lines would converge at the North and South poles and the equator would be in the form of a circle.

An observer's local meridian is easily located on the chart since it coincides with the vertical line whose GHA is equal to his longitude. GHA is not given directly on the chart but may be readily obtained from the SHA which is given. For any given instant GHA may be obtained from SHA by adding the GHA ♈ from the daily sheet:

GHA = SHA + GHA ♈

Conversely SHA = GHA − GHA ♈

Example: Locate on the chart the local meridian of an observer in longitude 110° W on Jan. 1, at 6^h10^m GCT. Since his longitude is 110° W, the GHA of his meridian will be 110°. From the daily sheet for Jan. 1 at 6^h10^m the GHA ♈ = 193°. The SHA of his meridian will therefore be 110° − 193° = −83° = 277°.

The identification of a star directly overhead; i. e., in the zenith, is easily made since the point overhead is on the local meridian and also has a Dec. equal to the observer's latitude.

Example: Assume that the observer in the above example is in latitude 40° N and that a star in the zenith is to be identified. The SHA of the star is exactly equal to the SHA of the local meridian and was found to be 277°. The star's Dec. is N 40° since the observer's latitude is equal to the Dec. of a point in the zenith.

Examination of the chart in the region of SHA = 277° and Dec. = N 40° shows the brightest star in the region to be the first magnitude star Capella. To verify, this region of the chart may be compared in detail with the sky. One finds the conspicuous triangle formed by the three first magnitude stars Pollux, Aldebaran, and Capella.

A star to the North or South of the zenith is easily identified because its angular distance from the zenith is equal to the difference between its declination and the observer's latitude. Thus, in the above example, Rigel (Dec. S 10°) would appear about 50° South of the zenith or at an altitude of 40°.

The chart may also be used with the "Star Identification Table" in H. O. 214. The table is used to convert an observed altitude and azimuth into Dec. and LHA. Since LHA + Long. = GHA, the problem becomes that of the above example; i. e., identification from SHA and Dec.

The Ecliptic, which if shown in the diagram on the daily sheet would be a straight line, is represented on the chart by a curved dotted line. The four bright stars of the diagram are easily found on the chart as they lie along the Ecliptic. The Sun, Moon, and planets may be plotted on the chart by means of their SHA and Dec.

Example: The daily sheet for Jan. 1, 0^h GCT gives GHA ♃ = 66°, GHA ♄ = 64°, GHA ♈ = 100°, and Dec. ♃ = Dec. ♄ = N12°. This gives SHA ♃ = 326°, SHA ♄ = 324°. Plotting SHA and Dec. places them on the Ecliptic about midway between the Vernal Equinox and Aldebaran, which agrees with the daily diagram.

x

POLARIS (1942)

LHA ♈	Corr.	LHA ♈	Corr.	LHA ♈	Corr.	LHA ♈	Corr.	LHA ♈	Corr.	LHA ♈	Corr.
° ′	′	° ′	′	° ′	′	° ′	′	° ′	′	° ′	′
357 55	−54	88 38	−27	128 21	+14	179 34	+55	270 37	+26	310 17	−15
0 00	−55	89 41	−26	129 19	+15	181 49	+56	271 40	+25	311 15	−16
2 16	−56	90 43	−25	130 18	+16	184 15	+57	272 42	+24	312 12	−17
4 45	−57	91 46	−24	131 18	+17	186 56	+58	273 42	+23	313 12	−18
7 15	−58	92 48	−23	132 18	+18	190 21	+59	274 42	+22	314 12	−19
10 25	−59	93 48	−22	133 16	+19	194 30	+60	275 43	+21	315 12	−20
14 34	−60	94 48	−21	134 13	+20	201 15	+61	276 46	+20	316 12	−21
21 15	−61	95 48	−20	135 12	+21	211 15	+60	277 47	+19	317 12	−22
31 15	−60	96 48	−19	136 15	+22	217 30	+59	278 45	+18	318 12	−23
37 30	−59	97 47	−18	137 17	+23	221 47	+58	279 42	+17	319 12	−24
41 25	−58	98 45	−17	138 18	+24	225 00	+57	280 42	+16	320 12	−25
44 41	−57	99 42	−16	139 18	+25	227 45	+56	281 41	+15	321 15	−26
47 15	−56	100 42	−15	140 19	+26	230 13	+55	282 41	+14	322 17	−27
49 45	−55	101 42	−14	141 24	+27	232 30	+54	283 39	+13	323 22	−28
52 02	−54	102 41	−13	142 30	+28	234 34	+53	284 37	+12	324 27	−29
54 03	−53	103 39	−12	143 32	+29	236 25	+52	285 34	+11	325 31	−30
55 57	−52	104 36	−11	144 34	+30	238 12	+51	286 32	+10	326 33	−31
57 49	−51	105 34	−10	145 40	+31	240 00	+50	287 30	+ 9	327 36	−32
59 30	−50	106 32	− 9	146 49	+32	241 40	+49	288 27	+ 8	328 45	−33
61 05	−49	107 30	− 8	147 57	+33	243 16	+48	289 25	+ 7	329 53	−34
62 39	−48	108 25	− 7	149 05	+34	244 50	+47	290 22	+ 6	331 01	−35
64 13	−47	109 21	− 6	150 13	+35	246 19	+46	291 17	+ 5	332 09	−36
65 44	−46	110 17	− 5	151 21	+36	247 47	+45	292 13	+ 4	333 20	−37
67 12	−45	111 15	− 4	152 30	+37	249 15	+44	293 10	+ 3	334 31	−38
68 33	−44	112 12	− 3	153 41	+38	250 39	+43	294 08	+ 2	335 42	−39
69 52	−43	113 08	− 2	154 52	+39	251 58	+42	295 03	+ 1	336 54	−40
71 11	−42	114 04	− 1	156 07	+40	253 17	+41	296 03	0	338 09	−41
72 30	−41	115 00	0	157 22	+41	254 36	+40	297 01	− 1	339 28	−42
73 48	−40	115 57	+ 1	158 41	+42	255 52	+39	297 57	− 2	340 47	−43
75 07	−39	116 55	+ 2	160 00	+43	257 07	+38	298 53	− 3	342 06	−44
76 18	−38	117 52	+ 3	161 23	+44	258 19	+37	299 48	− 4	343 28	−45
77 30	−37	118 47	+ 4	162 46	+45	259 31	+36	300 46	− 5	344 51	−46
78 41	−36	119 43	+ 5	164 10	+46	260 40	+35	301 44	− 6	346 19	−47
79 52	−35	120 38	+ 6	165 35	+47	261 49	+34	302 41	− 7	347 47	−48
81 01	−34	121 38	+ 7	167 03	+48	262 57	+33	303 36	− 8	349 15	−49
82 09	−33	122 35	+ 8	168 40	+49	264 05	+32	304 32	− 9	350 50	−50
83 15	−32	123 33	+ 9	170 19	+50	265 13	+31	305 28	−10	352 30	−51
84 20	−31	124 31	+10	172 00	+51	266 18	+30	306 26	−11	354 17	−52
85 26	−30	125 28	+11	173 45	+52	267 23	+29	307 24	−12	356 04	−53
86 31	−29	126 26	+12	175 34	+53	268 28	+28	308 21	−13	357 55	−54
87 36	−28	127 24	+13	177 30	+54	269 33	+27	309 19	−14	0 00	−55
88 38		128 21		179 34		270 37		310 17		2 16	

STARS

Alphabetical order				Order of SHA				
Name	Mag.	SHA	Dec.	SHA	Dec.	RA	Name	
		° ′	° ′	° ′	° ′	h m		
Acamar	3.4	315 59	S40 32	14 32	N14 54	23 02	Markab	
Achernar 1	0.6	336 07	S57 32	**16 24**	**S29 56**	**22 54**	**Fomalhaut**	
Acrux 2	1.1	174 09	S62 47	28 52	S47 15	22 05	Al Na'ir	
Adhara	1.6	255 55	S28 54	34 40	N 9 37	21 41	Enif	
Aldebaran 3	1.1	291 51	N16 23	**50 08**	**N45 04**	**20 39**	**Deneb**	
Alioth	1.7	167 08	N56 16	54 44	S56 55	20 21	Peacock	
Al Na'ir	2.2	28 52	S47 15	**63 01**	**N 8 43**	**19 48**	**Altair**	
Alnilam	1.8	276 41	S 1 14	77 05	S26 22	18 52	Nunki	
Alphard	2.2	218 49	S 8 25	**81 15**	**N38 44**	**18 35**	**Vega**	
Alphecca	2.3	126 56	N26 54	84 55	S34 25	18 20	Kaus Aust.	
Alpheratz 4	2.2	358 39	N28 46	91 11	N51 30	17 55	Etamin	
Al Suhail	2.2	223 32	S43 12	96 56	N12 36	17 32	Rasalague	
Altair 5	0.9	63 01	N 8 43	97 35	S37 04	17 30	Shaula	
Antares 6	1.2	113 32	S26 18	103 14	S15 39	17 07	Sabik	
Arcturus 7	0.2	146 45	N19 29	(109 22)	S68 55	16 43	α Tri. Aust.	
ε Argus	1.7	234 40	S59 20	**113 32**	**S26 18**	**16 26**	**Antares**	(*d*)
Bellatrix	1.7	279 30	N 6 18	120 46	S22 27	15 57	Dschubba	
Betelgeux 8	0.1–1.2	271 59	N 7 24	126 56	N26 54	15 32	Alphecca	
Canopus 9	−0.9	264 20	S52 40	(137 17)	N74 23	14 51	Kochab	
Capella 10	0.2	281 54	N45 56	**141 05**	**S60 36**	**14 36**	**Rigil Kent.**	
Caph	2.4	358 29	N58 50	**146 45**	**N19 29**	**14 13**	**Arcturus**	
θ Centauri	2.3	149 11	S36 05	149 11	S36 05	14 03	θ Centauri	
β Crucis	1.5	168 55	S59 22	**159 28**	**S10 52**	**13 22**	**Spica**	(*c*)
γ Crucis	1.6	173 00	S56 47	159 36	N55 14	13 22	Mizar	
Deneb 11	1.3	50 08	N45 04	167 08	N56 16	12 51	Alioth	
Deneb Kait.	2.2	349 50	S18 18	168 55	S59 22	12 44	β Crucis	
Denebola	2.2	183 28	N14 54	173 00	S56 47	12 28	γ Crucis	
Dschubba	2.5	120 46	S22 27	**174 09**	**S62 47**	**12 23**	**Acrux**	
Dubhe 12	2.0	194 57	N62 04	183 28	N14 54	11 46	Denebola	
Enif	2.5	34 40	N 9 37	194 57	N62 04	11 00	Dubhe	
Etamin	2.4	91 11	N51 30	**208 40**	**N12 15**	**10 05**	**Regulus**	(*b*)
Fomalhaut 13	1.3	16 24	S29 56	218 49	S 8 25	9 25	Alphard	
Hamal	2.2	329 02	N23 11	(221 50)	S69 29	9 13	Miaplacidus	
Kaus Aust.	2.0	84 55	S34 25	223 32	S43 12	9 06	Al Suhail	
Kochab	2.2	(137 17)	N74 23	234 40	S59 20	8 21	ε Argus	
Marfak	1.9	309 57	N49 39	**244 33**	**N28 10**	**7 42**	**Pollux**	
Markab	2.6	14 32	N14 54	**245 56**	**N 5 22**	**7 36**	**Procyon**	
Minplacidus	1.8	(221 50)	S69 29	255 55	S28 54	6 56	Adhara	
Mizar	2.4	159 36	N55 14	**259 21**	**S16 38**	**6 43**	**Sirius**	
Nunki	2.1	77 05	S26 22	**264 20**	**S52 40**	**6 23**	**Canopus**	
Peacock 14	2.1	54 44	S56 55	**271 59**	**N 7 24**	**5 52**	**Betelgeux**	
Polaris	2.1	(334 04)	N88 59	276 41	S 1 14	5 33	Alnilam	
Pollux 15	1.2	244 33	N28 10	279 30	N 6 18	5 22	Bellatrix	
Procyon 16	0.5	245 56	N ˙5 22	**281 54**	**N45 56**	**5 12**	**Capella**	
Rasalague	2.1	96 56	N12 36	**282 04**	**S 8 16**	**5 12**	**Rigel**	
Regulus 17	1.3	208 40	N12 15	**291 51**	**N16 23**	**4 33**	**Aldebaran**	(*a*)
Rigel 18	0.3	282 04	S 8 16	309 57	N49 39	3 20	Marfak	
Rigil Kent. 19	0.3	141 05	S60 36	315 59	S40 32	2 56	Acamar	
Ruchbah	2.8	339 30	N59 56	329 02	N23 11	2 04	Hamal	
Sabik	2.6	103 14	S15 39	(334 04)	N88 59	1 44	Polaris	
Shaula	1.7	97 35	S37 04	**336 07**	**S57 32**	**1 36**	**Achernar**	
Sirius 20	−1.6	259 21	S16 38	339 30	N59 56	1 22	Ruchbah	
Spica 21	1.2	159 28	S10 52	349 50	S18 18	0 41	Deneb Kait.	
α Tri. Aust.	1.9	(109 22)	S68 55	358 29	N58 50	0 06	Caph	
Vega 22	0.1	81 15	N38 44	358 39	N28 46	0 05	Alpheratz	

SHA★ = 360° − RA $\qquad$ GHA★ = GHA♈ + SHA★ $\qquad$ Jan – Apr., 1942

CONDENSED EXPLANATION

(See also explanation on white sheets)

1. Each daily sheet gives the Greenwich Hour Angle (GHA) and declination (Dec.) of the Sun, Vernal Equinox (♈), three planets, and Moon, at ten-minute intervals of Greenwich Civil Time (GCT). The values for A. M. are given on the front of the sheet and for P. M. on the back.

2. The diagram on the A. M. side of the sheet shows the positions of the Moon, planets, Vernal Equinox, and four stars, Aldebaran (*a*), Regulus (*b*), Spica (*c*), and Antares (*d*), with respect to the Sun. The two ends of the diagram are 180° from the Sun.

3. The column to the right of the heavy line on the A. M. side gives the correction (☾'s Par.) of an observed altitude of the Moon for parallax, the semidiameters of the Sun (SD ☉) and Moon (SD ☾) and the quantity "Corr. HA ☾" explained in §5.

4. The tables on the right of the heavy line on the P. M. side of the sheet give for various latitudes the times of rising and setting of the Sun and Moon, and the duration of Civil Twilight. The second column refers to the Sun. The third column (Twlt.) gives the duration of Civil Twilight. The fourth refers to the Moon, and the fifth (Diff.) gives the number of minutes later at which the Moon will rise or set on the succeeding day (interpolation for longitude).

5. Two identical tables are given to interpolate the tabulated GHA's for intermediate values of the GCT: one inside the front cover for use with the A. M. side of the sheet and one on the back of the flap for use with the P. M. side. This interpolation for the Moon may introduce an error of 1′, which may be eliminated by applying the correction given on the daily sheet, "Corr. HA ☾."

6. The inside back cover gives the Sidereal Hour Angle (SHA) and declination of stars. Two lists are given: one in alphabetical order, and the other in order of SHA which is also the order of GHA. Those values of the SHA which may be in error by more than 0′.5 are enclosed in parentheses. The GHA of a star is obtained by adding its SHA to the GHA ♈.

7. On the front of the flap is a star chart, and on the back of the flap is a table for converting an observed altitude of Polaris into latitude. Preceding the flap is a detailed explanation of the tables with examples (white sheets).

8. All observed altitudes must be corrected for refraction, and those of the Moon for parallax. If the upper or lower limb of the Sun or Moon is observed a correction must be included for semidiameter. Altitudes measured from the sea horizon must be corrected for dip. Tables for dip and refraction are given below. The corrections for semidiameter and parallax are given on the daily sheets (see § 3).

9. The tables, Interpolation of GHA, Dip, Polaris, ☾'s Par. and Corr. HA ☾, are the so-called "critical" type; i. e., the values of the argument are so chosen that the tabulated quantity increases by one unit.

10. The error of an interpolated GHA is never as great as 1′.8, and the average error is about 0′.5, except for those circumpolar stars whose SHA's are enclosed in parentheses.

REFRACTION

All observed altitudes must be corrected for refraction. *Subtract* the correction given below from the observed altitude.

Height in feet	Observed altitude						
	5°	10°	15°	20°	30°	45°	60°
	′	′	′	′	′	′	′
0	10	5	4	3	2	1	1
5, 000	8	5	3	2	1	1	0
10, 000	7	4	3	2	1	1	0
15, 000	6	3	2	2	1	1	0
20, 000	5	3	2	1	1	1	0
25, 000	4	2	2	1	1	0	0
30, 000	3	2	1	1	1	0	0
35, 000	3	2	1	1	1	0	0
40, 000	2	1	1	1	0	0	0

DIP

Altitudes measured from the *sea horizon* must be corrected for dip. *Subtract* the correction below from the observed altitude.

Height	Corr.	Height	Corr	Height	Corr.	Height	Corr.
Ft.	′	Ft.	′	Ft.	′	Ft.	′
0		160		620		1380	
2	1	180	13	670	25	1460	37
6	2	210	14	730	26	1540	38
12	3	250	15	780	27	1620	39
21	4	280	16	840	28	1700	40
31	5	310	17	900	29	1790	41
43	6	350	18	960	30	1870	42
58	7	390	19	1030	31	1960	43
75	8	430	20	1090	32	2060	44
93	9	480	21	1160	33	2150	45
114	10	520	22	1230	34	2250	46
137	11	570	23	1310	35	2340	47
162	12	620	24	1380	36	2440	48

STARS, 1938 JUNE 3

G.H.A. of First Point of Aries = G.H.A. ♈

Friday, June 3

G.M.T.	00m	10m	20m	30m	40m	50m
00	250 47	253 18	255 48	258 19	260 49	263 20
01	265 50	268 20	270 51	273 21	275 52	278 22
02	280 52	283 23	285 53	288 24	290 54	293 24
03	295 55	298 25	300 56	303 26	305 56	308 27
04	310 57	313 28	315 58	318 29	320 59	323 29
05	326 00	328 30	331 01	333 31	336 01	338 32
06	341 02	343 33	346 03	348 33	351 04	353 34
07	356 05	358 35	1 06	3 36	6 06	8 37
08	11 07	13 38	16 08	18 38	21 09	23 39
09	26 10	28 40	31 10	33 41	36 11	38 42
10	41 12	43 43	46 13	48 43	51 14	53 44
11	56 15	58 45	61 15	63 46	66 16	68 47
12	71 17	73 47	76 18	78 48	81 19	83 49
13	86 19	88 50	91 20	93 51	96 21	98 52
14	101 22	103 52	106 23	108 53	111 24	113 54
15	116 24	118 55	121 25	123 56	126 26	128 56

No.	Name	S.H.A.	Dec.
		° ′	° ′
11	ALTAIR	63 03	N. 8 42
14	ARCTURUS	146 47	N.19 30
38	*Polaris*	334 41	N.88 58
50	VEGA	81 17	N.38 43

PLANETS — Jupiter, −2·0

G.M.T.	G.H.A.	Dec.
	° ′	° ′
	June 3	
00	276 45	S. 11 43
06	6 59	11 42
12	97 13	11 42
18	187 27	11 42

STARS, SUN and PLANETS — 1939 APR. 24

G.H.A. of SUN

G.M.T.	00m	10m	20m	30m	40m	50m	SUN'S Dec.
h	° ′	° ′	° ′	° ′	° ′	° ′	° ′
00	180 24	182 54	185 24	187 54	190 25	192 55	N.12 30
01	195 25	197 55	200 25	202 55	205 25	207 55	12 31
14	30 26	32 56	35 26	37 56	40 26	42 56	12 41
15	45 26	47 56	50 26	52 56	55 26	57 56	N.12 42
16	60 26	62 56	65 26	67 56	70 26	72 56	12 43

STARS, SUN and PLANETS — 1940 JAN. 1

G.H.A. of First Point of ARIES = G.H.A. ♈

G.M.T.	00m	10m	20m	30m	40m	50m
h	° ′	° ′	° ′	° ′	° ′	° ′
00	99 30	102 01	104 31	107 02	109 32	112 03
05	174 43	177 13	179 44	182 14	184 44	187 15
06	189 45	192 16	194 46	197 16	199 47	202 17

STARS, SUN and PLANETS — 1940 FEB. 5

G.H.A. of First Point of ARIES = G.H.A. ♈

G.M.T.	00m	10m	20m	30m	40m	50m
h	° ′	° ′	° ′	° ′	° ′	° ′
00	134 00	136 31	139 01	141 32	144 02	146 32
01	149 03	151 33	154 04	156 34	159 04	161 35
02	164 05	166 36	169 06	171 37	174 07	176 37

STARS, 1940 JAN.—MAR.

No.	Name	Mag.	S.H.A.	Dec.
			° ′	° ′
7	ARCTURUS	0·2	146 46	N.19 29
16	PROCYON	0·5	245 57	N. 5 23
47	*Polaris*	2·1	334 18	. . 3 58

STARS, SUN and PLANETS — 1940 NOV. 28

G H A of SUN

G.M.T.	00m	10m	20m	30m	40m	50m	SUN'S Dec.
h	° ′	° ′	° ′	° ′	° ′	° ′	° ′
00	183 02	185 32	188 02	190 32	193 02	195 32	S.21 15
18	92 58	95 28	97 58	100 28	102 58	105 28	21 23
19	107 58	110 28	112 58	115 28	117 58	120 28	21 23
20	122 58	125 28	127 58	130 28	132 58	135 28	S.21 24

GREENWICH P. M. 1941 MARCH 3

GCT	☉ SUN GHA	Dec.	♈ GHA
h m	° ′	° ′	° ′
22 00	147 00	S6 42	131 17
10	149 30		133 47
20	152 00		136 18

GREENWICH A. M. 1941 MARCH 17

GCT	☉ SUN GHA	Dec.	♈ GHA	♂ MARS 1.1 GHA	Dec	☾ MOON GHA	Dec.
h m	° ′	° ′	° ′	° ′	° ′	° ′	° ′
10 00	327 52	S1 25	324 33	34 02	S22 46		
10	330 22		327 06	36 32			
20	332 52		329 36	39 02			
30	335 22	. .	332 06	41 33	. .		
12 00	357 53	S1 23	354 40	64 03	S22 46	123 55	S14 21
10	0 23		357 11	66 34		126 20	22
20	2 53		359 41	69 04		128 44	23
30	5 23	. .	2 11	71 34	. .	131 08	. 24
15 00	42 53	S1 20	39 48	109 05	S22 45		
10	45 23		42 18	111 35			
20	47 53		44 48	114 05			
30	50 23	. .	47 19	116 36	. .		
18 00	87 54	S1 17	84 55				
10	90 24		87 25				
20	92 54		89 56				
30	95 24	. .	92 26				
40	97 54		94 57				
50	100 24		97 27				
19 00	102 54	S1 16	99 57				

☾'s Par.

Alt. °	+ Corr. ′
0	60
7	59
12	58
16	57
19	56
22	

GREENWICH P. M. 1941 MARCH 21

GCT	☉ SUN GHA	Dec.	♈ GHA
h m	° ′	° ′	° ′
17 00	73 11	N0 16	73 49
10	75 41		76 19
20	78 11		78 50
30	80 41	. .	81 20
40	83 11		83 51
50	85 41		86 21
18 00	88 11	N0 17	88 51

STARS 1941 MARCH

Alphabetical order

Name	Mag.	SHA	Dec.
		° ′	° ′
Altair	0.9	63 01	N 8 42
Antares . (d) . .	1.2	113 33	S26 18
Arcturus	0.2	146 45	N19 29
ε Argus	1.7	234 40	S59 19
Bellatrix	1.7	279 30	N 6 18
Betelgeux	0.1–1.2	272 00	N 7 24
Canopus	−0.9	264 20	S52 40
Capella	0.2	281 55	N45 56
Polaris	2.1	(334 12)	N88 59
Pollux	1.2	244 34	N28 10
Procyon	0.5	245 56	N 5 22
Rasalague	2.1	96 57	N12 36
Regulus . . (b) . .	1.3	208 41	N12 18
Vega	0.1	81 16	N38 44

GREENWICH P. M 1942 APRIL 12

GCT	☉ SUN GHA	Dec.
h m	° ′	° ′
13 50	27 17	
14 00	29 47	N 8 35
10	32 17	
20	34 47	
30	37 17	. .
40	39 47	
50	42 17	
15 00	44 47	N 8 36
10	47 17	
20	49 47	
30	52 17	. .
40	54 47	
50	57 17	

GREENWICH A. M. 1941 AUGUST 8

GCT	☉ SUN GHA	Dec.	♈ GHA
h m	° ′	° ′	° ′
2 00	208 36	N16 18	346 12
10	211 06		348 42
20	213 36		351 12
30	216 06	. .	353 43
40	218 36		356 13
50	221 06		358 44
3 00	223 36	N16 17	1 14
10	226 06		3 44
20	228 36		6 15
30	231 06	. .	8 45
40	233 36		11 16
50	236 06		13 46
4 00	238 36	N16 17	16 16

Extracts from Air Almanac covering problems in text.

APPENDIX C

SOURCES OF INFORMATION

Write to	For
Civil Aeronautics Board, Department of Commerce, Washington, D. C.	General information. This organization has charge of the publication and distribution of aeronautical information, airway bulletins, weekly *Notices to Airmen*, the *Civil Aeronautics Journal*, and general promotion work for aviation. It also has charge of licensing and registering aircraft, etc. Civil Air Regulations (C.A.R.). Effective Nov. 1, 1937, the Bureau of Air Commerce promulgated the famous C.A.R. representing an enormous amount of work on the part of government officials. These regulations vitally affect all branches of the aviation industry and have become the airman's "bible." The C.A.R. are published in loose-leaf form with an efficient numerical index system and should be studied carefully by all persons affected by these new regulations. For the 1940–1941 program special bulletins have been prepared by the Civil Aeronautics Administration covering both primary and secondary courses under the C.P.T.P. as follows: Flight Instructor's Manual, *Civil Aeronautics Bulletin* 5. Digest of Civil Air Regulations for Pilots (*Civil Aeronautics Bulletin* 22). Civil Pilot Training Manual (*Civil Aeronautics Bulletin* 23). Practical Air Navigation (*Civil Aeronautics Bulletin* 24). Meteorology for Pilots (*Civil Aeronautics Bulletin* 25). Aerodynamics for Pilots (*Civil Aeronautics Bulletin* 26). Pilots' Airplane Manual (*Civil Aeronautics Bulletin* 27). Pilots' Powerplant Manual (*Civil Aeronautics Bulletin* 28).

	Pilots' Radio Manual (*Civil Aeronautics Bulletin* 29). Ground Instructor's Manual (*Civil Aeronautics Bulletin* 30). Upon request one's name will be added to the mailing list to be notified whenever new airway maps or new editions thereof are issued. The Coast and Geodetic Survey is the distributing agency for all aeronautical charts issued by the Air Corps and by the Civil Aeronautics Authority.
U. S. Weather Bureau, Washington, D. C.	Weather charts. Weather bulletins. The Weather Bureau cooperates with the Civil Aeronautics Authority in supplying weather data to U. S. airways.
National Advisory Committee for Aeronautics (NACA), Navy Building, Washington, D. C.	Technical reports. Technical notes. Over two hundred technical reports and a large number of technical notes have been published by the NACA covering the latest developments in all branches of aviation, including navigation. These are supplied free on request, except certain pamphlets which carry a nominal charge.
U. S. Coast and Geodetic Survey, Washington, D. C.	Department of Commerce Aeronautical Charts. Nautical charts covering the coasts and harbors of the United States. *Special Publication* 197, "Practical Air Navigation." Upon request, one's name will be added to the mailing list to be notified whenever new airway maps or new editions thereof are issued. The Coast and Geodetic Survey is the distributing agency for all aeronautical charts issued by the Air Corps and by the Civil Aeronautics Authority.
Naval Observatory, Washington, D. C.	Time service and time signals for the entire country. Radio time signals are broadcast hourly, except 9 A.M., 11 A.M., 9 P.M., and 11 P.M., E.S.T. or 75th-meridian time. Time signals are also broadcast by other stations at other times.
Bureau of Standards, Washington, D. C.	Continuous time service (See Chap. XII).
Hydrographic Office, Washington, D. C.	Navy strip maps. Navigation tables. "Radio Aids to Navigation." Charts (outside U. S.). Great-circle Course Charts. Pilot Charts of the Upper Air.

	Aircraft plotting sheets.
	Star Finder.
	Naval Air Pilots (restricted to naval aviators).
	Notices to Aviators.
Rand McNally & Company, Chicago, Ill.	"Standard Indexed Maps with Air Trails" for each state, and also for the United States.
Various instrument companies.	Data on the material they supply, including a description and directions for use.

Heretofore no single agency has supplied government charts and publications together with other required navigation equipment. Suppose a navigator plans a long flight requiring maps, drift indicator, navigation watches, tables, sextant, *Air Almanac*, radio equipment, meteorological data, special instruction, etc. Unless such a person has had considerable experience, he would not know from what sources to order the various items, and trouble and delay would result. In an effort to meet this need, the firm listed below, having official agencies, will supply any of the items listed above, in addition to the following items.

Write to	For
Weems System of Navigation, Annapolis, Md.	Aircraft plotter (Department of Commerce Type). Scales 1:1,000,000 and 1,500,000.
	Skeleton navigation charts of the world. Scale 1:5,000,000 (see Chap. IV).
	Universal plotting sheet to scale 1:1,000,000 for use with the Aircraft plotter.
	D. F. radio navigation charts of U. S.
	"Navigation Note Book" with universal plotting sheets facing ruled pages.
	"Star Altitude Curves" (new three-star edition).
	Star charts.
	Dalton Mark VII computer.
	Dalton Mark VIII computer. (Aluminum circular slide rule.)
	Dalton Type E1–A computer.
	Dalton Type E1–B computer.
	Dalton Model J computer.
	Mark II, III, and IV plotting boards.
	Sextants, second-setting watches, Gatty ground-speed and drift meter. Drift indicators, compasses, and any other item of navigation equipment.
	All government maps, charts, and publications.
	Foreign maps and charts.
	Instruction by resident and home-study courses.
	Consultant services.

Navigational and aviation texts.

"Radius of Action of Aircraft," Tornich. Covers wind and radius of action problems thoroughly as well as dead-reckoning navigation.

"Instrument Flying," Weems and Zweng. Covers requirements for instrument rating.

"Simplified Celestial Navigation," Weems and Link. Covers celestial navigation requirements for secondary CPTP course.

Index

B

K

L

N

V

W

* 9 7 8 0 2 8 2 4 9 4 1 5 5 *

ND - #0041 - 070324 - C0 - 229/152/23 - PB - 9780282494155 - Gloss Lamination